# Advances in Root Vegetables Research

*Edited by Prashant Kaushik*

Published in London, United Kingdom

Advances in Root Vegetables Research
http://dx.doi.org/10.5772/intechopen.100880
Edited by Prashant Kaushik

Contributors
Mirela Ahmadi, Ştefan A. Hulea, Ioan Peţ, Rashida Bashir, Samra Tabassum, Ayoub Rashid, Shafiqur Rehman, Ahmad Adnan, Rabia Ghaffar, Fachrur Rozi, Dian Adi Anggraeni Elisabeth, Ruly Krisdiana, Adri Adri, Yardha Yardha, Yanti Rina, Benjamin E. Ubi, Emmanuel O. Ajayi, Pamela E. Akin-Idowu, Olaide R. Aderibigbe, Dorcas O. Ibitoye, Gloria Afolayan, Oluyemesi M. Adewale, Esther A. Adesegun, Saleem Siddiqui, Naseer Ahmed, Chongtham Allaylay Devi, Puthem Robindro Singh, Bawitlung Lalramhlimi, Dóra Székely, Mónika Máté, Olalekan Onisuru, Oluwatayo Racheal Onisuru, Abukari Wumbei, Sopkoutie Kengni Nerlus Gautier, Joseph Kwowura Kwodaga, Djeugap Fovo Joseph, Yamdeu Joseph Hubert Galani, Nyamangyoku Ishibwela Obedi, Vethaiya Balasubramanian, Elijah K. Lelmen, Jacqueline K. Makatiani, Sharon Regan, Sagar S. Datir, Prashant Kaushik, Anand Kumar



First published in London, United Kingdom, 2023 by IntechOpen
IntechOpen is the global imprint of INTECHOPEN LIMITED, registered in England and Wales, registration number: 11086078, 5 Princes Gate Court, London, SW7 2QJ, United Kingdom

British Library Cataloguing-in-Publication Data
A catalogue record for this book is available from the British Library

Additional hard and PDF copies can be obtained from orders@intechopen.com

Advances in Root Vegetables Research
Edited by Prashant Kaushik
p. cm.
Print ISBN 978-1-80356-701-3
Online ISBN 978-1-80356-702-0
eBook (PDF) ISBN 978-1-80356-703-7

For EU product safety concerns:
IN TECH d.o.o., Prolaz Marije Krucifikse Kozulić 3, 51000 Rijeka, Croatia,
info@intechopen.com or visit our website at intechopen.com.

# Meet the editor

Dr. Prashant Kaushik holds Adjunct Faculty positions at three universities and is also a genome data engineer in a private company. He holds a BS (Hons) in agriculture, a PG diploma in floriculture and landscaping, an MS in vegetable science, and an MBA in production and operations management. After obtaining a Ph.D. in biotechnology from Universitat Politecnica de Valencia, Spain, Dr. Kaushik was a bioinformatician at Nagano University in Ueda, Japan and worked at Yokohama Ueki in Kikugawa City, Japan. Dr. Kaushik has published 110 SCI or SCI-E Q1, Q2, and Q3 research articles, and an equivalent number of non-indexed research publications. He has written six books and 14 book chapters. Among his research interests are molecular biology, plant breeding, genomics, and population genetics.

# Contents

# Preface

Root vegetables may be found in a wide variety of plant species; they include potatoes, carrots, beet, and turnips. Root vegetables often act as storage organs extended to store energy in the form of carbohydrates. They vary in terms of the quantities of starches, sugars, and a wide variety of other types of carbohydrates that they contain. Starchy root vegetables are important staple foods, especially in tropical countries. They have supplanted cereals over a significant area of Central Africa, West Africa, and Oceania, where they are consumed either directly or mashed to form dishes like *fufu* or *poi*. Those that have an exceptionally high concentration of carbohydrates in the form of starch are particularly valuable from an economic standpoint.

When preserved in root cellars, the vast majority of root vegetables are able to keep their quality and freshness for a number of months beyond harvesting, which is especially important at latitudes that are not tropical, since winter is typically a season when there is very little or no harvesting. A variety of methods may be used to lengthen the growing season, enabling harvesting to continue far into the winter. The use of polytunnels is essential to the success of these methods.

I hope that this book will act as a handbook for students, researchers, and practitioners working in the field of root vegetables research, as well as stimulating future study ideas by proposing relevant research topics. The book is arranged in 13 chapters, each presenting an overview of background material on the topic within each chapter and recent developments.

**Dr. Prashant Kaushik**
Universitat Politecnica de Valencia,
Valencia, Spain

CULTA Inc.,
Tokyo, Japan

Section 1

# Composition of Root Vegetables

Chapter 1

# Principles of Biophysical and Biochemical Characterization of Root Vegetables' Bioactive Proteins

*Olalekan Onisuru and Oluwatayo Racheal Onisuru*


## Abstract

The characterization of root vegetables' bioactive proteins biophysically and biochemically becomes imperative as they play an incredibly important role in the discovery, development, and formulation of bioactive proteins as biopharmaceutical products. This is because bioactive proteins differ in terms of size, molecular weights, structures, and physicochemical properties. Biophysical and biochemical characterization employs several techniques ranging from simple to complex procedures to give insight into proteins' high-order structures, functions, and biochemical activities. Owing to the increasing awareness and acceptance of the use of peptides and proteins of root vegetable origin as treatment agents against some debilitatingly chronic diseases, researchers are now exploring an eco-innovative approach to reduce their loss by getting to structurally and functionally characterizing them. Several biophysical and biochemical tools are employed routinely for protein characterization and some of which are ultraviolet-visual (UV-Vis) spectroscopy, high-performance liquid chromatography (HPLC), circular dichroism (CD), intrinsic tryptophan fluorescence (ITF), differential scanning calorimetry (DSC), thermal shift assay (TSA), among others.


**Keywords:** bioactive proteins, biophysical, biochemical, characterization techniques, biomolecules

## 1. Introduction

Proteins or bioactive active substance-made drugs have become an integral class of therapeutics serving as auspicious alternatives to treat many diseases that have till now proven recalcitrant to treatment [1]. The nutritional benefits of root vegetables are no longer in doubt nowadays as they are known to be bioactive proteins making them one of the heartiest and healthiest foods around with therapeutic benefits. Hence, they essentially serve as an alternative protein source rich in phytochemicals such as polyphenols, carotenoids, etc. This among other reasons which are environmental and physiological has made an increasing number of people now include one root vegetable or the other in their staple food [2, 3]. The burden of necessity is been placed on the increasing demand for proteins, especially those from plants and in particular root vegetables. This is owing to the growing world's population which


IntechOpen

stands at around 6.5 billion and is expected to double by the year 2063. This increased demand for root vegetable protein is further corroborated statistically as two-thirds of the planet's dietary protein comes from vegetables [4]. There exist a broad range of biological activities exhibited by bioactive proteins, and these activities are responsible for their application and interest in foods, supplements, and medicine [5].

The tremendous attention bioactive proteins have gained over the years is not unconnected with their disease prevention and treatment ability, which is owing to their multi-target health benefits. To this end, it becomes imperative to know their distinctive functionalities by separating them from their natural matrices and carrying out their characterization. This separation brings about the unfolding of their functional groups, which interact with target tissues [6]. For scientists to have a deeper understanding of such (potential or already established) therapeutic candidates' biomolecular mechanisms and interactions, biophysical and biochemical characterization of specific quality and functions becomes imperative. This is as characterization provides information in respect of identity (structure), purity, potency, safety, and stability. Also, these characterizations give insight and understanding into some of the compelling parameters essential for the maintenance of protein's activity as well as the conformation of the higher-order structure (HOS). These HOS include the tertiary structure (3-dimensional structure), secondary structure (protein's folding), and quaternary structure (sub-unit association) [7]. The increasing significance of biophysical analysis in the characterization of therapeutics such as bioactive proteins stems from the fact that it enhances the investigation and characterization of such biomolecules using physical techniques [8]. A typical characterization technique employed or carried out by researchers looked at the various structural make-up, functionalities, and stability of a bioactive protein. This is because they have varying degrees of implications as they affect the bioactivity of such bioactive proteins.

These physical techniques employed circular dichroism (CD), Fourier-transmission infrared (FTIR), spectroscopy, differential scanning calorimetry (DSC), intrinsic and extrinsic fluorescence, and dynamic light scattering, among others to elucidate their spectroscopic, thermodynamic, and hydrodynamic parameters [9–11]. The significance of a biochemical assay or characterization stems from the fact that this approach helps detect, quantify, and or study biological molecules such as bioactive protein's binding or activity [12].

## 2. Root vegetables bioactive proteins

Food proteins, particularly root vegetable bioactive proteins, generally now have scientific bases to be regarded as having nutritional and physiological functionalities, regulated by some encrypted peptides in their native protein sequence [13]. Carrot (*Daucus carota* L,), onion (*Allium cepa* L,), and lettuce (*Lactuca sativa*), among others, are globally seen as belonging to the most common root vegetables which are known to have specialized. Although these compounds or molecules are secondary metabolites that do not contribute to the root vegetables' vital process, they are however beneficial to many living organisms on health grounds [2]. These molecules or compounds which herein are referred to as bioactive proteins will at one time or the other need to be purified, compounded, and stored by targeting the biophysical properties of these bioactive proteins [14]. Root vegetables' bioactive proteins vary in the bioactive compound from one root vegetable to the other with some conferring characteristic color, taste, etc. For example, while carrot's (*D. carota* L,) characteristic orange color is largely due to the β-carotene presence in it, onion's (*A. cepa* L,) antioxidant, anti-inflammatory, and antimicrobial properties on the other hand have been linked to the presence of biologically active phytochemicals such as flavonoids, etc. [15]. Potato (*Solanum tuberosum* L,) as a root vegetable is also rich in bioactive phytochemicals such as β-carotene or carotenoids, ascorbic acids, polyphenols, and natural phenols among others, which determine the color of the potato's skin and pulp [16].

## 3. Protein characterization: principles and tools

Understanding bioactive proteins' mechanism of action is an integral component needed for their development as essential pharmaceutical targets and ultimately leading to their use therapeutically by man [17]. In addition to this, and with the advent of several biophysical tools and biochemical characterization techniques, much vital information embedded in root vegetable bioactive proteins can now be explored. This development is essential for both preliminary research and the commencement of the drug discovery process employing root vegetables' bioactive proteins [17]. Owing to the increasing awareness and acceptance of the use of peptides and bioactive proteins of plant origin such as root vegetables' bioactive compounds as treatment agents for some debilitatingly chronic diseases, manufacturing sectors are now exploring an eco-innovative approach to reducing the loss of these bioactive proteins (**Table 1**) [18, 19].

Hence, biophysical tools come in handy to provide critical information in respect of the characterization and behavior of these root vegetable proteins. Generally, protein characterization is not only an incredible aspect of the manufacturing of biopharmaceutical products, but it also plays a significant role in the discovery and development of such pharmaceuticals into ready-to-use therapeutics. This is because proteins or bioactive compounds differ in respect of size, molecular structure, and physicochemical properties. Hence, characterizing proteins allows researchers to have information through the protein identification, profiling, and quantification of the protein's major and minor components. Also, because a typical bioactive protein must possess a unique three-dimensional structure for it to elicit its beneficial biological activities, and due to the possibility of substantial molecular structure changes, characterizing bioactive protein then becomes a necessity [5]. The various biophysical and biochemical characterization techniques available are broadly tabulated as seen in **Figure 1** below.

## 4. Biophysical characterization of root vegetables bioactive proteins

Biophysical techniques though eclectic is a veritable tool that gives insightful information in respect of biological molecules' electronic structure (size and shape), dynamics, polarity, as well as a mode of interaction [17]. Physical sciences techniques, principles, application, and study of biological systems have progressively been on the front burner of biological research for the determination of protein's structural and dynamic properties. This has undoubtedly led to expanding the understanding of their nature, mechanism, and functional roles [21]. Biophysical methods of characterization include several techniques that directly measure the structure, properties, dynamics, or function of biomolecules such as those of bioactive proteins from root

| Target | A biophysical or biochemical approach | Typical information that shows the acceptability |
|---|---|---|
| 1. Identity | • Amino acid analysis & sequencing<br>• LC-MS (liquid chromatography-mass spectrometry)<br>• Peptide mapping to identify post-translational modifications (PTMs) (eg phosphorylation) | • Exact, correct sequence identified<br>• Correct relative molecular mass (Mr) within instrument error<br>• Number & sites of phosphorylation; extent of phosphorylation |
| 2. Purity | • SDS-PAGE (sodium dodecyl sulfate-polyacryl-amide electrophoresis)/native PAGE<br>• Dynamic laser light scattering (DLS)<br>• Analytical gel filtration<br>• Analytical ultracentrifugation (AUC) | • Single band on a gel; still a single band at high loading<br>• Monodisperse, Mr. ± 20% expected<br>• Defined a single Gaussian peak for a monomer<br>• Indicates homogeneity & correct M |
| 3. Concentration | • Ultraviolet (UV) spectrum<br>• Bradford assay | • Peak at 280 nm; Peak at 205 nm; No peaks above ~340 nm; Test for light scattering (look into ratio at different wavelengths eg A280/A230); concentration calculated using ε<br>• Linearity with BSA standards |
| 4. Functionality | • Functional assay<br>• Isothermal calorimetry (ITC)<br>• Surface plasmon resonance (SPR)<br>• Functional comparison between protein batches<br>• Validity of construct | • Functional activity observed with expected parameters (eg $k_{cat}$, $K_m$, $k_{cat}/K_m$)<br>• With known tool ligand: n ± 15% of expected; $K_d$ within 2-fold of reference value; $\Delta H$ within 1 kcal/Mol<br>• Direct binding assay (DBA): $K_d$ within 2-fold of reference value; Expected theoretical $R_{max}$; Inhibition in solution assay (ISA): [Protein] within ±15% of two different concentration measures (Bradford & A280); competition observed between target definition compound (TDC) and TDC in solution<br>• Compare $K_d$, $\Delta H$, stoichiometry, $K_m$, $k_{cat}$, $k_{cat}/K_m$ (usually >$10^6$ $s^{-1}$ $M^{-1}$), $K_i$; Single phase kinetics<br>• Compare $K_d$, $K_m$, $K_i$, $\Delta H$ with full-length protein; compare structure-activity relationship (SAR) |
| 5. Stability | • Differential scanning calorimetry (DSC)<br>• Differential scanning fluorimetry (DSF)<br>• Selwyn's test | • Good pre-transition baseline; visible Tm (above 37°C); good post-transition baseline<br>• Good pre-transition baseline; visible Tm (above 37°C); good post-transition baseline<br>• Overlay of plots of [P] vs. [E].t for different combinations of [E] and t |

**Table 1.**
*Potential biochemical and biophysical approaches for protein quality control checks [20].*

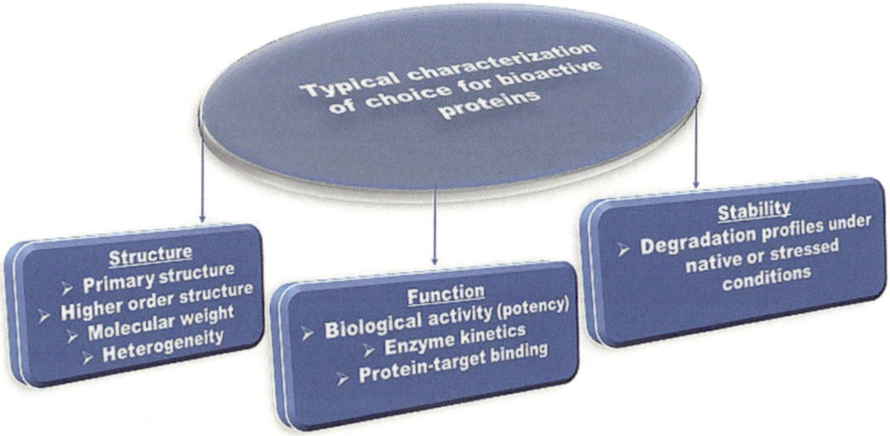


**Figure 1.**
*Checklist of a typical characterization of choice looked into a bioactive protein.*

vegetables [20]. Some available biophysical tools are suitably handy to be employed to assess root vegetable bioactives' protein information, and data, as well as interpret data regarding their structure, solubility, size, etc. [17, 22]. Ultraviolet-visible (UV-Vis) and fluorescence spectroscopy, dynamic light scattering (DLS), differential scanning calorimetry (DSC), intrinsic tryptophan fluorescence (ITF), thermal shift assay (TSA), and size exclusion chromatography (SEC) are examples of simple biophysical methods [14]. These biophysical characterizations among others look at the bioactive protein's higher-order structure such as secondary, tertiary, quaternary, and oligomeric structure, stability, and solubility.

Differential scanning calorimetry is perhaps the only technique of characterization that provide complete thermodynamic parameters of a substance such as a bioactive protein. This is as DSC, a technique that measures the thermal molecular stability and structure of a protein by quantifying the enthalpy ($\Delta H$), transition temperature ($T_m$), and changes in heat capacity ($\Delta Cp$), which are parameters protein's primary structure cannot reveal, are thus elucidated employing this technique [23]. Bioactive proteins like typical protein, interacts, lives, function, and die in a highly crowded environment. This protein interacts with other proteins or molecules called binding is essential for its biological functionalities. Hence, isothermal titration calorimetry (ITC) as a biophysical technique measures the energetic changes that occur as a result of binding between two proteins with heat either released or absorbed. This technique, therefore, estimates these binding affinity $K_A$ (which may either be favorable or unfavorable), enthalpy ($\Delta H$), entropy ($\Delta S$), and stoichiometry [24].

## 5. Biochemical characterization of root vegetables bioactive proteins

Root vegetables' bioactive protein biochemical characterization involves the estimation of the bioactive content and molecular weight determination using sodium dodecyl sulfate polyacrylamide gel electrophoresis (SDS-PAGE) techniques [25, 26], centrifugation, two-dimensional electrophoresis mass spectrometry, and circular dichroism [26–29]. Biochemical characterization involves a determination of the biochemical properties of a sample or biological molecule such as root vegetable bioactive proteins while investigating their enzymatic activities in terms of activation or inhibition [17, 22, 30]. These biochemical characterization techniques give insight into the biochemical functionalities of macromolecules such as root vegetable bioactive proteins. Root vegetables' bioactive protein can also be elucidated to give their structural information in terms of their precise molecular mass, and N-terminal sequence, among others [28]. There are a couple of experimental techniques that are utilized for this characterization and they include assays that allow for the detection, isolation, and purification of proteins [27, 31]. Effect of pH and temperature assay alongside enzymatic activity, solubility at physiological pH, etc., are some of the techniques employed in the course of biochemical characterization. Other biochemical characterizations that may also be carried out include carrying out the protein's enzymatic activities and terms of its activation or inhibition [30]. However, owing to the laborious nature of traditional biochemical characterization and despite the introduction of automation-enhanced next-generation sequencing (NGS) technology, biochemical characterization remains a low throughput technique. This then explains why researchers have switched to an alternative means of elucidating protein's biochemical component, and this involves the use of a computational approach [29, 32, 33]. All of these characterization techniques and those of biophysical characteristics are indispensable to the development of root vegetables' bioactive proteins as human therapeutics [22].

## 6. Conclusion

The imperativeness of characterizing bioactive protein from root vegetables has been scientifically established beyond the acknowledgement of their therapeutic benefit. With the continued advancement in the available biophysical and biochemical techniques, more elucidation and insight into the factors that contribute to or affect their functionalities, structural orientation, and stability will continue to be available for researchers, with the intent they can be further improved on and developed as biopharmaceuticals. The various characterization principles and techniques that researchers have employed so far are with the view of ascertaining the authenticity of the nutritional and physiological content cum characteristics of root vegetables' bioactive protein, and this cannot be complete without biophysical and biochemical characterization. This is so as protein's higher order structure which can either be affected by environmental changes in a particular way, or when it interacts with other molecules such as ligands, denaturant, glycosylation, and oxidation states, remains a critical parameter in the development of bioactive protein into therapeutics.

## Conflict of interest

The authors declare that they have no financial or relationship conflicting interests.

## Author details

Olalekan Onisuru[1]* and Oluwatayo Racheal Onisuru[2]

1 Protein Structure and Function Research Unit, School of Molecular and Cell Biology, University of Witwatersrand, South Africa

2 Chemical Sciences Department, University of Johannesburg, South Africa

*Address all correspondence to: 2430435@students.wits.ac.za

## References

[1] Bolje A, Gobec S. Analytical techniques for structural characterization of proteins in solid pharmaceutical forms: An overview. Pharmaceutics. 2021;**13**(4):534

[2] Soto VC, González RE, Galmarini CR. Bioactive compounds in vegetables, is there consistency in the published information? A systematic review. The Journal of Horticultural Science and Biotechnology. 2021;**96**(5):570-587. DOI: 10.1080/14620316.2021.1899061

[3] Multari S, Neacsu M, Scobbie L, Cantlay L, Duncan G, Vaughan N, et al. Nutritional and phytochemical content of high-protein crops. Journal of Agricultural and Food Chemistry. 2016;**64**(41):7800-7811

[4] Ahmed N, Ali A, Riaz S, Ahmad A, Aqib M. Vegetable proteins: Nutritional value, sustainability, and future perspectives. In: Vegetable Crops-Health Benefits and Cultivation. London, United Kingdom: IntechOpen; 2021. pp. 43-56

[5] McClements DJ. Encapsulation, protection, and delivery of bioactive proteins and peptides using nanoparticle and microparticle systems: A review. Advances in Colloid and Interface Science. 2018;**253**:1-22

[6] Okoye CO, Ezeorba TPC, Okeke ES, Okagu IU. Recent findings on the isolation, identification and quantification of bioactive peptides. Applied Food Research. 2022:100065

[7] Rodríguez-Mendieta I. Biophysical Analysis. Bioprocess International. 2015;**13**:8

[8] Chaudhuri R, Cheng Y, Middaugh CR, Volkin DB. High-throughput biophysical analysis of protein therapeutics to examine interrelationships between aggregate formation and conformational stability. The AAPS Journal. 2014;**16**(1): 48-64

[9] Kelly SM, Jess TJ, Price NC. How to study proteins by circular dichroism. Biochim Biophys Acta (BBA)-Proteins Proteomics. 2005;**1751**(2):119-139

[10] Kong J, Yu S. Fourier transform infrared spectroscopic analysis of protein secondary structures. Acta Biochimica et Biophysica Sinica Shanghai. 2007;**39**(8):549-559

[11] Themistou E, Singh I, Shang C, Balu-Iyer SV, Alexandridis P, Neelamegham S. Application of fluorescence spectroscopy to quantify shear-induced protein conformation change. Biophysical Journal. 2009;**97**(9):2567-2576

[12] Young VR, Pellett PL. Plant proteins in relation to human protein and amino acid nutrition. The American Journal of Clinical Nutrition. 1994;**59**(5):1203S-1212S

[13] Chakrabarti S, Guha S, Majumder K. Food-derived bioactive peptides in human health: Challenges and opportunities. Nutrients. 2018;**10**(11):1-17

[14] Houde DJ, Berkowitz SA. Biophysical characterization and its role in the biopharmaceutical industry. In: Biophysical Characterization of Proteins in Developing Biopharmaceuticals. Elsevier B.V.; 2015. pp. 23-47. DOI: 10.1016/B978-0-444-59573-7.00002-6

[15] De Ancos B, Colina-Coca C, González-Peña D, Sánchez-Moreno C.

Bioactive compounds from diverse plant, microbial, and marine sources. In: Biotechnology of Bioactive Compounds: Sources and Applications. No. 1. Hoboken, New Jersey: Wiley-Blackwell; 2015. pp. 3-36

[16] Stojkovic D, Smiljkovic M, Kostic M, Sokovic M. Root vegetables as a source of biologically active agents-lesson from soil. In: Barros L, editor. Phytochem Veg A Valuab source Bioact Compd Petropoulos. SA, Ferreira: ICFR; 2018. pp. 1-39

[17] Kwan TOC, Reis R, Siligardi G, Hussain R, Cheruvara H, Moraes I. Selection of biophysical methods for characterisation of membrane proteins. International Journal of Molecular Sciences. 2019;**20**(10):1-26

[18] Ruiz JCR, Ancona DAB, Campos MRS. Proteínas y peptidos de origen vegetal en la reduccion de lipidos; potencial nutracéutico. Nutrición Hospitalaria. 2014;**29**(4):776-784

[19] Usman I, Hussain M, Imran A, Afzaal M, Saeed F, Javed M, et al. Traditional and innovative approaches for the extraction of bioactive compounds. International Journal of Food Properties. 2022;**25**(1):1215-1233. DOI: 10.1080/10942912.2022.2074030

[20] Holdgate G, Embrey K, Milbradt A, Davies G. Biophysical methods in early drug discovery. Admet DMPK. 2019;**7**(4):222-241

[21] Dobson CM. Biophysical techniques in structural biology. Annual Review of Biochemistry. 2019;**88**:25-33

[22] Arakawa T, Philo JS. Biophysical and biochemical analysis of recombinant proteins. In: Pharmaceutical Biotechnology. USA: CRC Press, Informa Healthcare, Inc.; 2016. pp. 39-64

[23] Patel. Guide To Biophysical Methods of Macromolecular Characterization. 2019. pp. 9-25

[24] Velazquez-Campoy A, Leavitt SA, Freire E. Characterization of protein-protein interactions by isothermal titration calorimetry. In: Protein-Protein Interactions. New Jersey: Springer, Humana Press Inc; 2015. pp. 183-204

[25] Raghavendra R, Neelagund SE. Biochemical characterization of novel bioactive protein from silkworm (Bombyx mori L) fecal matter. Applied Biochemistry and Biotechnology. 2012;**167**(5):1002-1014

[26] Kaur H, Bhagwat SR, Sharma TK, Kumar A. Analytical techniques for characterization of biological molecules - Proteins and aptamers/oligonucleotides. Bioanalysis. 2019;**11**(2):103-117

[27] Brent MM, Iwata A, Carten J, Zhao K, Marmorstein R. Structure and biochemical characterization of protein acetyltransferase from Sulfolobus solfataricus. The Journal of Biological Chemistry. 2009;**284**(29):19412-19419

[28] Tubbs CE. The Biochemical Characterization of Protein DE and its Interaction with Rat Epididymal Sperm. USA: North Carolina State University ProQuest Dissertations Publishing; 2000

[29] Hon J, Borko S, Stourac J, Prokop Z, Zendulka J, Bednar D, et al. EnzymeMiner: Automated mining of soluble enzymes with diverse structures, catalytic properties and stabilities. Nucleic Acids Research. 2020;**48**(W1):W104-W109

[30] Lindenthal C, Klinkert M-Q. Identification and biochemical characterisation of a protein phosphatase 5 homologue from Plasmodium

falciparum. Molecular and Biochemical Parasitology. 2002;**120**(2):257-268

[31] Boobathy S, Soundarapandian P, Prithivraj M, Gunasundari V. Biochemical characterization of protein isolated from seaweed, Gracilaria edulis. Current Research Journal of Biological Sciences. 2010;**2**(1):35

[32] Perbandt M, Eberle R, Fischer-Riepe L, Cang H, Liebau E, Betzel C. High resolution structures of Plasmodium falciparum GST complexes provide novel insights into the dimer-tetramer transition and a novel ligand-binding site. Journal of Structural Biology. 2015;**191**(3):365-375

[33] Beneyton T, Thomas S, Griffiths AD, Nicaud J-M, Drevelle A, Rossignol T. Droplet-based microfluidic high-throughput screening of heterologous enzymes secreted by the yeast Yarrowia lipolytica. Microbial Cell Factories. 2017;**16**(1):1-14

Chapter 2

# Root Vegetables: Biology, Nutritional Value and Health Implications

*Mirela Ahmadi, Ștefan A. Hulea and Ioan Peț*


## Abstract

Plants served as main staple for humanity since time immemorial. Plant roots science is a fascinating domain that offers a window to the complex world of plants-microorganisms relationship. Plant roots were used throughout human history both as a food source particularly in times of food scarcity as well as for medicinal purposes aid in the treatment of various human disorders. Root vegetables are excellent sources of fiber and antioxidants and are low in calories and lipids—being indispensable in human diet. There is an increasing interest in the biochemical processes occurring in the rhizosphere between root tissues and the bacterial/fungal colonizers especially in soils where there is a deficiency in minerals such as iron, phosphorus and selenium or there is higher load of toxic metals such as aluminum, cadmium, nickel and lead. That interest stems from the need to improve crop yields in hostile environmental conditions such as drought and low nutrient availability in soils. In this chapter, we will focus on the typical edible plant roots as well as bulbs (are not proper roots) looking at their nutrient content as well as their use as health enhancers.


**Keywords:** edible roots vegetables, health enhances

## 1. Introduction

As roots grow and search for nutrients they evolve in a very complex environment called rhizosphere. This is defined as the area around plant root that is populated by a variety of different microorganism species, which "cooperate" with the plant for the benefit of both. However, not all bacteria in rhizosphere are beneficial and plants have developed mechanisms to protect themselves against harmful bacteria. It has been estimated that there are over 10,000 bacterial species in the rhizosphere, not all with "good intentions" toward the plant.

The rhizosphere comprises two main compartments: ecto-rhizosphere and endo-rhizosphere. The former is the outermost zone that extends from the rhizoplane out into the bulk soil. The latter includes parts of the cortex and endodermis between which bacteria find a "home." As McNear wrote in 2013: "the rhizosphere is not a region of definable size and shape but instead, consists of a gradient in chemical, biological and physical properties, which change both radially and longitudinally along the root" [1].

Roots are in constant "touch" with their surroundings seeking water and nutrients and also shedding root cap and border cells, mucilage and exudates. The latter comprises part of the carbon fixed via photosynthesis, namely inorganic carbon, i.e. $HCO_3^-$ and organic carbon, such as organic acids and polyphenols. The exchange of material is influenced by the plant species, climate, presence of insects that feed on plants, nutrient availability, soil moisture and its physicochemical properties. Out of all organic compounds released from roots the low molecular weight compounds are the most studied because they serve as nutrients for the bacteria in the rhizosphere. The organic compounds also serve as chemo-attractants for the soil microbial population. For example, the exudates of leguminous plant roots attract rhizobium bacteria such as *Rhizobium leguminosarum*. This bacterium colonizes the root and helps the plant by converting atmospheric nitrogen into $NH_4^+$ that is further used in amino acid synthesis [2]. The enzyme complex involved is called nitrogenase and it catalyzes the reaction:

$$\mathrm{N_2} + 8\,\mathrm{e^-} + 8\,\mathrm{H^+} + 16\,ATP\,2\,NH_3 + \mathrm{H_2} + 16\,ADP + 16\,\mathrm{P_i}$$

The nitrogenase complex consists of two enzymes: dinitrogenase reductase (a dimeric Fe-protein) and dinitrogenase (a tetrameric FeMo-protein). The nitrogenase is rapidly inactivated by atmospheric oxygen. That is why the root nodules provide for a low oxygen environment, so that the enzyme is kept active.

Leguminous plants, which provide the largest simple source of vegetable protein in human diet and livestock feed have evolved signaling systems when under nitrogen deprivation. Legumes possess specific flavonoids that under nitrogen scarcity are released near the root tips, close to the emerging root hair zone that is the site of infection by rhizobium bacteria.

Plant flavonoids are secondary metabolites derived from the phenylpropanoid pathway and include chalcones, flavonols, flavones, anthocianins among others [3]. Flavonoids accumulate in the dividing cells of roots and some of them act as chemo-attractants for the rhizobium bacteria. The rhizobial signaling molecules are called nodulation factors and include lipo-chitooligosaccharides having a N-acetylglucosamine backbone, N-acetylated on the terminal non-reducing sugar. The substitutions on the oligosaccharide moiety determine the specificity of the symbiosis. Some plant flavonoids such as luteolin-7-O-glucoside and quercetin-3-O-galactoside can act as growth regulators of rhizobium bacteria.

One of the bacterial phylum present in the rhizosphere of legumes is *Firmicutes*. These are beneficial bacterial that colonize human gastrointestinal (GI) tract and as such they produce butyric acid that lowers gut pH and limits the growth of harmful bacteria. Many microbes in the rhizosphere, including *Firmicutes* have developed mechanisms for physically interacting with plant roots and through complex processes can reach other parts of the plant, e.g. stem and leaves. The ingested part plants housing colonizing bacteria help these microbes settle in the colon and in so doing help keep in check pathogens.

The symbiosis between nitrogen-fixing bacteria and leguminous plants is one way by which plants cope with limited availability of nitrogen in the soil. Besides this root exudates promote nutrient acquisition by changing the pH within the rhizosphere or chelating ions in soil solution. The root exudates contain organic acids such as malic and citric acids that acidify the soil and solubilize phosphate bound in soil minerals. Moreover, in case of chemical fertilizers plants respond differently depending on the chemical form of nitrogen in the soil. An excess of ammonium ion ($NH_4^+$) leads to a more alkaline environment whereas an excess of nitrate results in a lower pH in the rhizosphere. The pH fluctuations influences the availability of minerals such as zinc, calcium and magnesium. In addition, plant-bacteria cooperation can broaden immune functions of the plant host [4]. Accumulating evidence suggests that the chemical composition of root exudates is of paramount importance in selecting beneficial bacteria, which in turn leads to healthier and more productive plants [5].

Iron (Fe) is an essential mineral for plant growth and development. It is well known that in alkaline soils (representing some 30% of the world's arable land) plants do not grow well because at higher pH, Fe is trapped in Fe oxides ($Fe_2O_3$). So plants have developed strategies for getting hold of iron. Thus, the root exudates contain a mixture of organic acids and phenols that reduce the pH in the rhizosphere, hence allowing for the reduction of Fe(III) to Fe(II), which is then taken up by the root epidermal cells (strategy I). Another strategy for Fe uptake is based on the solubilization of $Fe_2O_3$ by strong Fe(III)-chelating agents called phytosiderophores. They belong to the mugineic acid family and are released into rhizosphere by efflux transporters. The mechanisms of Fe uptake by plant roots have been extensively studied in the weed *Arabidopsis thaliana*. Its habitat includes side roads, railway tracks and disturbed habitats. It has been shown that secretion of phenolic compounds by this plant is critical for Fe acquisition from soils with low Fe availability. In an elegant study on the mechanism of Fe mobilization by *A. thaliana*, Schmidt and colleagues demonstrated that polyphenols such as coumarins are essential for Fe uptake by the plant [6]. Coumarins act both as reductants of Fe(III) and as ligands of Fe(II).

Two other minerals are in the attention of plant scientists, namely inorganic phosphorus ($P_i$) and aluminum (Al). In acidic soils (that occupy a sizable portion of arable lands worldwide) low $P_i$ availability and high Al toxicity limit plant growth and productivity. Work on *A. thaliana* revealed that organic acid, phytohormones and Fe homeostasis are critical factors in plant's response to nutritional stressors such as low $P_i$ and Al toxicity [7]. In acidic soils with high concentrations of Fe, Al, Mn, $P_i$ is easily fixed in the form of insoluble salts. Moreover, when pH is below 5.5, Al becomes soluble and toxic to plant roots, impairing the absorption of water and nutrients. Recent studies suggested that Al-tolerant phosphobacteria isolated from ryegrass could assist plants to deal with $P_i$ shortage and Al toxicity [8]. An active area of research deals with ways to activate genes involved in changing root system architecture (RSA) in conditions dictated by limited nutrient availability and metal toxicity. Changing RSA involves the inhibition of primary root growth and promoting the growth of lateral roots and hair.

## 2. Nutritional value and health benefits of edible root vegetables

Wild plant roots have been eaten by humans since ancient times, especially during periods of food scarcity or famine. With the advent of agriculture in settled communities the roots of cultivated plants became permanent fixtures on the panoply of human diet. In this chapter we will focus on cultivated edible plants, whose roots are routinely used as foods and consumed either raw or cooked. Besides being nutritious due to their macro- and micronutrients content, they also contain numerous phytochemicals that are increasingly sought after by the food and pharma-/nutraceuticals industries both as food quality enhancers and promoters of health and disease prevention, respectively.

Roots can be broadly classified in:

- edible taproots – consist of a main thick root from which other thin secondary roots grow laterally (ex.: carrots, radishes);
- edible tuberous roots – consist of lateral thick roots that serve main as nutrient stores (ex.: sweet and regular potatoes).

### 2.1 Beetroot

Nutritionally, beetroot is a food source rich in proteins, carbohydrates, amino acids, phytosterols, vitamins and minerals, fibers, as well as nitrates. It also contains many phytochemicals such as polyphenols, flavonoids, betalains: betacyanins and betaxanthins [9].

Betalains are water-soluble nitrogen-containing pigments exhibiting red-violet and yellow-orange colors. Due to glycosylation and acylation of the hydroxyl groups in the molecule betalains have a great structural diversity. Betanin (betanidin-5-O-β-glucoside) is the most represented betacyaninin plants. It is also one of the few natural compounds that were approved for use as colorant in the food industry, cosmetics and pharmaceuticals (trade name: E162). Betanin is a strong reactive oxygen species (ROS) scavenger and exhibits gene-regulatory activity via Nfr2 (nuclear factor erythroid-derived 2)-like-dependent signaling pathway that triggers the induction of phase II enzymes synthesis and antioxidant defense mechanisms. It has been suggested that betanin may also prevent LDL oxidation and DNA damage [10].

The type of beetroot processing has a considerable influence on the antioxidant power displayed by this vegetable. Thus, it was found that fresh, dried and pureed beetroot exhibited the highest antioxidant power, as expressed by the total phenolic content. Moreover, the liquid nitrogen method of beetroot processing resulted in the highest bioavailability of biologically active compounds. Beetroot active compounds have shown antitumor activity in vitro cell culture and animal model experiments.

Among all vegetables beetroot has the highest amount of nitrate (2.8 g/100 g wet weight). Nitrate as such has no biological effects but its metabolization product nitric oxide ($NO^{\bullet}$) has. Nitrate is absorbed in the upper part of duodenum but some 25% ends up in entero-salivary cycle where bacteria in the mouth convert it to nitrite. This nitrite is further reduced in the GI tract by several reductases to nitric oxide. Nitric oxide is a vasodilator (relaxes the smooth muscle cell in the vasculature) causing the vessels to widen hence prevent an increase in blood pressure. A decrease in nitric oxide supply leads to endothelial dysfunction, which is the primary risk for cardiovascular diseases (CVD). Clinical studies on healthy subjects demonstrated that beetroot juice intake was protective against endothelial dysfunction induced by an acute ischemic insult caused by brachial artery occlusion [11]. Beetroot juice supplementation significantly reduced systolic and diastolic blood pressure. Accumulating evidence suggest that several conditions such as hypertension, atherosclerosis, T2D and inflammation (chronic or acute) benefit from beetroot consumption. Betalains appear to interfere with the proinflammatory signaling cascade in which NF-κB plays a critical role by activating the transcription genes that regulate and amplify the inflammatory response.

Animal model experiments indicated that beetroot supplementation had a protective effect against drug-induced liver and kidney injury. The mechanisms likely involved are the anti-inflammatory, antioxidant and anti-apoptotic activities. In humans, beetroot supplementation was shown to improve hemoglobin status in adolescent anemic girls [12]. In another study involving healthy subjects it was shown that daily consumption of a 10% beetroot juice beverage resulted in a 34% decrease in plasma glucose level after 4 weeks of supplementation suggesting an improved glucose metabolism [13].

### 2.2 Carrots

Carrots are a food source rich in micronutrients, phytochemicals and fiber. The main macronutrients are represented by carbohydrates (6.6–7.7 g/100 g), protein (0.8–1.1 g/100 g) and lipids (0.2–0.5 g/100 g). Carrots contain several B group vitamins (thiamine, niacin, folic acid, in sub milligram range), vitamin C (21-775 mg/100 g between cultivars) and minerals Na, K, Ca, Mg, Cu, Zn in milligram range). The fiber is represented by insoluble fiber (cellulose, hemicellulose and lignin) whereas the soluble fiber consists of pectin, gums and mucilage [14]. Carrots are a significant source of phenolic compounds, carotenoids and polyacetylenes [15].

Phenolic compounds are widely present in the plant kingdom and include phenolic acids, flavonoids, tanins, lignans, curcuminoids and stilbenoids. The concentration of phenolic compounds increases in the direction xylem toward peel (periderm) as they are released through the exudate into the surrounding medium of the root. Phenolic compounds play an important role in the acquisition of metal ions and the facilitation of microbes – root interactions. There is a large body of evidence that phenolics exert a host of health benefits including antioxidant, anti-inflammatory and antiproliferative properties. In so doing they decrease the risk of cardiovascular diseases, diabetes, inflammatory conditions and slow down the aging process. Anthocyanins from black carrot root were found to possess anti-proliferative activity in cell culture and animal model experiments. Unfortunately, epidemiological studies in humans failed to demonstrate clear cut benefits in cancer patients [16]. Most clinical studies however, had a short duration, insufficient to draw a definite conclusion on the subjects. It is worth mentioning here that in 2014 in U.S. some 40% of all cancer cases were attributable to risk factors such as smoking, alcohol consumption, bad diet, low physical activity and the rest of 60% were caused by DNA replicative errors that led to gene mutations [17]. It is therefore of paramount importance to pay attention to a diet rich in phytochemicals such as fresh fruits and vegetables in order to reduce the risk of cancer.

Isoprenoid precursors through a series of reactions yield lycopene, which is further processed to yield α- and β-carotene. α-carotene is converted to lutein whereas β-carotene is turned into zeaxanthin. In humans, conversion of β-carotene into vitamins A occurs mainly in the gut and liver and much less in other tissues. The efficiency factor for the conversion of dietary β-carotene to vitamins A is 12:1 by weight. Epidemiological data indicate that diets rich in carotenoid-containing foods are associated with a reduced risk of developing chronic diseases such as CVD, diabetes, cancer and age-related macular degeneration [18, 19]. Retinol in vitamins A has been shown to play a central role in these processes. The major factors affecting bioavailability of carotenoids and their conversion to vitamin A are food matrices, food preparation and the fat content of meals.

The third group of phytochemicals in carrots are polyacetylenes. They are non-volatile compounds comprising at least two conjugated triple C-C bonds [15]. There is evidence to suggest that these compounds have the potential to improve human health due to their antifungal, antibacterial and anti-inflammatory properties.

### 2.3 Celery

Celery plant parts (leaves, stalk, and root) have been used since Antiquity for medicinal purposes in the treatment of conditions such as joint pain, gout, and fever cause by bacterial infection, constipation, heartburn, etc. Celery is rich in vitamins (A, C, D, E, K, B group), minerals (K, Mg, Ca, Zn, Fe, Cu, Se) and phytochemicals (carotenes, phenolic acids (ferulic acid, caffeic acid, chlorogenic acid, *p*-hydroxy-benzoic acid), flavonoids (apigenin, luteolin, kaempferol), anthocyanins fiber and volatile oils (in seeds). Mannitol, glucose and fructose are the major monosaccharides in celery. In a study on the nutrient, biomass, minerals and antioxidant distribution in different plant parts between six celery genotypes belonging to leafy, stalk and root types it was found that all celery types exhibited the highest level of antioxidants in leaves [20]. Polyphenols and flavonoids content in roots was slightly lower than that in leaves and stalks. Minerals show a selective distribution among celery parts. For instance, Se, Cu, Zn and K were found in slightly lower amount in roots than in leaves. Both celery stalks and root are consumed either raw or cooked. They share the same type of nutrients in slightly different proportions. The nutritional value is the same for stalks and root.

Celery seeds and root extracts exhibited anti-proliferative and pro-apoptotic activity against several human cancer cell lines. These extracts also reduced the ability of dendritic cells to proliferate during lipopolysaccharide stimulation. As a result, there was no decrease in the pro-inflammatory TNF-α and IL-6 levels but a reduced production of the anti-inflammatory IL-10. Interestingly, it has been recently shown that during severe cases of COVID-19 infections there was a spike in the production of anti-inflammatory IL-10 but for some unknown reason IL-10 failed to suppress COVID-associated cytokine storm that causes increased inflammation. The IL-10 level in COVID patients has been linked to disease severity and prognosis [21]. Regular consumption of celery has been consistently shown that it leads to decreased inflammation, oxidative stress and reduced risk of developing hypertension and coronary heart disease. Apigenin in celery was shown in animal model experiments to help improve liver function by increasing the antioxidant power and hepatoprotective activity [22].

### 2.4 Ginger

Ginger has been used for a long time as a spice rather than a food staple. It is consumed raw or pickled. It is rich in polyphenols (gingerols, shogaols, paradols), flavonoids and several terpenoids, which are the main constituents of ginger essential oils. Ginger roots also contain polysaccharides, lipids, organic acids and fiber. Due to its high content in polyphenols ginger possesses a strong antioxidant activity. Dried ginger appears to have the highest antioxidant power. Cell culture experiments revealed that ginger extracts protected against oxidative stress as it stimulated the expression of antioxidant enzymes that reduce ROS generation and lipid peroxidation. The antioxidant activity was mediated through the Nrf2 signaling pathway [23].

Ginger extracts were shown, in animal model experiments to alleviate the severity of inflammatory bowel disease. The polyphenols in these extracts inhibit the inflammation signaling pathways represented by the NF-κB and MAPK pathways.

Cell culture and mice model experiments using ginger-derived nanoparticles have shown that this novel therapeutic approach could be a promising way for the treatment/prevention of inflammatory bowel diseases [24, 25].

Ginger extracts were also shown to be cytotoxic to breast, cervical, colorectal and prostate cancer cells. The 6-gingerol from ginger may exert its effect on cancer cells via the inhibition of proliferation and the induction of apoptosis in these cells. Apoptosis is induced by the decreased expression of genes involved in the Ras/ERK and PI3K/Akt signaling pathways. Interestingly, a natural ginger extract exhibited a 2.4-fold higher inhibition of tumor growth than a mixture of 6-shogaol, 6-gingerol, 8-gingerol and 10-gingerol [26].

In a cross-sectional observational study involving 4628 participants, age 18 to77, it has been found that daily ginger consumption was associated with a decreased risk of hypertension and atherosclerosis [27].

Animal model experiments indicated that administration of ginger extracts to rats fed a high-fat diet resulted in improved plasma lipid profiles and increased plasma HDL-C level, thus reducing the risk of atherosclerosis. Ginger treatment decreased the activity of angiotensin-1 converting enzyme and increased $NO^{\bullet}$ in hypertensive rats. Mice fed a high-fat diet treated with 6-paradol from ginger exhibited a significantly lower blood glucose level. 6-gingerol treatment of diabetic rats improved glucose tolerance by increasing glucagon-like peptide-1 expression and increased the transport of GLUT4 to cell membrane. In a clinical observational study ginger intake led to reduced levels of fasting plasma glucose, HbA1c, insulin, triglycerides in T2D patients [28]. In a RCT on gestational diabetes mellitus women it was found that ginger tablet supplementation for 6 weeks resulted in reduced fasting blood glucose, fasting insulin and HOMA index. However, there was no change in the 2 hours post-prandial blood glucose level [29]. Ginger intake, in line with a long tradition to treat respiratory disorders was shown to possess bronchodilating activity and anti-hyperactivity. This effect was due probably to the relaxation of smooth muscle cells of the airways as animal model experiments demonstrated. Ginger phytochemicals could also improve symptoms of allergic asthma by reducing allergic airway inflammation [30].

A recent systematic review of over 100 randomized controlled clinical trials on the effects of ginger consumption on a host of human disorders found that conditions such as inflammation, metabolic syndrome, irritable bowel disease, some cancers and pain relief for arthritis, chemotherapy-induced nausea and vomiting showed clear health benefits while others such as diabetes, cardiovascular disease and neurological disorders were not significantly helped by ginger intake [31]. In the realm of pain relief it was reported that ginger intake was helpful in alleviating primary dysmenorrhea pain and was as effective as medications such as ibuprofen and mefenamic acid. Ginger's mode of action involves the suppression of cyclooxygenase and lipoxygenase responsible for the production of pro-inflammatory prostaglandins and leukotrienes, respectively.

Like in all RCT (randomized clinical trials) to date, the wide spectrum of trial designs, number of participants, dosage of active compounds administered, standardization of methodology, etc. makes it difficult to fully assess the effectiveness of natural products from plants in the treatment/prevention of human diseases. Better designed trials will be able to address the shortcomings encountered so far.

### 2.5 Curcuma longa (turmeric)

Turmeric has been long used by the folk medicine as adjuvant for liver obstruction and jaundice, ulcers and inflammation as well as a host of other ailments such as cold, digestive problems, skin infections, wound healing, asthma, tumors and others. Turmeric was also used as a spice, food preservative and coloring agent. The plant part most important for health is the tuberous rhizome from which turmeric is formed. The main bioactive compound in turmeric is curcumin, also known as diferuloylmethane [32]. Chemically, it is a diarylheptanoid, which is a phenolic pigment responsible for the yellow color of turmeric. Besides the macronutrients (proteins, lipids and carbohydrates), turmeric rhizome contains minerals (Mn, Ca, Mg, Cu, Fe, Zn, P, Na, K), vitamins (B, A, E, K, C), polyphenols, terpenoids, alkaloids, fiber and resins. Monoterpenes are predominant in the essential oils of flowers and leaves whereas sesquiterpenes are predominant in the oils of roots and rhizome. Curcumin makes up 0.3–5.4% of raw turmeric and is the most investigated compound from this plant to date.

Animal model experiments and human clinical studies showed that curcumin may be an important adjuvant therapy in conditions such as gastrointestinal and respiratory disorders, inflammatory disorders, diabetes, CVD and cancer (colorectal, pancreatic and lung cancer). In T2D patients curcumin improved insulin sensitivity, enhanced adiponectin secretion and lowered leptin, resistin, IL-6, IL-1β and TNF-α levels. A meta-analysis of RCT found that curcumin supplementation led to lower blood lipid profiles in CVD patients [33]. One type of cancer that currently has a poor prognosis and survival rate is glioblastoma (GBM). Several cellular signaling pathways such as p53, MAPK, PI3K/Akt, JAK/STAT and NF-κB were found to be dysregulated in GBM.

Curcumin appears to modulate these pathways as in vitro and in vivo experiments suggested. Unfortunately, there was only one clinical trial on the possible anti-tumor effect of curcumin in GBM patients. Using the micellar curcumin formulation it was found that the intra-tumoral curcumin concentration was too low to cause a short-term anti-tumor effect. This was likely due to the small dose of curcumin administered to patients [34]. Due to its chemical structure curcumin has low solubility at neutral and acidic pH, hence reduced bioavailability. In the duodenum curcumin undergoes rapid metabolization via the formation of glucuronides and sulfates that are excreted. Free curcumin was not detected in the serum of GBM patients. It is important therefore to increase the solubility and absorption of this bioactive compound. To that effect efforts are underway to encapsulate curcumin in micelles, nanoparticles, liposomes to make sure it is delivered to the target tissue. There is evidence that curcumin-loaded nanoparticles could suppress the viability, proliferation and migration of glioma stem cells through the induction of cell cycle arrest and apoptosis [34].

In an effort to improve the efficacy of conventional drugs for the treatment of amyotrophic lateral sclerosis (ALS) curcumin nanoparticles were added to standard therapy for this condition. It was found that curcumin was safe and well tolerated [35]. In another RCT curcumin reduced oxidative stress, improved aerobic metabolism and slowed down the progression of the disease [36].

Since the antiviral activity of curcumin is well documented it has been recently speculated that curcumin might be used as adjuvant therapy in the treatment of SARS-CoV-2 infections [37].

### 2.6 Horseradish

Horseradish (*Armoracia rusticana* L.) is a perennial crop of the Brassicaceae family, indigenous to Eastern Europe and Mediterranean. Upon cutting or shredding the root releases a strong pungent, burning and lachrymatory odor. Horseradish roots (HR) have been used since ancient times as a culinary spice recognized for its nutritional value. Its other uses include food preservative, antiseptic and fermenter. Its folk medicinal use includes the treatment of gout, kidney stones, asthma, bladder infections as well as a possible remedy against rheumatic pain, headache and high blood pressure. The root contains small amounts of macronutrients, fiber, vitamins (C, folate), minerals (Mg, Ca, K, Zn, Mn, Se). It is rich in sulfur-containing glycosides known as glucosinolates [37]. Upon grating glucosinolates (GLS) are hydrolyzed by the enzyme myrosinase to volatile oils such as isothiocyanates, nitriles and thiocyanates. Isothiocyanate sulforaphane is the predominant volatile oil in broccoli. The sharp taste and odor of HR is mainly due to sinigrin-derived allyl isothiocyanate (AITC). AITC is a well-known safe and non-toxic food-borne antimicrobial agent. HR essential oils were shown to possess potent antifungal activity. Thus, essential oils in the vapor phase could control a honey bee larvae disease caused by the fungus *Ascosphaera apis* [38].

There is evidence to suggest that HR oils have potential anti-cancer effects against lung, colorectal, ovarian, oral, prostate cancer as well as glioblastoma. Recent cell culture experiments using cisplatin-resistant oral cancer cells demonstrated that AITC can inhibit Akt/mTOR proliferation signaling pathway and promote mitochondria-dependent apoptotic pathway via AITC-enhanced activity of caspase-3 and caspase-9 in these cells [39].

A study back in 2007 using a HR preparation on patients with acute sinusitis, acute bronchitis and acute urinary tract infection showed that HR was just as effective as the standard treatment with antibiotics and displayed a significantly lower potential for adverse events [40]. In another cell culture experiment, *E. coli* LPS-treated murine macrophages were incubated with HR extracts. It was found that there was a significant reduction in the inflammatory markers TNF-α and IL-6 through the inhibition of NF-κB and p65 activation [41]. HR extracts also reduced ROS production and increased heme oxygenase-1 expression, which meant an enhanced cellular protection during inflammation.

Despite promising results from in vitro and in vivo studies of the health benefits of HR oils there are no RCT to date on the use of HR extracts in clinical settings.

### 2.7 Radish

Like in the case of horseradish, radish roots have been used for centuries for the treatment of conditions such as stomach pain, constipation, fever, urinary tract infections, liver inflammation, ulcers and cardiac disorders. More recently, in vitro and animal model experiments reported antibacterial, antioxidant and anxiety lowering effects [42].

Radish contains carbohydrates, protein, fiber, vitamins (B group and C) and minerals (Ca, Mg, Fe, Zn, Mn, K, P). Besides the macro-and micronutrient arsenal radish contains secondary metabolites such as polyphenols, isothiocyantes (sulforaphane, sulforaphene, indole-3-carbinol) and glucosinolates (GSL) similar in composition with those in HR. GSL are found exclusively in cruciferous vegetables. They can be classified in three major classes: aliphatic GSL (derived from Met, Ileu, Leu, Val), aromatic GSL (derived from Phe, Tyr) and indolic GSL (derived from Trp). They are sulfur-rich secondary metabolites involved in plant's defense mechanisms against herbivores and pathogens. GSL occur in pungent plants of the Brassicaceae order. To date more than 200 types of GSL have been identified [43].

Selenium (Se) has long been recognized as being essential to animal and human nutrition. Although Se is not considered essential to plants is nevertheless thought as a beneficial element. Low Se soil levels translate in low Se levels in crops used for human consumption. Se as selenate ($Na_2SeO_4$) is absorbed by plants via sulfur transporters and is incorporated into selenocysteine (SeCys) and seleno-methionine (SeMet). SeCys is part of the seleno-glutathione peroxidase, a powerful antioxidant enzyme. SeMet and methyl-selenocysteine were found to have anticarcinogenic properties.

Because of the importance of Se for human health efforts were made to increase Se uptake by root veggies like radish. Schiavon et al. [44] have shown that Se biofortification in radish resulted in enhanced nutritional value through accumulation of methyl-selenocysteine and secondary metabolites such as glucosinolates, polyphenols and amino acids. The method of Se fertilization and the dosage of selenate applied to plants is important as excess Se may interfere with cysteine and methionine biosynthesis and can also affect negatively glucosinolate accumulation in plants. This is because Se and sulfur share the same uptake pathway. Nitrogen assimilation may also be affected by a high Se concentration in the fertilizer as Se may interfere with molybdenum (Mo) uptake. Mo is a cofactor in the enzyme nitrate reductase that converts nitrate to nitrite, the first reaction in nitrate assimilation by plants. Nitrite is further reduced to ammonia by nitrite reductase. At higher Se dosage the concentration of GSH in roots was lower than at low Se dosage because of a lower entry of sulfate into the sulfur assimilation pathway. The foliar Se spray of radishes grown in soil yielded a higher production of cysteine and GSH in roots. The study of Schiavon et al. [44] also revealed that Se foliar spray resulted in higher levels of all types of glucosinolates in roots including the glucoraphanin, a powerful anticarcinogen. The authors concluded that Se foliar fertilization is a better way to achieve a higher Se and other bioactive compounds in the roots than the hydroponic method.

In the following we will examine some of the properties that make radish such a valued vegetable in terms of nutrition and health enhancement promoter. Turmeric has been long used by the folk medicine as adjuvant for liver obstruction and

a. Antioxidant activity

Radish contains a large selection of phytochemicals that includes carotenoids, GSL, isothiocyanates, phenolic acids, polyphenols, flavanol, flavanone and anthocyanins. Some are present in only one tissue (GSL and carotenoids in sprouts) while others occur in more than one tissue, e.g. anthocyanin, in root and leaves. The leaves have a higher amount of polyphenols than the root. The leafy part constitutes an excellent source compounds with antioxidant power. Hence, it is recommended that all parts of the radish plant be consumed, including the sprouts. The flavonoids in radish have the ability to chelate iron, thus blocking iron-catalyzed generation of reactive oxygen species (ROS).

b. Detoxification activity

Cell culture and animal model experiments revealed that radish extracts attenuated the chemically-induced rat liver injury by decreasing lipid peroxidation caused by oxidative stress (OS). There is evidence to suggest that administration of radish extracts to rats upregulated the expression of cytochrome P450, Nrf-2/HO-1 signaling pathway, which are known to activate the expression of antioxidant enzymes. Besides fresh extracts some studies employed fermented radish in presence of *Lactobacillus spp*. Apparently, fermentation helps disassociate fibers in vegetables and that facilitate the release of bioactive compounds. Thus, fermented Spanish black radish had a hepatoprotective effect in CCl4-induced rat liver injury [45]. In another paper from the same group it was shown that the administration of fermented black radish to mice challenged with methionine and choline deficient diet significantly attenuated the increase in serum enzymes associated with hepatic cell injury such as alanine aminotransferase and aspartate aminotransferase [46]. Liver fibrosis and inflammation were also mitigated by treatment with fermented black radish.

Radish extracts proved helpful in the detoxification of xenobiotics. Thus, a single center, open label, pilot study investigated black radish supplementation to healthy males who received a controlled dose of acetaminophen (ibuprophen). The results showed that changes over a 4 week period of the ibuprophen metabolite and estradiol-17β suggested that there was an upregulation of phase I and phase II detoxification enzymes [47].

c. Anticancer activity

There is some evidence to suggest that indole-3-carbinol and its metabolite 3,3′-diindolyl-methane could suppress the growth and proliferation in tumor cell lines. These compounds target several features of cancer cell metabolism such as cell cycle regulation and survival including NF-κB/Akt signaling pathway, estrogen receptor signaling, caspase activation and endoplasmic reticulum stress [48]. Several studies reported the anti-cancer properties of GSL and isothiocyanates [42]. Thus, extracts of Spanish black radish inhibited the proliferation of HepG2 human tumor cells in vitro by up-regulating the phase I and II detoxification system. Apparently, the anti-cancer effect was due to the GSL compounds glucoraphasatin and 4-methylthio-3-butenyl isothiocyanate [49]. Sulforaphane and sulforaphene in Thai rat-tailed radish extract exhibited a strong cytotoxic effect against colon tumor cell line HCT116 [50]. The mechanism of action involved the increased production of ROS in these cells and the disruption of microtubule polymerization, hence affecting cell cycle regulation. A prospective EPIC-Heidelberg cohort study comprising 11,405 subjects and a mean follow-up time of 9.4 years found that a high GSL intake from cruciferous vegetables including radish was inversely associated with prostate cancer risk.

A very interesting use of radish extracts with the goal of obtaining cytotoxic agents against cancer cells is the reduction of graphene oxide (GO) in the presence of mild reductants such as those in radish extracts. GO is generated by the exfoliation of graphene from graphite in the presence of strong acids and bases. For biomedical applications GO is reduced by eco-friendly compounds such as polyphenols and flavonoids in plants such as radish. It has been found that reduced GO could significantly inhibit the proliferation of human breast and lung cancer cell lines [51].

d.Potential anti-diabetic properties

Unlike other medical conditions discussed above the potential of radish extracts to exert an anti-diabetes activity has been investigated so far only by using in vitro or in vivo (animal model experiments) systems. Thus, radish extracts administered to streptozotocin-induced diabetic rats caused a significant reduction in blood glucose, insulin and triglycerides levels [52]. Radish extracts also reduced the starch-induced postprandial glycemic load suggesting that is has a potent anti-diabetic activity.

It has been speculated that the hypoglycemic effect was due to an improved insulin sensitivity rather than increased insulin output. The phenolic compounds in radish may also assist in reducing oxidative stress via production of ROS, which are known to be elevated in diabetes. In addition, isothiocyanates in radish were shown to induce phase II antioxidant enzymes such as glutathione transferase, heme oxygenase-1, NAD(P)H-quinone reductase and UDP-glucuronosyl transferase. In vitro studies, demonstrated that aqueous radish extracts inhibited the activity of α-amylase and α-glucosidase, hence a decreased absorption of poly-and oligosaccharides in the GI tract.

## 2.8 Parsnip

Like the other root vegetables discussed above parsnip has been a staple for humans since ancient times. Parsnip is rich in vitamins and minerals (particularly potassium), phytochemicals (polyphenols, flavonoids, polyacetylenes, terpenes), essential oils and fiber. 100 g of parsnip provide about 75 kcal. A typical parsnip root contains 80% water, 5% carbohydrates, 1% protein, 0.3% lipids and 5% fiber.

Traditionally, parsnip has been used by folk medicine, particularly in the old Persian medical practice for topical and oral treatment of headaches, stomatitis, dermatitis, kidney stones and fever as well as recommended as gastric tonic, laxative and diuretic [53].

We cannot emphasize strongly enough the importance of fiber intake for a healthy life because of the proven health benefits of a fiber-rich diet. The dietary fiber comprises cellulose, hemicellulose, lignin, pectin and β-glucans. According to WHO and FAO dietary fiber consists of ten or more monomeric units that are neither digested nor absorbed in the small intestine and they are labeled as complex carbohydrates. Fruits and vegetables contain soluble and insoluble fiber. The former includes pectins, gums, insulin-type fructans and some hemicellulose. The latter comprises lignin, cellulose, some hemicellulose, resistant starches and analogous carbohydrates such as methyl cellulose. The content of dietary fiber in parsnip is 30% of the dry matter, composed mainly by neutral sugars (18%), pectic compounds (10%) and Klason lignin (1.92%). Klason lignin represents the insoluble residue portion left after removing the ash by acid hydrolysis of the plant tissue [54, 55].

When designing a healthy meal one should bear in mind the potential interplay between soluble fiber and fat. Experiments on mice indicated that mice fed a high fat diet that included soluble fiber exhibited a weight gain [56]. This outcome might be due to increased short chain fatty acids production after fermentation in the colon and subsequent increase in energy absorption.

Regular intake of soluble fiber has been associated with lower cholesterol and glucose levels, increased mass of friendly gut bacteria and a lower risk of developing metabolic syndrome, T2D and CVD. An observational study comprising healthy subjects found an inverse association between fiber intake and the concentration of serum C reactive protein. In another study on T2D patients it was shown that a higher fiber intake led to a decrease in the levels of circulating pro-inflammatory cytokine IL-18. It is well documented that high levels of circulating pro-inflammatory cytokines are associated with an increased risk for diabetes and CVD so a diet rich in fiber lowers the risk of getting T2D or CVD. Dietary fibers also decrease blood glucose excursions and lower insulin response.

The phytochemicals in parsnip root have been shown to possess a wide spectrum of pharmacological properties, which made them useful in tackling conditions such as neurological, respiratory, gastrointestinal, liver, skin, heart and urogenital disorders [54]. In vitro cell culture experiments have also shown cytotoxic effects of parsnip phytochemicals on cancer cell lines. A parsnip furanocoumarin such as xanthotoxin was shown to prevent memory impairment induced by injection of scopolamine in mice suggesting that xanthotoxin has neuroprotective effects on the cholinergic neurotransmission and also reduced oxidative stress in the brain [57].

### 2.9 Garlic

Garlic (*Allium sativum* L.) has been used for centuries both as food and as remedy in many cultures for a number of ailments such as cold, influenza, snake bites, bacterial and fungal infections and hypertension. Traditional medicine has also used garlic for the treatment of indigestion, respiratory and urinary tract infections and heart disorders. Epidemiological data indicated that Allium species extracts reduce the risk of diabetes and heart disease, activate the immune response when challenged by microbial and fungal pathogens and also show anti-aging, anti-diabetic and anti-cancer properties. Bulbs of garlic contain hundreds of phytochemicals including sulfur-containing compounds, polyphenols (alliin, allicin, diallyl disulfide, S-allylcysteine, S-allylmercaptocysteine, etc) such as quercetin, luteoline and apigenin, saponins, tanins and polysaccharides. When garlic is chopped or crushed alliin (allyl thiosulfinate), the main cysteine sulfoxide is converted into allicin by the enzyme allinase. The enzyme has pyridoxal phosphate as cofactor and it turns alliin into aminoacrylic acid and allyl sulfenic acid. Two molecules of allyl sulfenic acid (highly reactive) form allicin (2-propenylthiosulfinate). Allicin readily crosses cell membranes and reacts with thiol groups [58].

There have been numerous observational and clinical trials in the last two decades assessing the therapeutic effects of garlic preparations on pathologies such as diabetes, CVD, hypertension, metabolic syndrome, skin disorders, cancer, bacterial and fungal infections due mainly to the antioxidant, anti-inflammatory and lipid lowering effects shown by these preparations [59]. Garlic compounds were shown to elicit a number of biological responses such as the modulation of several cell signaling pathways (Akt/mTOR, MAPK, Nrf2, protein kinase B, 5'-AMP-activated protein kinase) as well as the activity of cytokines, intercellular adhesion molecules, cyclooxygenase, inducible NO synthase and others). Many studies were hampered by the low bioavailability and fast metabolization of garlic compounds in the human body and that affected the interpretation of the results. For example, the effect of garlic treatment on people with elevated blood pressure showed mixt results. One study indicated that the treatment resulted in slight improvement in cases of mild hypertension whereas another study showed no effect. Moreover, aged garlic extracts contain mainly water-soluble organosulfur compounds such as S-allyl cysteine and S-allylmercaptocysteine, which show other pharmacokinetics properties than oil-soluble S-containing compounds and that may influence the outcome of garlic supplementation.

Table headings list study design, medical condition examined, number of patients, type of intervention, duration of study and outcome. The clinical trials on T2D patients receiving garlic preparations with or without standard medication indicated that in general there was a significant reduction in blood glucose and HbA1c levels as well as an improved plasma lipid profile. Patients with gastric lesions supplemented with garlic preparations over a long period of time showed a decreased risk of developing gastric cancer incidence and mortality. On the other hand, patients with liver, prostate and colon cancer supplemented with 4 capsules of garlic preparation daily for 6 months did not show an improvement of their condition and the quality of life. Garlic preparations were found useful in reducing the level of oxidative stress and the production of pro-inflammatory cytokines such as IL-6 and CRP commonly associated with most human pathologies. Garlic preparations were found helpful for combating microbial infections. There was an inverse association between oral bacteria level and a lime-containing garlic extracts mouth wash in children with severe early caries. The inhibitory effect of garlic may include morphological alterations in bacterial cell wall and inhibition of microbial adherence to the epithelial cells of the host.

In general, garlic therapy yielded mixed results suggesting that not all pathologies are alike and not all patients respond in the same way to garlic supplementation. Better garlic formulations together with the standard therapy and a healthy diet and lifestyle should improve the outcome of treatment and reduce the risk of developing the conditions in the first place.

### 2.10 Onions

Onion (*Allium cepa*) has been used for centuries as food and spice as well as traditional medicine for a host of conditions such as fever, headache, dropsy, dyspnea, chronic bronchitis, cough and arthritis. The health benefits of eating onions are thought to be due to their antioxidant, immunomodulatory and anti-inflammatory activity [60–62].

Onions contain vitamins ($B_1$, $B_2$, $B_6$, folate, vitamin C), minerals (Mg, Ca, K, P), phenolic acids (gallic acid, ferulic acid, protocatechuic acid), flavonoids (flavanones, flavonols, flavanonols, kaempferol, anthocyanins), sulfur-containing compounds (diallyl sulfide, diallyl disulfide, S-methyl cysteine sulfoxide, etc), organic acids (citric, tartric, malic, oxalic, succinic), monosaccharides (glucose, fructose), fructooligosaccharides, phytoalexins and saponins. The main flavonol in onion is qercetin, in free form and as glucoside.

In an elegant randomized double-blind, placebo-controlled cross-over clinical trial it was found that supplementation with 162 mg/d quercetin from onion skin extract powder to overweight-to-obese patients for 6 weeks resulted in a modest drop in blood pressure (BP) in hypertensive but not in pre-hypertensive individuals [63]. These findings suggest that a threshold of higher BP might be necessary in order to detect a BP-lowering effect of quercetin. In addition, in the clinic's office where the BP measurements were performed no significant effects of quercetin supplementation on systolic BP were recorded and only about 50% of the participants showed a decrease in systolic BP. In contrast to animal model studies showing that quercetin attenuated hypertension and vascular dysfunction in a $NO^•$-dependent fashion, in the human trial above the biomarkers of endothelial function such as plasma endothelin-1, soluble vascular cell adhesion molecule-1, reactive hyperemia index were not affected by quercetin supplementation. The marker of inflammation CRP and angiotensin converting enzyme activity were also unaffected by quercetin treatment. The authors of the study concluded that for the hypertensive patients quercetin could decrease the 24 h systolic BP but without affecting the markers associated with inflammation and endothelial function. For the time being the molecular mechanisms underlying the BP-lowering effect of quercetin remain unclear.

In another RCT study it was found that onion extracts containing 50 mg quercetin increased the circulating endothelial progenitor cells and improved the flow-mediated dilation while BP and blood lipid profile were not affected.

It is also worth mentioning here that a meta-analysis of several RCTs on the effect of quercetin on BP indicated that a significant anti-hypertensive effect was only apparent at doses above 500 mg/day taken for longer than 8 weeks.

Platelet aggregation constitutes an aggravating factor in atherosclerosis. In vitro experiments using rat platelets have demonstrated that methanol extracts of onion skins were able to inhibit platelet aggregation. Quercetin and quercetin glucosides as well as organosulfur compounds appear to be involved in this inhibition. Allicin in onions was found to be a potent inhibitory factor toward ADP, arachidonic acid and collagen-induced platelet aggregation. It has been proposed that quercetin and organosulfur compounds from onion may be included in a preparation to be used for the prevention/management of atherosclerosis.

Most studies on the effect of onion compounds on cancer have been carried out on cancer cell lines and animal models. The polyphenols and S-containing compounds were mainly responsible for the observed effects. Quercetin glucosides in onion extracts were shown to possess antiproliferative activity against human breast, colorectal and prostate cancer cell lines. One possible mechanism is the inhibition of the PI3K/Akt signaling pathway, which results in apoptosis. Diallyl-trisulfide from onion was shown to trigger cancer cell cycle arrest at G2/M phase and the release of ROS that promote apoptosis and restrict tumor cell formation and development.

Onion extracts have been investigated in animal model experiments in relation to their potential use as anti-diabetic agents. For instance, STZ-induced diabetic rats treated with *A. cepa* bulb juice exhibited a 50% reduction in the fasting blood glucose level. C57BL/6 J mice treated with red onion extracts showed a decreased high fat diet-induced mass accumulation and insulin resistance. Flavonoids such as quercetin and S-methylcysteine in onions are mainly responsible for the hypoglycemic activity. These compounds were shown to decrease blood glucose levels, plasma lipids, oxidative stress and lipid peroxidation as well as modulating insulin secretion. Preliminary clinical trials demonstrated the blood glucose lowering effect of onion extracts.

Onion constituents show clear benefits against respiratory and allergic disorders. Experiments with allergic asthma guinea pigs indicated that onion quercetin significantly alleviated asthma symptoms. The mechanism of action includes β2-adrenoreceptors stimulation, inhibition of Ca channel blocking, histamine H1 receptors and phosphodiesterase activity. Protective effects of onion extracts were demonstrated by epidemiological and population case–control studies. Onion extracts as well as purified thiosulfinates and kaempferol acted by relaxing tracheal smooth muscles hence, improving clinical symptoms and reducing the severity of asthmatic attacks.

Biological active compounds in onions were shown in vitro experiments to possess antimicrobial activity. Red onion extracts were more effective antimicrobial agents than those from white and yellow onions. The bacterial species tested included *E. coli*, *Salmonella typhimurium*, *Klebsiella* spp. Onion extracts in ethanol, chloroform or water were also effective against fungi and molds. In a vivo study on broiler chickens it was found that onion powder-containing chicken feed had an effect on the gut microflora in that the number of *E. coli* was reduced whereas the number of *Lactobacillus* spp. was increased, which has a positive effect on chickens' health.

## 3. Conclusions

The last 30 years or so have been marked by an impressive progress in our knowledge about the chemical composition and the mode of action of biologically active compounds in fruits and vegetables, both wild and cultivated. In the present chapter we focused on the biochemistry and the potential benefits of compounds occurring in the roots and bulbs of some cultivated vegetables that have been for thousands of years part of human staple in all cultures. These chemicals are synthesized by plants to help attract friendly bacteria and/or ward off pathogens as well as increasing the absorption of vital nutrients including minerals.

All of the vegetable roots discussed in this chapter contain a wealth of bioactive compounds such as phenolic acids, polyphenols, sulfur-containing compounds, isothiocyanates, glucosinolates, mono- and polysaccharides, phytosterols, saponins, fiber and others. These chemicals have been extensively studied by using a variety of in vitro and in vivo experimental systems. Numerous clinical trials tried to assess the usefulness of either isolated compounds from roots and bulbs or whole extracts from these tissues for the treatment of major diseases such as diabetes, cardiovascular disease, cancer, allergies as well as bacterial infections either alone or in conjunction with standard therapies. In most cases there was a significant improvement in the condition of patients and quality of life. If total cure could not be achieved at least these natural compounds can assist in the prevention of many diseases in the first place. It is hoped that based on the knowledge accumulated so far the nutraceutical industry will come up with better product formulation regarding these bioactive compounds so there will be a better outcome for the patients. Besides supplementation with plant-based products, a diet rich in fruits and vegetables, an active lifestyle with physical activity and stress management will ensure a good health and a happy life.

## Author details

Mirela Ahmadi[1]*, Ştefan A. Hulea[2] and Ioan Peţ[1]*

1 Banats' University of Agriculture Sciences and Veterinary Medicine "King Michael the 1st of Romania" from Timisoara, Timisoara, Romania

2 Nutrition Consultant, Toronto, ON, Canada

*Address all correspondence to: mirelaahmadi@usab-tm.ro and ioanpet@usab-tm.ro

## References

[1] McNear DH Jr. The rhizosphere – Roots, soil and everything in between. Nature Education Knowledge. 2013;**4**(3):1

[2] Liu A, Contador CA, Fan K, Lam HM. Interaction and regulation of carbon, nitrogen and phosphorus metabolisms in root nodules of legumes. Frontiers in Plant Science. 2018;**9**:1860. DOI: 10.3389/fpls.2018.01860

[3] Abdel-Latif K, Bogusz D, Hocher V. The role of flavonoids in the establishment of plant roots endosymbioses with arbuscular mycorrhiza, rhizobia and Frankia bacteria. Plant Signaling & Behavior. 2012;7(6):636-641. DOI: 10.4161/psb.20039

[4] Vannier N, Agler M, Hacquard S. Microbiota-mediated disease resistance in plants. PLoS Pathogens. 2019;**15**: e1007740. DOI: 10.1371/journal.ppat.1007740

[5] Pascale A, Proietti S, Pantelides IS, Stringlis IA. Modulation of the root microbiome by plant molecules: The basis for targeted disease suppression and plant growth promotion. Frontiers in Plant Science. 2019;**10**:1741. DOI: 10.3389/fpls.2019.01741

[6] Schmidt H, Günther C, Weber M, Spörlein C, Loscher S, Böttcher C, et al. Metabolome analysis of *Arabidopsis thaliana* roots identifies a key metabolic pathway for iron acquisition. PLoS One. 2014;**9**(7):e102444. DOI: 10.1371/journal.pone.0102444

[7] Chen W, Tang L, Eang J, Zhu H, Jin J, Yang J, et al. Research advances in the mutual mechanisms regulating response of plant roots to phosphate deficiency and aluminum toxicity. International Journal of Molecular Science. 2022;**23**(3):1137. DOI: 10.3390/ijms23031137

[8] Parra-Almuna L, Diaz-Cortez A, Ferrol N, Luz MM. Aluminum toxicity and phosphate deficiency activates antioxidant systems and up-regulates expression of phosphate transporters in ryegrass (*Lolium perenne* L.) plants. Plant Physiology and Biochemistry. 2018;**130**:445-454. DOI: 10.1016/j.plaphy.2018.07.031

[9] Chen L, Zhu Y, Hu Z, Wu S, Jin C. Beetroot as a functional food with huge health benefits: Antioxidant, antitumor, physical function and chronic metabolomics activity. Food Science & Nutrition. 2021;**9**(11):6406-6420. DOI: 10.1002/fsn3.2577

[10] Esatbeyoglu T, Wagner AE, Schini-Kerth VB, Rimbach G. Betanin – A food colorant with biological activity. Molecular Nutrition & Food Research. 2015;**59**(1):36-47. DOI: 10.1002/mnfr.201400484

[11] Webb AJ, Patel N, Loukogeorgakis S, Okorie M, Aboud Z, Misra S, et al. Acute blood pressure lowering, vasoprotective and anti-platelet properties of dietary nitrate via bioconversion to nitrite. Hypertension. 2008;**51**:784-790. DOI: 10.1161/HYPERTENSIONAHA.107.103523

[12] Priya NG, Malarvizhi M, Jothi AJ. Beetroot juice on hemoglobin among adolescent girls. Journal of Nursing and Health Science. 2013;**2**:9-13

[13] Olumese FE, Oboh HA. Effects of daily intake of beetroot juice on blood glucose and hormones in young healthy

subjects. Nigerian Quarterly Journal of Hospital Medicine. 2016;**26**:455-462

[14] Sharma KD, Karki S, Thankur NS, Attri S. Chemical composition, functional properties and processing of carrot – A review. Journal of Food Science and Technology. 2012;**49**(1):22-32. DOI: 10.1007/s13197-011-0310-7

[15] Ahmad T, Cawood M, Iqbal Q, Arino A, Batool A, Tarid RMS, et al. Phytochemicals in *Daucus carota* and their health benefits – Review article. Food. 2019;**8**(9):424. DOI: 10.3390/foods8090424

[16] Wang LS, Stoner GD. Anthocyanins and their role in cancer prevention. Cancer Letters. 2008;**269**(2):281-290. DOI: 10.1016/j.canlet.2008.05.020

[17] Islami F, Sauer AG, Miller KD, Siegel RL, Fedewa SA, Jacobs EJ, et al. Proportion and number of cancer cases and deaths attributable to potentially modifiable risk factors in the United States. CA: A Cancer Journal for Clinicians. 2017;**68**(1):31-54. DOI: 10.3322/caac.21440

[18] Del Rio D, Rodriguez-Mateos A, Spencer JPE, Tognolini M, Borges G, Crozier A. Dietary (poly)phenolics in human health: Structures, bioavailability and evidence of protective effects against chronic diseases. Antioxidants & Redox Signaling. 2013;**18**(14):1818-1892. DOI: 10.1089/ars.2012.4581

[19] Tang G. Bioconversion of dietary provitamin a carotenoids to vitamin a in humans. The American Journal of Clinical Nutrition. 2010;**91**(5):1468S-1473S. DOI: 10.3945/ajcn.2010.28674G

[20] Golubkina NA, Kharchenko VA, Moldovan AI, Koshevarov AA, Zamana S, Nadezhkin S, et al. Yield, growth, quality, biochemical characteristics and elemental composition of plant parts of celery leafy, stalk and root types grown in the northern hemisphere. Plants (Basel). 2020;**9**(4):484. DOI: 10.3390/plants9040484

[21] Islam H, Chamberlain TC, Mui AL, Little JP. Elevated interleukin 10 levels in COVID-19: Potentiation of pro-inflammatory responses or impaired anti-inflammatory action? Frontiers in Immunology. 2021;**12**:677008. DOI: 10.3389/fimmu.2021. 677008

[22] Wang E, Chen F, Hu X, Yuan Y. Protective effects of apigenin against furan-induced toxicity in mice. Food & Function. 2014;**5**(8):1804-1812. DOI: 10.1039/c4fo00038b

[23] Mao QQ, Xu XY, Cao SY, Gan EY, Corke H, Beta T, et al. Bioactive compounds and bioactivities of ginger (*Zingiber officinale* roscoe). Food. 2019;**8**(6):185. DOI: 10.3390/foods8060185

[24] Bologa C. Foods that promote health. In: Hulea SA, Ahmadi M, editors. Foods That Harm, Foods That Promote Health: A Biochemical and Nutritional Perspective in Health and Disease Prevention. Irvine, CA, USA: Universal Publishers Inc.; 2021. pp. 181-213

[25] Zhang M, Viennois E, Prasad M, Zhang Y, Wang L, Zhang Z, et al. Edible ginger-derived nanoparticles: A novel therapeutic approach for the prevention and treatment of inflammatory bowel disease and colitis-associated cancer. Biomaterials. 2016;**101**:321-340. DOI: 10.1016/j.biomaterials.2016.06.018

[26] Gundala SR, Mukkavilli R, Yang C, Yadav P, Tandon V, Vangala S, et al. Enterohepatic recirculation of bioactive ginger phytochemicals is associated with enhanced tumor growth

inhibitory activity of ginger extract. Carcinogenesis. 2014;**35**(6):1320-1329. DOI: 10.1093/carcin/bgu011

[27] Wang Y, Yu H, Zhang X, Feng Q, Guo X, Li S, et al. Evaluation of daily ginger consumption for the prevention of chronic diseases in adults: A cross-sectional study. Nutrition. 2017;**36**:79-84. DOI: 10.1016/j.nut.2016.05.009

[28] Arablou T, Aryaeian N, Valizadeh M, Sharifi F, Hosseini AF, Djalali M. The effect of ginger consumption on glycemic status, lipid profile and some inflammatory markers in patients with type 2 diabetes mellitus. International Journal of Food Sciences and Nutrition. 2014;**65**:515-520. DOI: 10.3109/09637486.2014.880671

[29] Hajimoosayi F, Sadatmahalleh SJ, Kazemnejad A, Pirijani R. Effect of ginger on the blood glucose level of women with gestational diabetes mellitus with impaired glucose tolerance test (GTT): A randomized double-blind placebo-controlled trial. BMC Complementary Medicine and Therapies. 2020;**20**(1):116. DOI: 10.1186/s12906-020-02908-5

[30] Khan AM, Shahzad M, Asim MBR, Imran M, Shabbir A. *Zingiber officinale* ameliorates allergic asthma via suppression of Th2-mediated immune response. Pharmaceutical Biology. 2015;**53**(3):359-367. DOI: 10.3109/13880209.2014.920396

[31] Anh NH, Kim SJ, Long NP, Min JE, Yoon YC, Lee EG, et al. Ginger on human health: A comprehensive systematic review of 109 randomized controlled trials. Nutrients. 2020;**12**(1):157. DOI: 10.3390/nu12010157

[32] Fuloria S, Mehta J, Chandel A, Sekar M, Rani NNIM, Begum MY, et al. A comprehensive review on the therapeutic potential of *Curcuma longa* Linn. in relation to its major active constituent curcumin. Frontiers in Pharmacology. 2022;**13**:820806. DOI: 10.3389/fphar.2022.820806

[33] Qin S, Huang L, Gong J, Shen S, Huang J, Ren H, et al. Efficacy and safety of turmeric and curcumin in lowering blood lipid levels in patients with cardiovascular risk factors: A meta-analysis of randomized controlled trials. Nutrition Journal. 2017;**16**(1):68. DOI: 10.1186/s12937-017-0293-y

[34] Wong SC, Kamarudin MNA, Naidu R. Anticancer mechanism of curcumin on human glioblastoma. Nutrients. 2021;**13**(3):950. DOI: 10.3390/nu13030950

[35] Ahmadi M, Agah E, Nafissi S, Jaafari MR, Harirchian MH, Sarraf P, et al. Safety and efficacy of nanocurcumin as add-on therapy to riluzole in patients with amyotrophic lateral sclerosis: A pilot randomized clinical trial. Neurotherapeutics. 2018;**15**(2):430-438. DOI: 10.1007/s13311-018-0606-7

[36] Chico L, Ienco EC, Bisordi C, Gerfo AL, Petrozzi L, Petrucci A, et al. Amyotrophic lateral sclerosis and oxidative stress: A double-blind therapeutic trial after curcumin supplementation. CNS & Neurological Disorders Drug Targets. 2018;**17**(10):767-779. DOI: 10.2174/1871527317666180720162029

[37] Lunz K, Stappen I. Back to roots – An overview of the chemical composition and bioactivity of selected root essential oils. Molecules. 2021;**26**(11):3155. DOI: 10.3390/molecules26113155

[38] Kloucek P, Smid J, Flesar J, Havlik J, Titera D, Rada V, et al. In vitro inhibitory activity of essential oil vapors against

*Ascosphaera apis*. Natural Product Communications. 2012;7:253-256. DOI: 10.1177/1934578X1200700237

[39] Chang PY, Tsai FJ, Bau DT, Hsu YM, Yang JS, Tu MG, et al. Potential effect of allyl isothiocyanate on inhibiting cellular proliferation and inducing apoptotic pathway in human cisplatin-resistant oral cancer cells. Journal of the Formosan Medical Association. 2021;**120**(1): 515-523. DOI: 10.1016/j.jfma.2020.06.025

[40] Goos KH, Alberecht U, Schneider B. On-going investigations on efficacy and safety of a herbal drug containing nasturtium herb and horseradish root in acute sinusitis, acute bronchitis and acute urinary tract infection in children in comparison with other antibiotic treatments. Article in German. Arzneimittel-Forschung. 2007;**57**(4): 238-246. DOI: 10.1055/s-0031-1296611

[41] Marzocco S, Calabrone L, Adesso S, Larocca M, Franceschelli S, Autore G, et al. Anti-inflammatory activity of horseradish (*Armoracia rusticana*) root extracts in LPS-stimulated macrophages. Food & Function. 2015;**6**(12):3778-3788. DOI: 10.1039/C5FO00475F

[42] Manivannan A, Kim JH, Kim DS, Lee ES, Lee HE. Deciphering the nutraceutical potential of *Raphanus sativus* – A comprehensive overview. Nutrients. 2019;**11**(2):402. DOI: 10.3390/nu11020402

[43] Wang J, Qui Y, Wang X, Yue Z, Yang X, Chen X, et al. Insights into the species-specific metabolic engineering of glucosinolates in radish (*Raphanus sativus* L.) based on comparative genomic analysis. Scientific Reports. 2017;7:16040. DOI: 10.1038/s41598-017-16306-4

[44] Schiavan M, Berto C, Malagoli M, Trentin A, Sambo P, Dall'Acqua S, Pilon-Smits EAH. Selenium biofortification in radish enhances nutritional quality via accumulation of methyl-selenocysteine and promotion of transcripts and metabolites related to glucosinolates, phenolics and amino acids. Frontiers in Plant Science. 2016;7:1371. DOI: 10.3389/fpls.2016.01371

[45] Kim J, Ahn M, Kim SE, Lee HS, Kim HK, Kim GO, et al. Hepatoprotective effect of fermented black radish (*Raphanus sativus* L. var. niger) in CCl4-induced liver injury in rats. Journal of Preventive Veterinary Medicine. 2017;**41**(4):143-149. DOI: 10.13041/jpvm.2017.41.4.143

[46] Ahn M, Kim J, Choi Y, Ekanayake P, Chun JY, Yang DW, et al. Fermented black radish (*Raphanus sativus* L. var. *niger*) attenuates methionine and choline deficient diet-induced nonalcoholic fatty liver disease in mice. Food Science & Nutrition. 2019;7(10):3327-3337. DOI: 10.1002/fsn3.1200

[47] Evans M, Paterson E, Barnes DM. An open label pilot study to evaluate the efficacy of Spanish black radish on the induction of phase I and phase II enzymes in healthy male subjects. BMC Complementary and Alternative Medicine. 2014;**14**:475. DOI: 10.1186/1472-6882-14-475

[48] Weng JR, Tsai CH, Kulp SK, Chen CS. Indol-3-carbinol as a chemopreventive and anti-cancer agent. Cancer Letters. 2008;**262**(2):153. DOI: 10.1016/j.canlet.2008.01.033

[49] Hanlon PR, Webber DM, Barnes DM. Aqueous extract from Spanish black radish (*Raphanus sativus* L. var. *niger*) induces detoxification enzymes in the HepG2 human hepatoma cell line. Journal of Agricultural and

Food Chemistry. 2007;**55**:6439-6446. DOI: 10.1021/jf070530f

[50] Pocasap P, Weerapreeyakul N, Barusrux S. Cancer preventive effect of Thai rat-tailed radish (*Raphanus sativus* L. var. *caudatus Alef*). Journal of Functional Foods. 2013;**5**:1372-1381. DOI: 10.1016/j.jff.2013.05.005

[51] Punniyakotti P, Aruliah R, Angaiah S. Facile synthesis of reduced graphene oxide using *Acalypha indica* and *Raphanus sativus* extracts and their in vitro cytotoxicity activity against human breast (MCF-7) and lung (A549) cancer cell lines. 3. Biotech. 2021;**11**(4):157. DOI: 10.1007/s13205-021-02689-9

[52] Banihanni SA. Radish (*Raphanus sativus*) and diabetes. Nutrients. 2017;**9**(9):1014. DOI: 10.3390/nu9091014

[53] Lattimer JM, Haub MD. Effects of dietary fiber and its components on metabolic health. Nutrients. 2010;**2**(12):1266-1289. DOI: 10.3390/nu2121266

[54] Kenari HM, Kordafshari G, Moghimi M, Eghbalian F, Taher KD. Review of pharmacological properties and chemical constituents of *Pastinaca sativa*. Journal of Pharmacopuncture. 2021;**24**(1):14-23. DOI: 10.3831/KPI.2021.24.1.14

[55] Castro A, Bergenståhl B, Tornberg E. Parsnip (*Pastinaca sativa* L.): Dietary fiber composition and physicochemical characterization of its homogenized suspensions. Food Research International. 2012;**48**(2):598-608. DOI: 10.1016/j.foodres.2012.05.023

[56] Isken F, Klaus S, Osterhoff M, Pfeiffer AFH, Weickert MO. Effects of long-term soluble vs insoluble dietary fiber intake on high fat diet-induced obesity in C57BL/6J mice. The Journal of Nutritional Biochemistry. 2010;**21**:278-284. DOI: 10.1016/j.jnutbio.2008.12.012

[57] Skalicka-WozniakK,BudzynskaB,BialaG, Boguszewska-Czubara A. Scopolamine-induced memory impairment is alleviated by xanthotoxin: Role of acetylcholinesterase and oxidative stress processes. ACS Chemical Neuroscience. 2018;**9**(5):1184-1194. DOI: 10.1021/acschemneuro.8b00011

[58] Batiha GE, Beshbishy AM, Wasef LG, Yha E, Al-Sagan A, MEA E-H, et al. Chemical constituents and pharmacological activities of garlic (*Allium sativum* L.): A review. Nutrients. 2020;**12**(3):872. DOI: 10.3390/nu12030872

[59] Ansary J, Forbes-Hernandez TY, Gil E, Cianciosi D, Zhang J, Elexpuru-Zabelata M, et al. Potential health benefits of garlic based on human intervention studies: A brief overview. Antioxidants (Basel). 2020;**9**(7):619. DOI: 10.3390/antiox9070619

[60] Chakraborty AJ, Uddin TM, Zidan BMRM, Mitra S, Das R, Nainu F, et al. *Allium cepa*: A treasure of bioactive phytochemicals with prospective health benefits. Evidence-based Complementary and Alternative Medicine. 2022;**2022**:4586318. DOI: 10.1155/2022/4586318

[61] Marrelli M, Amodeo V, Statti G, Conforti F. Biological properties and bioactive components of *Allium cepa* L.: Focus on potential benefits in the treatment of obesity and related comorbidities. Molecules. 2019;**24**(1):119. DOI: 10.3390/molecules24010119

[62] Beigoli S, Behrouz S, Zia AM, Ghasemi SZ, Boskabady M, Marefati N, et al. Effects of *Allium cepa* and its constituents on respiratory and allergic disorders: A comprehensive

review of experimental and clinical evidence. Evidence-based Complementary and Alternative Medicine. 2021;**2021**:5554259. DOI: 10.1155/2021/5554259

[63] Brüll V, Burak C, Stoffel-Wagner B, Wolffram S, Nickenig G, Muller C, et al. Effects of a quercetin-rich onion skin extract on 24h ambulatory blood pressure and endothelial function in overweight-to-obese patients with (pre-) hypertension: A randomized double-blinded placebo-controlled cross-over trial. The British Journal of Nutrition. 2015;**114**(8):1263-1277. DOI: 10.1017/S0007114515002950

Chapter 3

# Bioactive Components of Root Vegetables

*Rashida Bashir, Samra Tabassum, Ayoub Rashid, Shafiqur Rehman, Ahmad Adnan and Rabia Ghaffar*

## Abstract

Health and nutrition values force the lifestyle to embrace functional food which accommodates health-promoting nutrients. Root vegetables are an excellent source of health-promoting phytoconstituents, including phenolic acids, flavonoids, essential oils, proteins, and bioactive pigments. These bioactive compounds impart broad-spectrum pharmacological activities, including anti-hepatotoxicity, anti-hyperlipidemia, anti-inflammatory, anti-hypertension, anti-depressant, and anti-hypoglycemia. In this context, quantification via a compatible extraction technique is essential. However, these bioactive compounds are sensitive to heat processing, growth conditions, pre-extraction treatments, and extraction techniques. The recovery of bioactive compounds and their health benefits can be further enhanced by suitable processing, storage, and proper supplementation. The present review aims to comprehensively discuss the bioactive compounds of root vegetables along with factors influencing these compounds and the involvement of root vegetables in oxidative stress reduction, as reported in the literature (2001–2022).

**Keywords:** bioactive compounds, phenolic acids, flavonoids, essential oils, proteins and bioactive pigments, anti-hepatotoxicity, anti-hyperlipidemia, anti-inflammatory, anti-hypertension, anti-depressant, anti-hypoglycemia, ant-carcinogenic activities

## 1. Introduction

Vegetable-rich diets are highly recommended owing to their health-promoting functions. Whereas, vegetables with modified roots (edible roots) possess bioactive compounds with diverse biological activities, most prominently antioxidant properties [1]. Analysis has indicated a significant association between vegetable consumption with haemorrhagic and ischaemic stroke protection. This association is a major reason behind the increased consumption of vegetables in the past few years [2]. However, worldwide dietary intake of fruits and vegetables is still low, which may increase the risk of cardiovascular diseases and cancer. According to a survey, 2.635 million deaths per year are linked to the insufficient consumption of fruits and vegetables. It is important to mention that 600 g per day per individual consumption of fruits and vegetables can result in a 1.8% reduction of the worldwide disease burden [3]. Root vegetables such as


IntechOpen

**Figure 1.**
*A: Turnip (Brassica rapa), B: Rutabaga (Brassica nupus), C: Carrots (Daucas carota), D: Sweet potato (Ipomoea batatas), E: Taro (Colocasia esculenta), F: Beetroot (Beta vulgaris), G: Cassava (Manihot esculenta), H: Parsnip (Pastinaca sativa), I: Radish (Raphanus sativus), J: Purple yam (Dioscorea alata), K: Mustard root (Brassica juncea).*

carrot, sweet potato, turnip, radish, rutabaga, beetroot, etc. (**Figure 1**) tend to possess bioactive compounds at varying extents as different factors can influence the accumulation and recovery of these bioactive compounds. The present review summarizes the contents of major natural products and bioactive compounds of commonly consumed root vegetables, along with the factors influencing these bioactive compounds and their role in oxidative stress management. Electronic databases are used for data collection. Authentic databases such as Web of Science, ScienceDirect, Pubmed, and Scopus were preferred for reviewing appropriate and quality publications (2001–2022).

## 2. Phenolic compounds

Phenolic compounds constitute a major class of biologically active plant metabolites (see **Table 1**, bioactive phenolic compounds) involving phenols with one or more hydroxyl groups. These phenolic compounds are divided into two main subclasses, phenolic acids, and polyphenolic compounds. Phenolic acids involve hydroxycinnamic acid and hydroxybenzoic acid derivatives [24]. Polyphenolic compounds constitute a diverse subclass, further divided into different groups such as flavonoids, stilbenes, coumarins, and lignans, depending upon the number of phenolic rings and structural diversity [25]. Root vegetables serve as a major source of phenolic compounds, and the phenolic content of some root vegetables is reported to be higher than non-root vegetables [26]. Among root vegetables, radish has a high phenolic content of 1315.83 μg gallic acid equivalent/g of extract, even higher than potatoes and kohlrabi [26]. The phenolic compounds tend to vary in different parts of the plants [27]. Carrots peels are reported to have more phenolic acids than vascular tissues. 5′-Caffeoylquinic acid is a major compound in different carrot varieties with the

| Vegetable | Bioactive compound | Bioactivities | Ref. |
|---|---|---|---|
| Radish | Quercetin, ferulic acid, caffeic acid | Antioxidation | [4, 5] |
| Turnip | Isorhamnetin-3,7-di-O-glucoside, kaempferol-3-O-(feruloyl)sophoroside-7-O-glycoside, sinapic acid | Antioxidation | [5, 6] |
| Rutabaga | 3-caffeoylquinic acid, 5-caffeoylquinic acid, 5-feruloylquinic acid, 4-*p*-coumaroylquinic acid | Antioxidant | [7, 8] |
| Carrot | 5′-caffeoylquinic acid, cis-5′-caffeoylquinic acid | Antioxidation, anti-inflammatory | [9, 10] |
| Beetroot | Gallic acid, naringenin, myricertin, catechol | Antioxidation, anti-inflammatory, anti-hypertensive | [11–14] |
| Parsnip | Qucercetin 3,7-O-diglucoside, quercetin 3-O-rutinoside, 5-O-caffeoylshikimic acid | Antioxidation | [15] |
| Taro | Catechin, 1-O-feruloyl-D-glucoside, 3,5-dicaffeoylquinic acid | Antimetastatic | [16, 17] |
| Sweet potato | 5-caffeoylquinic acid, 3,5-dicaffeoylquinic acid, 3,4- dicaffeoylquinic acid | Antioxidant, hypoglycemic | [18, 19] |
| yam | Sinnapic acid, ferulic acid | Antimicrobial | [20, 21] |
| Cassava | Quercetin, gallic acid, ellagic acid, scopoletin, isovanillin | Antioxidant | [22, 23] |

**Table 1.**
*Major Bioactive phenolic compounds and bioactivties of some common root vegetables.*

highest content value in peels (15.04 mg/100 g FW), followed by cis-5′-chlorogenic isomer (3.51 mg/100 g FW) of Caffeoylquinic acid [9]. Another study has reported higher phenolic content in rutabaga sprouts (reaching up to 125.7 mg GAE/g DW) as compared to seeds (6.9 mg GAE/g DW) and roots (5.1 mg GA/g DW) [28]. Phenolic accumulation also differs depending upon the varieties, as a study has indicated higher phenolic content in black radish (13.7 mg GAE/g DW) compared to white and red varieties [29]. Higher phenolic content of plants correlates with antioxidant activity, such as parsnip, a carrot-related root vegetable rich with phenolic glycosides that positively correlate with antioxidant activity [15]. Beetroot with high phenolic content of 56.65 mg GAE/100 ml of juice, encapsulated as soybean proteins reaching up to 150 mg GAE/100 ml [11]. Turnip is reported to contain flavonoids and hydroxy-cinnamic derivatives. HPLC analysis of turnip tops has identified flavonoid glycosides and total phenolic content reaching up to mg 191.39 mg/100 g fresh weight [30]. Because turnip is usually consumed in processed form, phenolic compounds are negatively impacted upon applying heating procedures for processing. Franciso et al. [31] have studied the impact of different heat treatments on the turnip. Flavonoid content decreased to 4.58 μmol/g dry weight upon conventional boil initially at 13.85 μmol/g DW in raw samples. Likewise, high-pressure cooking also deteriorated flavonoids (5.00 μmol/g DW). A post-processed vegetable, Taro has high flavonoid content such as luteolin (44.5%), apigenin (52.7%), and chrysoeriol glycosides (2.63%) in dry land growth conditions [32]. On the other hand, sweet potato contains phenolic acid derivatives, including isomers of caffeoylquinic acid and dicaffeoylquinic acid, whose content is also reduced by applying processing techniques. Boiling is the least favored technique for preserving phenolic compounds in sweet potatoes [18]. Cassava roots are widely consumed in African countries. Milled flour of dried cassava is used in many delicacies; however, drying impacts the phenolic composition of cassava. A study has indicated that roasting reduces the phenolic content (51.35 mg/g in roasted flour) as compared to sun drying (64.82 mg/g in sun-dried flour) [22]. However, other than processing techniques, extraction methods and parameters greatly impact the recovery of phenolic compounds from a vegetable source. A study has compared different techniques for black carrot extraction, reporting optimized microwave-assisted extraction (power = 348.06, time = 9.86 min, solvent: solid ratio = 19.3 g/mL, solvent = 19.8% ethanol) to be best suited for phenolic recovery with the content value of 264.9 mg GAE/100 ml while conventional extraction to be least suited with approximately 2.5 times less phenolic yield than former technique [33].

Vegetables stored at room temperature lose sensory and composition quality due to the action of polyphenol oxidase and peroxidase enzymes. A study has recommended "refrigerate storing" to prolong the shelf life of vegetables [34]. Peels of vegetables are often discarded during food processing; however, incorporation of peels in food can be beneficial, as a study carried out on rutabaga has indicated higher phenolic content in peels (18.14 mg/GAE/g) as compared to a pulp (11.57 mg GAE/g) using ultrasound-assisted extraction [8].

## 3. Glucosinolates

Glucosinolates are sulfur and nitrogen-containing compounds predominantly occurring in brassica vegetables (see **Table 2**). Depending on their structure, these compounds are further divided into three subclasses, aliphatic, aromatic, and indole glucosinolates [43]. Glucosinolates, well-known cancer-preventing bioactive compounds, and isothiocyanates (**Figure 2**) provide characteristic flavors to brassica vegetables [44]. Turnip is an excellent source of glucosinolates with 100–130 mg /100 g FW total glucosinolate content [45]. When compared with other Chinese leafy vegetables (cabbage, pakchoi, cai-Tai, choy-sum) of the brassica family, the turnip has shown the highest glucosinolate content, with gluconapin being the highest accumulated glucosinolate (65.84 mg/100 g FW) [45]. UPLC/MS analysis has indicated progoitrin as another high accumulated glucosinolate in turnip [38]. Rutabaga is a hybrid plant of cabbage and turnip with a considerable amount of glucosinolates (7.34 μmol/g DW), with progoitrin as major glucosinolate; however,

| Plant | | Glucosinolate | Bioactivity | Ref. |
|---|---|---|---|---|
| Radish | White | Glucorasphatin, glucoraphenin, glucobrassicanapin | Detoxification enzyme induction | [35–37] |
| | Black | Glucoraphenin, glucosisymbrin, glucosisaustricin | | |
| Turnip | | Progoitrin, glucoalyssin, gluconasturtiin,4-hydroxyglucobrassicin, glyconapin | Bone formation | [38] |
| Rutabaga | | Progoitrin, sinigrin, glucoerrucin | Antimicrobial | [39, 40] |
| Mustard roots | | 2-propenyl glucosinolate (Sinigrin), 2-hydroxy-2-phenylethyl glucosinolate (glucobarbarin), | Nematicidal | [41, 42] |

**Table 2.**
*Glucosinolates of root vegetables of the Brassicaceae family.*

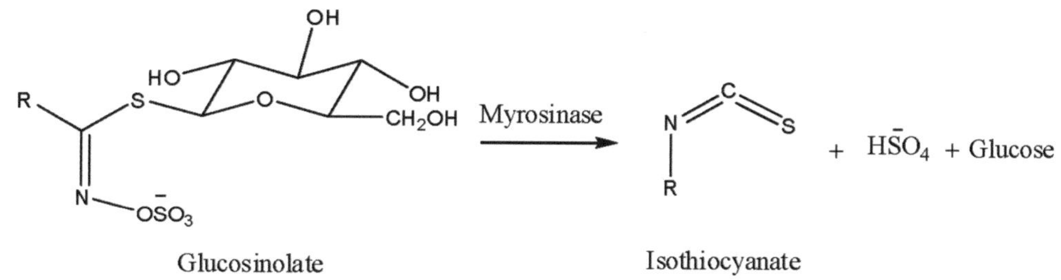


**Figure 2.**
*Glucosinolate conversion into isothiocayanate by the action of myrosinase.*

total glucosinolate content is reduced upon cooking(6.86 μmol/g DW) [46]. Boiling and high-pressure cooking can reduce glucosinolate content by three times [31].

Glucosinolate content of radish varies depending upon the radish cultivar; however, glucoraphanin is a major constituent of the glucosinolate profile of all radish cultivars. Red radish has reported glucosinolate content of 163.91 mg sinigrin equivalent/100 g FW [47]. 4-methylthio-3-butenyl glucosinolate is prominent in white radish, which is converted into 4-methylthio-3-butenyl isothiocyanate by the action of myrosinase enzyme. The resulting metabolite is responsible for the specific pungent flavor of radish. 4-methylthio-3-butenyl glucosinolate comprises 90% of the total glucosinolate profile in Japanese varieties [48]. Spanish black radish also has identified radish-specified glucosinolates, which induce detoxification enzymes in HepG2 cell lines. However, the glucoraphanin (4-methylthio-3-butenyl glucosinolate), a major glucosinolate of radish, is not preserved in black radish-based dietary supplements [35, 36]. Tissue damage during cutting, chewing, cooking, and fermentation leads to glucosinolate degradation into isothiocyanates, thiocyanates, and cyanides. The major degradation product of radish glucosinolates other than erucin is 5-(methylthio)-4-pentenenitrile [49]. Mustard is also reported in the literature for high glucosinolates levels and isothiocyanate byproducts. A study on mustard seeds has indicated higher levels of aliphatic glucosidase in all mustard varieties, with total glucosidase content ranging from 51.96 to 64.36 umol/g FW in root mustard [50]. Glucosinolates are highly sensitive to extraction techniques and pre-extraction treatments. A study conducted on brassica vegetables has indicated cold methanolic extraction with conventional wet tissue freeing is suitable for glucosinolate recovery compared to hot methanolic extraction. Also, freeze-drying was found to be avoidable for short time storage [51].

## 4. Color imparting bioactive compounds

### 4.1 Anthocyanins

Root vegetables possess natural-colored pigments that give them their characteristic colors (see **Table 3**). Anthocyanins are water-soluble pigments generated from glycosylation of the anthocyanidin class of flavonoids. Till now, 670 such compounds have been isolated and identified from colored plants and flowers [52]. These pigments cover a wide spectrum of colors, from red to purple, depending on their structure [63]. Red radish contains anthocyanins that have similar properties as synthetic Red No. 40 and is stable to be used commercially [64]. The extraction and post-extraction concentration procedure affects the recovery of these natural

| Plants | Compounds | Colour | Ref. |
|---|---|---|---|
| **Anthocyanins** | | | |
| Radish | Pelargonidin-3-caffeoyl-duglycoside-5-glucoside, Pelargonidin-3-feruloyl-duglycoside-5-glucoside, Pelargonidin-3-*p*-coumaroyl-duglycoside-5-glucoside | Red | [47] |
| | Cyanidin 3- caffeoyl sophoroside-5-glucoside, cyanidin 3- feruloyl sophoroside-5-diglucoside, cyanidin 3-(glucosyl-p coumaroyl) sinapolysophoroside-5-Malonylglucoside etc. | Purple | [52] |
| Sweet potato | Cyanidin 3-caffeoyl-*p*-hydroxybenzoyl sophoroside-5-glucoside, peonidin 3-caffeoyl-*p*-hydroxybenzoyl sophoroside-5-glucoside | Purple | [53] |
| Turnip | Peonidin O-hexoside, Rosindin-O-hexoside, Delphinidin-O-malonylhexoside, cyaniding-3-O-glucoside, malvidin-3,5-diglucoside | Purple | [54] |
| Carrot | Cyanidin 3-xylosyl-galactoside, cyaniding 3-xylosyl-feruloylglucosylgalactoside, cyaniding 3-xylosyl-p-coumaroylglucosylgalactoside | Black | [55] |
| | Ferulic acid cyanidin 3-xylosyl-glucosylgalactoside, *p*-coumaric acid cyanidin 3-xylosyl-glusosylgalactose, cyanidin 3-xylosylglucosyl-galactoside | Purple | [56] |
| **Carotenoids** | | | |
| Carrot | β-Carotene, α-carotene, lutein | Orange | [57] |
| | | purple | [57] |
| | Lycopene, β-carotene, α-carotene, lutein | Red | [57] |
| Sweet potato | β-Carotene | Orange | [58] |
| Cassava | β-Carotene, 9-Z- β-carotene, 13-Z- β-carotene | Cream colour | [58] |
| Radish | Lutein, β-carotene | White | [59] |
| | Lutein, β-carotene, neoxanthin, zeaxanthin | Purple | [60] |
| **Betalains** | | | |
| Beetroot | Betanin, isobetanin, 2,17′-bidecarboxy-neobetanin, miraxanthin II, vulgxanthin I | Deep red | [61] |
| | Vulgaxanthin I, indicaxanthin, miraxanthin | Yellow | [62] |

**Table 3.**
*Bioactive color pigments of root vegetables.*

colorants. Hydroethanolic and acidified water extraction system with membrane pertraction concentration is suitable for high anthocyanin recovery and led to 62.58 mg/100 ml anthocyanin recovery from red radish [65]. Hydrogen-rich water is reported to influence anthocyanin accumulation in the Yanghua cultivar of radish [66]. A study has indicated radish growth under UV-A light with hydrogen-rich water, and calcium chloride treatment promotes the accumulation of anthocyanins (reaching a relative anthocyanin content value of 42.06, which was at 9.72 in control plants) [67]. Purple yam is a rich source of anthocyanins with antimicrobial activity [21]. HPLC-DAD analysis has quantified 31 mg/100 g DW anthocyanin content in fresh yams. Cyanidin and peonidin glycosides acylated with hydroxycinnamic acids are leading anthocyanins of purple yam; however, processing of yams by blanching leads to a total anthocyanin loss of 60% [20]. Purple radish has also identified the presence of acylated cyaniding sophoroside glycosides and diglycosides [52]. Within the biological system, anthocyanins are generated under the influence of light in the presence of chalcone synthase enzymes. Light-responsive genes are expressed in turnip at different levels in anthocyanin biosynthesis [68]. Among other sweet potato varieties, purple-fleshed sweet potatoes have indicated high amounts of peonidin anthocyanins (1039 mg//100 g DW), which is almost three times higher than cyanidin-based anthocyanins [53]. Purple Carrots are found to be a good source of anthocyanin, with content value reaching up to 33,876 mg/kg DM; however, air or freeze-drying of fresh carrots leads to anthocyanins loss [56]. A study on black carrots also has identified a considerable amount of anthocyanins [69].

### 4.2 Carotenoids

Carotenoids represent a class of natural products with lipid-soluble vibrant color pigments. These pigments are responsible for the yellow to red spectrum of colors in plants and other organisms [70]. Unlike flavonoid-based anthocyanins, carotenoids are tetraterpenoid molecules containing eight isoprenoid units that generate a C40 carbon skeleton [71]. In green tissues, carotenoids are located in chloroplasts, while in colored vegetables, accumulation occurs in chromoplasts imparting different colors to the plant [72]. Among root vegetables, carrot is the most prominent source of carotenoids. Carotene content varies depending upon the carrot cultivars. It is reported to be in the range of 58.15–102.02 mg/kg FW) with β-carotene constituting the major portion of total carotenoid content [73]. Studies have indicated the immense impact of light on carotenoids [74]. In carrots, the biosynthetic pathway of carotenoids is influenced by light, as it can induce repression of gene expression of β-carotene and α-carotene biosynthesis [75]. Parsnip, on the other hand, despite being a carrot-like root vegetable, only has minor amounts of carotenoids [76]. β-Carotene is also dominant in sweet potato varieties with orange flesh. The carbon chain of the majority of carotenoids contains trans double bonds. In sweet potatoes, 127 μg/g of total β-carotene contains 123 μg/g of all-E-β-carotene [58]. Turnip leaves are also reported to possess carotenoids with content values reaching up to 250 μg/g [77]; however, a study on white radish leaves has shown higher carotenoid content (486.95 μg/g DW) as compared to turnip leaves. Moreover, despite being a white-fleshed variety, a minute amount of carotenoids is also identified in radish roots [59].

### 4.3 Betalains

Betalain pigments are betalamic acid derivatives divided into two subclasses: betacyanin (red-violet pigments) and betaxanthin (yellow pigments), depending upon cyclo DOPA and amine condensation on betalamic acid, respectively [61]. Black radish peel extracts appears yellowish and is reported to contain 22.5 mg/100 g DW of betaxanthin while only 7.7 mg/100 g DW of betacyanin pigments [78]. Beetroot is the richest vegetable source of betalains (17.24 mg/g DW), most prominently betacyanins responsible for its bright red color. A study has indicated approximately three times higher betacyanin content than betaxanthins in peels and other root sections of beetroot [61]. Beetroot peels are also betalain rich. An optimized study on betalain extraction from beetroot peels has identified 1.5% citric acid, 50% ethanol, and 52.52°C temperature and 49.9 min extraction time to be best suited for betalain recovery with the content value of 1.20 mg/g DW [79].

## 5. Vitamins

Vitamins are necessary for normal growth, and their deficiency can cause health Complications (see **Table 4**). Vitamin C or ascorbic acid is a natural antioxidant that helps prevent different diseases by reducing oxidative stress [93]. Radish is a major source of vitamin C with 38.83 10–2 g/kg of vitamin C as compared to turnips and carrots; however slight change in vitamin C accumulation is seen in plants grown in high $CO_2$ conditions [86]. The ascorbic acid content of sweet potato is comparable to radish. Different cultivars, including white, orange, and purple sweet potato, have indicated vitamin C content ranging from 17 to 37 mg/100 g DW [94]. Turnip greens and turnip tops are good sources of vitamin C; however, the heat processing leads to a significant loss of vitamin C. Steaming can result in approximately 60% vitamin C loss, while boiling and high-pressure cooking can completely diminish vitamin C content from a turnip [31]. Other than heat treatment, the vitamin C content is also sensitive to long-term storage. A study on carrot cultivars reported 88–132 mg/kg vitamin C content in fresh carrots, which decreased by 58% (on average) upon 30 days of storage [95]. Vitamin E refers to a group of lipid-soluble antioxidants that play an essential role in cell signal regulation and proliferation [96]. Among different types of vitamin K, α and γ tocopherol are the most prominent types [89]. Carrots serve as an excellent source of vitamin E; however, orange and purple carrots varieties have higher vitamin E content than white carrots [90]. Green leafy plant sections contain higher vitamin E content than roots, as determined in radish leaves containing up to 48.5 μg/g DW of vitamin E with α-tocopherol as major vitamin E. At the same time, roots only contain up to 0.17 μg/g DW with γ-tocopherol in the highest quantity; however, vitamin C and E are also sensitive to post-harvest storage. Vitamin K is also found in root vegetables but is predominant in green leafy portions. Phylloquinone, commonly known as vitamin K1, is the predominant form of vitamin K and is reported in considerable amounts in radish leaves [89]. As discussed in the previous section, carrots are rich in carotenoids, among which α-carotene and β-carotene are major pro-vitamin A compounds that are converted into vitamin A (retinol) in a biological system. According to the US National academy of science, 24 μg of pro-vitamin, A carotenoid is equivalent to 1 μg of vitamin A (retinol) [80, 81].

| Vitamin/ provitamins | Plant source | Recommended intake/day | Deficiency | Refs. |
|---|---|---|---|---|
| Vitamin A/ Provitamin A (β-carotene) | Carrot | 700 μg | Night blindness, xerophthalmia | [80–82] |
| Vitamin B (folates) | Beetroot | 400 μg dietary folate equivalent | Anemia | [83–85] |
| Vitamin C (ascorbic acid) | Turnip, Radish | 75–90 mg | Skin-related complications, low wound healing | [86–88] |
| Vitamin E (tocopherols) | Radish, carrot | 0.67–15 mg Tocopherol equivalent | Weak immune system | [89–92] |

**Table 4.**
*Major root vegetable sources of vitamins and health conditions related to deficiency.*

## 6. Other bioactive compounds

Phytosterols are plant-based steroid molecules with cardioprotective and antitumor activities. Seeds are a rich source of bioactive phytosterols, such as radish seeds with good phytosterol profiles mainly consisting of brassicasterol, campesterol, and sitosterol [97]; however, considerable quantities of phytosterols are also reported in edible roots. A study has shown 366.16 mg/100 g phytosterol content in carrots with campesterol and sitosterol as major constituents [98]. Radish sprouts also have shown campesterol (947 μg/g) and β-sitosterol (899 μg/g) in considerable amounts [99]. Sweet potato contains anticancer phytosterol, daucosterol linolenate, daucosterol linoleate, and daucosterol palmitate that regulate gut microbiota [100]. Saponins compounds such as steroidal saponins are reported to possess anti-proliferative activities. Yams are an excellent source of saponins with a content range of 37.36–129.97 mg/g. While individual steroidal saponins include dioscin, gracilllin, protodioscin, and protoracillin [101]. Dioscin, a steroidal saponin, has been extracted from wild yams and has shown a promoting effect on GATA 3 expression, a tumor suppressor in breast cancer [102]. Diosgenin is also a major steroidal saponin in yam with high antioxidant activity [103]. Other than steroidal saponins, root vegetables have reported triterpenoid saponins with antioxidant activity. Beetroot is a rich source of triterpenoid saponins. Triterpenoid aglycones of saponins include oleanolic acid, hederagenin, akebonoic acid, and glycogenin, which were linked with hexose, uronic, deoxyhexose, and pentose sugar to generate terpenoid saponins [104]. Other than beetroot, sweet potatoes also have reported triterpene saponin with antioxidant properties. Sandrosaponin is a major triterpene saponin of sweet potatoes with a content value of 161.20 mg/100 g, constituting approximately 81% of the total saponins of sweet potatoes [105]. Volatile fractions of plants contain useful alcohols, terpenes, and hydrocarbons, among which terpenes are the most important due to their significant role in providing aroma and the flavor to vegetables. The accumulation of terpenes in root vegetables varies depending upon the color variation. Orange carrots have a high accumulation of β-caryophyllene, α-humulene, and bornyl acetate, while the yellow variety contains β-bisabolene and γ-bisabolene in higher amounts [106]. Likewise, hydrodistillation of radish has also identified phytol as a major terpene with the highest abundance (69.7% of total compounds identified) in the white variety, followed by neophytadiene (1.5%) in the black variety and β-damascone (1.4%) in the red variety [49].

## 7. Vegetable supplementation against oxidative stress

Oxidative stress can lead to many health complications such as cancer, inflammatory diseases, cardiovascular diseases, neurodegenerative diseases, and aging [107]. As vegetables' secondary metabolites act as natural antioxidants, dietary supplementation of vegetable products can help reduce oxidative stress. The following section addresses the effect of vegetable supplementation on oxidative stress. A study on Wister rats under cadmium-induced oxidative stress has indicated that five days of intake of carrot juice as drinking water before stress induction can lower cadmium concentration in both liver and kidney along with the oxidative stress as determined by a lower concentration of malondialdehyde (MDA) (see **Table 5**) in pretreated rats [108]. Similar results, i.e., malondialdehyde reduction, were seen with beetroot juice and radish supplementation, proving the positive effect of raw consumed root vegetables on oxidative stress management [110, 112]. A study has shown betanin, a major

| Supplement | Subject | Study design | MDA level | Refs. |
|---|---|---|---|---|
| Carrot juice | Wister rats | Pretreatment with carrot juice for five days followed by two days cadmium treatment | 59.11*,40.26[a] μg/g | [108] |
| | Human | Daily intake of 16 fl oz. carrot juice 3 months | 42*,18[a] μM in plasma samples | [109] |
| Beetroot juice | Wister rats | Supplementation of 3 ml of juice and carcinogen for aberrant crypt foci induction for 30 days | 118.07*,100.41[a] nmol/g kidney tissue | [110] |
| | Human | Daily intake of 250 ml of beetroot juice for 30 days before exercise | 10.69*, 5.75[a] nmol/ml in plasma samples | [111] |
| Radish extract | Balb/c mice | 10 days treatment with 40 mg/g body weight (BW) zearalenone for oxidative stress induction and 5 mg / kg b.w of radish extract. | 4.51*, 0.73[a] nmol/mg liver tissue | [112] |
| Sweet potato jelly | Wister rats | 14 days treatment (diabetic rats) with jelly containing 400 g of purple sweet potato. | 8.95*–2.39[a] nmol/ml | [113] |
| Yam powder | Spraque-Dawley rats | 12 weeks treatment with 5 g/kg of BW/day of yam extract to methionine induced hyperhomocysteinemic rats. | 1.2*,b–0.6[a,b] μM serum | [114] |

[a]*Values with supplementation.*
**Values without supplementation.*
[b]*Approximate value.*

**Table 5.**
*Carrot, beetroot and radish supplementation effect on oxidative stress as determined by malondialdehyde (MDA) level.*

beetroot compound, as an efficient antioxidant against oxidative stress in rats with acute kidney damage [115]. In hepatotoxic conditions, the oxidative stress on the cells increases; however, black radish extract treatment to human hepatocyte carcinoma (HepG2) cells and rats with liver injury has indicated a dose-dependent increase in hepatic proteins expressions along with radical scavenging by 3-(E)-(methylthio) methylene-2-pyrrolidinethione, a compound isolated from black radish and lipid accumulation prevention which collectively produce a hepatoprotective effect [116]. Commercially available purple sweet potato pigments are also suitable for lowering oxidative stress resulting from hepatic injury in mice [117]. Antioxidants such as vitamin E and anthocyanins of purple carrots act as oxidative damage protectors in rat organs, and when provided in a combination, MDA level drops significantly [118]. A study has indicated 2, 4-di-tert-butylphenol as a prominent antioxidant of sweet potato, lowing oxidative stress in neuronal cell damage in mice [119]. Leave supplementation also effectively reduces oxidative stress in a dose-dependent manner. A study on rats (fed on a cholesterol-rich diet) has indicated an improvement in cholesterol profile along with lower MDA levels upon turnip leaves supplementation, compared to positive control [120]. A similar kind of study on cholesterol-fed hamsters has also shown a reduction in oxidative stress upon supplementing sweet potato leaves [121]. Yams extract has also shown a reduced hyperhomocysteinemic-induced oxidative stress in rats [114].

## 8. Concluding remarks and future perspectives

This chapter has comprehensively and effectively addressed the bioactive compounds in commonly consumed root vegetables. Root vegetables are rich in phenolic compounds, glucosinolates, bioactive pigments, and vitamins, as well as saponins, phytosterols, and volatile aromatic components. The content value of these compounds varied in different root vegetables to a great extent; however, glucosinolates were majorly confined to brassica vegetables. Likewise, anthocyanins and carotenoids were dominant in purple and orange to red-fleshed root vegetables, respectively, and betalains were hardly reported in any vegetable other than beetroot. Moreover, saponins, phytosterols, and terpenes are also reported in considerable amounts. The accumulation of these bioactive compounds was found to be dependent on plant varieties and plant parts, and growth conditions. Also, the post-harvest treatments (heat processing, storage, extraction techniques and parameters, pre-extraction drying) greatly influenced the bioactive compound's recovery. Notably, phenolic compounds, glucosinolate, and vitamin C were highly deteriorated upon heating, while carotenoids were found to be extremely light sensitive. As oxidative stress is related to many health complications, root vegetable supplementation against oxidative stress was reported in this chapter as well. Root vegetables were found to be suitable against oxidative stress in hepatic injury, kidney damage, neuronal cell damage, hyperlipidemic, hyperhomocysteinemic, diabetic, and pre-cancerous conditions due to their antioxidation properties. In the light of all the stated information in the present review, fresh root vegetables can serve as an essential part of the routine diet with only a gentle heat processing if necessary. However, factors affecting the bioactive compounds of root vegetables need to be further studied to improve their accumulation, recovery, and stability to attain more nutraceutical benefits.

## Acknowledgements

A special thanks to INTECHOPEN LIMITED, for funding this book chapter entitles "Bioactive Components of Root Vegetables" as well as their Author Services Manager for time-to-time discussion and solutions.

## Conflict of interest

The authors declare no conflict of interest.

## Abbreviations

| | |
|---|---|
| TPC | total phenolic content |
| TFC | total flavonoid content |
| TTC | total tannin content |
| TCC | total carotenoid content |
| DW | dry weight |
| FW | fresh weight |
| GAE | gallic acid equivalent |
| HPLC-DAD | high performance liquid chromatography-diode-array detection |

## Author details

Rashida Bashir[1*], Samra Tabassum[1], Ayoub Rashid[2], Shafiqur Rehman[3], Ahmad Adnan[2] and Rabia Ghaffar[4]

1 Department of Chemistry, Division of Science and Technology, University of Education Lahore, Lahore, Pakistan

2 Department of Chemistry, GC University Lahore, Lahore, Pakistan

3 Faculty of Sciences, Department of Chemistry, University of Central Punjab, Lahore, Pakistan

4 Department of Botany, Division of Science and Technology, University of Education Lahore, Lahore, Pakistan

*Address all correspondence to: rashida.bashir@ue.edu.pk

## References

[1] Marić L, Malešić E, Jurinjak Tušek A, Benković M, Valinger D, Jurina T, et al. Effects of drying on physical and chemical properties of root vegetables: Artificial neural network modelling. Food and Bioproducts Processing. 2020;**119**:148-160

[2] He FJ, Nowson CA, MacGregor GA. Fruit and vegetable consumption and stroke: Meta-analysis of cohort studies. The Lancet. 2006;**367**(9507):320-326

[3] Lock K, Pomerleau J, Causer L, Altmann DR, McKee M. The global burden of disease attributable to low consumption of fruit and vegetables: Implications for the global strategy on diet. Bulletin of the World health Organization. 2005;**83**:100-108

[4] Li Z, Lee HW, Liang X, Liang D, Wang Q, Huang D, et al. Profiling of phenolic compounds and antioxidant activity of 12 cruciferous vegetables. Molecules. 2018;**23**(5):1139

[5] Chihoub W, Dias MI, Barros L, Calhelha RC, Alves MJ, Harzallah-Skhiri F, et al. Valorisation of the green waste parts from turnip, radish and wild cardoon: Nutritional value, phenolic profile and bioactivity evaluation. Food Research International. 2019;**126**:108651

[6] Francisco M, Moreno DA, Cartea ME, Ferreres F, García-Viguera C, Velasco P. Simultaneous identification of glucosinolates and phenolic compounds in a representative collection of vegetable *Brassica rapa*. Journal of Chromatography A. 2009;**1216**(38):6611-6619

[7] Jakopic J, Veberic R, Slatnar A. Changes in quality parameters in rutabaga (*Brassica napus* var. *napobrassica*) roots during long term storage. LWT. 2021;**147**:111587

[8] Stefanucci A, Zengin G, Llorent-Martinez EJ, Dimmito MP, Della Valle A, Pieretti S, et al. Chemical characterization, antioxidant properties and enzyme inhibition of Rutabaga root's pulp and peel (*Brassica napus* L.). Arabian Journal of Chemistry. 2020;**13**(9):7078-7086

[9] Zhang D, Hamauzu Y. Phenolic compounds and their antioxidant properties in different tissues of carrots (*Daucus carota* L.). Journal of Food Agriculture and Environment. 2004;**2**:95-100

[10] Metzger BT, Barnes DM. Polyacetylene diversity and bioactivity in orange market and locally grown colored carrots (*Daucus carota* L.). Journal of Agricultural and Food Chemistry. 2009;**57**(23):11134-11139

[11] Tumbas Šaponjac V, Čanadanović-Brunet J, Ćetković G, Jakišić M, Vulić J, Stajčić S, et al. Optimisation of beetroot juice encapsulation by freeze-drying. Polish Journal of Food and Nutrition Sciences. 2020;**70**(1):25-34

[12] El-Beltagi HS, Mohamed HI, Megahed BMH, Gamal M, Safwat G. Evaluation of some chemical constituents, antioxidant, antibacterial and anticancer activities of beta Vulgaris L. root. Fresenius Environmental Bulletin. 2018;**27**(9):6369-6378

[13] Shepherd AI, Costello JT, Bailey SJ, Bishop N, Wadley AJ, Young-Min S, et al. Beet the cold: Beetroot juice supplementation improves peripheral blood flow, endothelial function and

anti-inflammatory status in individuals with Raynaud's phenomenon. Journal of Applied Physiology. 2019;**127**(5):1478-1490

[14] Siervo M, Shannon O, Kandhari N, Prabhakar M, Fostier W, Kochl C, et al. Nitrate-rich beetroot juice reduces blood pressure in tanzanian adults with elevated blood pressure: A double-blind randomized controlled feasibility trial. The Journal of Nutrition. 2020;**150**(9):2460-2468

[15] Nikolić NČ, Lazić MM, Karabegović IT, Stojanović GS, Todorović ZB. A characterization of content, composition and scavenging capacity of phenolic compounds in parsnip roots of various weight. Natural Product Communications. 2014;**9**(6):1934578X1400900620

[16] Kumar V, Sharma HK. Process optimization for extraction of bioactive compounds from taro (*Colocasia esculenta*), using RSM and ANFIS modeling. Journal of Food Measurement and Characterization. 2017;**11**(2):704-718

[17] Kundu N, Campbell P, Hampton B, Lin C-Y, Ma X, Ambulos N, et al. Antimetastatic activity isolated from *Colocasia esculenta* (taro). Anti-Cancer Drugs. 2012;**23**(2):200

[18] Jung J-K, Lee S-U, Kozukue N, Levin CE, Friedman M. Distribution of phenolic compounds and antioxidative activities in parts of sweet potato (*Ipomoea batata* L.) plants and in home processed roots. Journal of Food Composition and Analysis. 2011;**24**(1):29-37

[19] Guo H, Kong F, Yan C. Optimization of polysaccharide ultrasonic extraction conditions using purple sweet potato tubers based on free radical scavenging and glycosylation inhibitory bioactivities. Pharmacognosy Magazine. 2017;**13**(51):504

[20] Fang Z, Wu D, Yü D, Ye X, Liu D, Chen J. Phenolic compounds in Chinese purple yam and changes during vacuum frying. Food Chemistry. 2011;**128**(4):943-948

[21] Mahmad N, Taha RM, Othman R, Abdullah S, Anuar N, Elias H, et al. Anthocyanin as potential source for antimicrobial activity in *Clitoria ternatea* L. and *Dioscorea alata* L. Pigment & Resin Technology. 2018;**47**(6):490-495

[22] Adefegha SA, Okeke BM, Oyeleye SI, Oboh G. Effects of processing on starch composition, glycemic indices, phenolic profile, and possible antidiabetic properties of cassava (*Manihot esculenta*) flours. Journal of Food Processing and Preservation. 2021;**45**(7):e15586

[23] Yi B, Hu L, Mei W, Zhou K, Wang H, Luo Y, et al. Antioxidant phenolic compounds of cassava (*Manihot esculenta*) from Hainan. Molecules. 2011;**16**(12):10157-10167

[24] LI L, Shewry PR, Ward JL. Phenolic acids in wheat varieties in the HEALTHGRAIN diversity screen. Journal of Agricultural Food Chemistry. 2008;56:9732-9739.

[25] Dinelli G, Segura-Carretero A, Di Silvestro R, Marotti I, Arraez-Roman D, Benedettelli S, et al. Profiles of phenolic compounds in modern and old common wheat varieties determined by liquid chromatography coupled with time-of-flight mass spectrometry. Journal of Chromatography A. 2011;**1218**(42): 7670-7681

[26] Farahnaky A, Kamali E, Golmakani MT, Gavahian M, Mesbahi G, Majzoobi M. Effect of ohmic and microwave cooking on some bioactive compounds of kohlrabi, turnip, potato, and radish. Journal of Food Measurement and Characterization. 2018;**12**(4):2561-2569

[27] Goyeneche R, Roura S, Ponce A, Vega-Gálvez A, Quispe-Fuentes I, Uribe E, et al. Chemical characterization and antioxidant capacity of red radish (*Raphanus sativus* L.) leaves and roots. Journal of Functional Foods. 2015;**16**:256-264

[28] Pasko P, Bukowska-Strakova K, Gdula-Argasinska J, Tyszka-Czochara M. Rutabaga (*Brassica napus* L. var. *napobrassica*) seeds, roots, and sprouts: A novel kind of food with antioxidant properties and proapoptotic potential in Hep G2 hepatoma cell line. Journal of Medicinal Food. 2013;**16**(8):749-759

[29] Bors MD, Semeniuc CA, Socaci S, Varva L, Moldovan O, Vlaic R, et al. Total phenolic content and antioxidant capacity of radish as influenced by the variety and vegetative stage. Bulletin UASVM Food Science and Technology. 2015;**72**(1):77-81

[30] Romani A, Vignolini P, Isolani L, Ieri F, Heimler D. HPLC-DAD/MS characterization of flavonoids and hydroxycinnamic derivatives in turnip tops (*Brassica rapa* L. subsp. *sylvestris* L.). Journal of Agricultural and Food Chemistry. 2006;**54**(4):1342-1346

[31] Francisco M, Velasco P, Moreno DA, García-Viguera C, Cartea ME. Cooking methods of *Brassica rapa* affect the preservation of glucosinolates, phenolics and vitamin C. Food Research International. 2010;**43**(5):1455-1463

[32] Gonçalves RF, Silva AM, Silva AM, Valentão P, Ferreres F, Gil-Izquierdo A, et al. Influence of taro (*Colocasia esculenta* L. Shott) growth conditions on the phenolic composition and biological properties. Food Chemistry. 2013;**141**(4):3480-3485

[33] Kumar M, Dahuja A, Sachdev A, Kaur C, Varghese E, Saha S, et al. Valorisation of black carrot pomace: Microwave assisted extraction of bioactive phytoceuticals and antioxidant activity using Box-Behnken design. Journal of Food Science and Technology. 2019;**56**(2):995-1007

[34] de Brito FAL, Rãbia K, Jãnior LFC, Coelho DG, da Silva Vieira MR, do Nascimento Simãµes A. Changes in phenol metabolism of minimally processed baby cassava under different temperatures: An alternative to commercialization. African Journal of Biotechnology. 2017;**16**(17):928-936

[35] Hanlon PR, Webber DM, Barnes DM. Aqueous extract from Spanish black radish (*Raphanus sativus* L. Var. *niger*) induces detoxification enzymes in the HepG2 human hepatoma cell line. Journal of Agricultural and Food Chemistry. 2007;**55**(16):6439-6446

[36] Ediage EN, Di Mavungu JD, Scippo M-L, Schneider Y-J, Larondelle Y, Callebaut A, et al. Screening, identification and quantification of glucosinolates in black radish (*Raphanus sativus* L. *niger*) based dietary supplements using liquid chromatography coupled with a photodiode array and liquid chromatography-mass spectrometry. Journal of Chromatography A. 2011;**1218**(28):4395-4405

[37] Yi G, Lim S, Chae WB, Park JE, Park HR, Lee EJ, et al. Root glucosinolate profiles for screening of radish (*Raphanus sativus* L.) genetic resources. Journal of Agricultural and Food Chemistry. 2016;**64**(1):61-70

[38] Jeong J, Park H, Hyun H, Kim J, Kim H, Oh HI, et al. Effects of glucosinolates from turnip (*Brassica rapa* L.) root on bone formation by human osteoblast-like MG-63 cells and in normal young rats. Phytotherapy Research. 2015;**29**(6):902-909

[39] Paśko P, Okoń K, Krośniak M, Prochownik E, Żmudzki P, Kryczyk-Kozioł J, et al. Interaction between iodine and glucosinolates in rutabaga sprouts and selected biomarkers of thyroid function in male rats. Journal of Trace Elements in Medicine and Biology. 2018;**46**:110-116

[40] Dastan K, Tehranifard A, Vazirian M. Investigation of Anti-pseudomanal Activity of *Brassica napus* L. International Conference on Environmental and Computer Science (IPCBEE). Singapore, Singapore: IACSIT Press; 16 Sepmtember 2011. pp. 144-146

[41] Gimsing A, Kirkegaard J. Glucosinolate and isothiocyanate concentration in soil following incorporation of Brassica biofumigants. Soil Biology and Biochemistry. 2006;**38**(8):2255-2264

[42] Oliveira RD, Dhingra OD, Lima AO, Jham GN, Berhow MA, Holloway RK, et al. Glucosinolate content and nematicidal activity of Brazilian wild mustard tissues against Meloidogyne incognita in tomato. Plant and Soil. 2011;**341**(1):155-164

[43] Li S, Schonhof I, Krumbein A, Li L, Stützel H, Schreiner M. Glucosinolate concentration in turnip (*Brassica rapa* ssp. *rapifera* L.) roots as affected by nitrogen and sulfur supply. Journal of Agricultural and Food Chemistry. 2007;**55**(21):8452-8457

[44] Zhang H, Schonhof I, Krumbein A, Gutezeit B, Li L, Stützel H, et al. Water supply and growing season influence glucosinolate concentration and composition in turnip root (*Brassica rapa* ssp. *rapifera* L.). Journal of Plant Nutrition and Soil Science. 2008;**171**(2):255-265

[45] Chen X, Zhu Z, Gerendás J, Zimmermann N. Glucosinolates in Chinese Brassica campestris vegetables: Chinese cabbage, purple cai-tai, choysum, pakchoi, and turnip. HortScience. 2008;**43**(2):571-574

[46] Kapusta-Duch J, Kusznierewicz B, Leszczyńska T, Borczak B. Effect of cooking on the contents of glucosinolates and their degradation products in selected Brassica vegetables. Journal of Functional Foods. 2016;**23**:412-422

[47] Jing P, Zhao S-J, Ruan S-Y, Xie Z-H, Dong Y, Yu L. Anthocyanin and glucosinolate occurrences in the roots of Chinese red radish (*Raphanus sativus* L.), and their stability to heat and pH. Food Chemistry. 2012;**133**(4):1569-1576

[48] Ishida M, Kakizaki T, Morimitsu Y, Ohara T, Hatakeyama K, Yoshiaki H, et al. Novel glucosinolate composition lacking 4-methylthio-3-butenyl glucosinolate in Japanese white radish (*Raphanus sativus* L.). Theoretical and Applied Genetics. 2015;**128**(10):2037-2046

[49] Blažević I, Mastelić J. Glucosinolate degradation products and other bound and free volatiles in the leaves and roots of radish (*Raphanus sativus* L.). Food Chemistry. 2009;**113**(1):96-102

[50] Zhang C, Di H, Lin P, Wang Y, Li Z, Lai Y, et al. Genotypic variation of glucosinolates and their breakdown products in mustard (*Brassica juncea*) seeds. Scientia Horticulturae. 2022;**294**:110765

[51] Doheny-Adams T, Redeker K, Kittipol V, Bancroft I, Hartley SE. Development of an efficient glucosinolate extraction method. Plant Methods. 2017;**13**(1):1-14

[52] Lin L-Z, Sun J, Chen P, Harnly JA. LC-PDA-ESI/MSn identification of new anthocyanins in purple bordeaux radish

(*Raphanus sativus* L. variety). Journal of Agricultural and Food Chemistry. 2011;**59**(12):6616-6627

[53] Kim HW, Kim JB, Cho SM, Chung MN, Lee YM, Chu SM, et al. Anthocyanin changes in the Korean purple-fleshed sweet potato, Shinzami, as affected by steaming and baking. Food Chemistry. 2012;**130**(4):966-972

[54] Zhuang H, Lou Q, Liu H, Han H, Wang Q, Tang Z, et al. Differential regulation of anthocyanins in green and purple turnips revealed by combined de novo transcriptome and metabolome analysis. International Journal of Molecular Sciences. 2019;**20**(18):4387

[55] Polat S, Guclu G, Kelebek H, Keskin M, Selli S. Comparative elucidation of colour, volatile and phenolic profiles of black carrot (*Daucus carota* L.) pomace and powders prepared by five different drying methods. Food Chemistry. 2022;**369**:130941

[56] Macura R, Michalczyk M, Fiutak G, Maciejaszek I. Effect of freeze-drying and air-drying on the content of carotenoids and anthocyanins in stored purple carrot. Acta Scientiarum Polonorum Technologia Alimentaria. 2019;**18**(2):135-142

[57] Wang Y-H, Li T, Zhang R-R, Khadr A, Tian Y-S, Xu Z-S, et al. Transcript profiling of genes involved in carotenoid biosynthesis among three carrot cultivars with various taproot colors. Protoplasma. 2020;**257**(3):949-963

[58] Kimura M, Kobori CN, Rodriguez-Amaya DB, Nestel P. Screening and HPLC methods for carotenoids in sweetpotato, cassava and maize for plant breeding trials. Food Chemistry. 2007;**100**(4):1734-1746

[59] Park CH, Park S-Y, Park YJ, Kim JK, Park SU. Metabolite profiling and comparative analysis of secondary metabolites in Chinese cabbage, radish, and hybrid xBrassicoraphanus. Journal of Agricultural and Food Chemistry. 2020;**68**(47):13711-13719

[60] Mamatha BS, Sangeetha RK, Baskaran V. Provitamin-A and xanthophyll carotenoids in vegetables and food grains of nutritional and medicinal importance. International Journal of Food Science & Technology. 2011;**46**(2):315-323

[61] Sawicki T, Bączek N, Wiczkowski W. Betalain profile, content and antioxidant capacity of red beetroot dependent on the genotype and root part. Journal of Functional Foods. 2016;**27**:249-261

[62] Stintzing FC, Schieber A, Carle R. Identification of Betalains from Yellow beet (*Beta vulgaris* L.) and Cactus Pear [*Opuntia ficus-indica* (L.) Mill.] by High-Performance Liquid Chromatography−Electrospray Ionization Mass Spectrometry. Journal of Agricultural and Food Chemistry. 2002;**50**(8):2302-2307

[63] Wiczkowski W, Szawara-Nowak D, Topolska J. Red cabbage anthocyanins: Profile, isolation, identification, and antioxidant activity. Food Research International. 2013;**51**(1):303-309

[64] Otsuki T, Matsufuji H, Takeda M, Toyoda M, Goda Y. Acylated anthocyanins from red radish (*Raphanus sativus* L.). Phytochemistry. 2002;**60**(1):79-87

[65] Patil G, Madhusudhan MC, Ravindra Babu B, Raghavarao KSMS. Extraction, dealcoholization and concentration of anthocyanin from red radish. Chemical Engineering and Processing: Process Intensification. 2009;**48**(1):364-369

[66] Su N, Wu Q, Liu Y, Cai J, Shen W, Xia K, et al. Hydrogen-rich

water reestablishes ROS homeostasis but exerts differential effects on anthocyanin synthesis in two varieties of radish sprouts under UV-A irradiation. Journal of Agricultural and Food Chemistry. 2014;**62**(27):6454-6462

[67] Zhang X, Wei J, Huang Y, Shen W, Chen X, Lu C, et al. Increased cytosolic calcium contributes to hydrogen-rich water-promoted anthocyanin biosynthesis under UV-A irradiation in radish sprouts hypocotyls. Frontiers in Plant Science. 2018;**9**:1020

[68] Zhou B, Wang Y, Zhan Y, Li Y, Kawabata S. Chalcone synthase family genes have redundant roles in anthocyanin biosynthesis and in response to blue/UV-A light in turnip (*Brassica rapa*; Brassicaceae). American Journal of Botany. 2013;**100**(12):2458-2467

[69] Ersus S, Yurdagel U. Microencapsulation of anthocyanin pigments of black carrot (*Daucus carota* L.) by spray drier. Journal of Food Engineering. 2007;**80**(3):805-812

[70] Elvira-Torales LI, García-Alonso J, Periago-Castón MJ. Nutritional importance of carotenoids and their effect on liver health: A review. Antioxidants. 2019;**8**(7):229

[71] Jaber BA, Majeed KR, Al-Hashimi AG. Isolation and identification of new strains of bacteria producing carotenoids pigments which isolated from a different sources in Basrah, Iraq. Basrah Journal of Agricultural Sciences. 2022;**35**(1):81-93

[72] Clotault J, Peltier D, Berruyer R, Thomas M, Briard M, Geoffriau E. Expression of carotenoid biosynthesis genes during carrot root development. Journal of Experimental Botany. 2008;**59**(13):3563-3573

[73] Koca Bozalan N, Karadeniz F. Carotenoid profile, total phenolic content, and antioxidant activity of carrots. International Journal of Food Properties. 2011;**14**(5):1060-1068

[74] Antonious GF, Kasperbauer MJ. Color of light reflected to leaves modifies nutrient content of carrot roots. Crop Science. 2002;**42**(4):1211-1216

[75] Fuentes P, Pizarro L, Moreno JC, Handford M, Rodriguez-Concepcion M, Stange C. Light-dependent changes in plastid differentiation influence carotenoid gene expression and accumulation in carrot roots. Plant Molecular Biology. 2012;**79**(1):47-59

[76] Fratianni A, Niro S, Messia MC, Panfili G, Marra F, Cinquanta L. Evaluation of carotenoids and furosine content in air dried carrots and parsnips pre-treated with pulsed electric field (PEF). European Food Research and Technology. 2019;**245**(11):2529-2537

[77] Iqbal S, Younas U, Chan KW, Saeed Z, Shaheen MA, Akhtar N, et al. Growth and antioxidant response of *Brassica rapa* var. *rapa* L. (turnip) irrigated with different compositions of paper and board mill (PBM) effluent. Chemosphere. 2013;**91**(8):1196-1202

[78] Yücetepe A, Altin G, Özçelik B. A novel antioxidant source: Evaluation of in vitro bioaccessibility, antioxidant activity and polyphenol profile of phenolic extract from black radish peel wastes (*Raphanus sativus* L. var. *niger*) during simulated gastrointestinal digestion. International Journal of Food Science & Technology. 2021;**56**(3):1376-1384

[79] Lazăr S, Constantin OE, Stănciuc N, Aprodu I, Croitoru C, Râpeanu G. Optimization of betalain pigments extraction using beetroot

by-products as a valuable source. Inventions. 2021;**6**(3):50

[80] Coe KM, Ellison S, Senalik D, Dawson J, Simon P. The influence of the Or and Carotene Hydroxylase genes on carotenoid accumulation in orange carrots [*Daucus carota* (L.)]. Theoretical and Applied Genetics. 2021;**134**(10):3351-3362

[81] Tang G, Qin J, Dolnikowski GG, Russell RM, Grusak MA. Spinach or carrots can supply significant amounts of vitamin A as assessed by feeding with intrinsically deuterated vegetables. The American Journal of Clinical Nutrition. 2005;**82**(4):821-828

[82] Cordeiro A, Bento C, Matos A, Ramalho A. Vitamin A deficiency is associated with body mass index and body adiposity in women with recommended intake of vitamin A. Nutrición Hospitalaria. 2018;**35**:1072

[83] Jastrebova J, Witthöft C, Grahn A, Svensson U, Jägerstad M. HPLC determination of folates in raw and processed beetroots. Food Chemistry. 2003;**80**(4):579-588

[84] Bailey LB. Folate and vitamin B 12 recommended intakes and status in the United States. Nutrition Reviews. 2004;**62**(suppl_1):S14-S20

[85] Odewole OA, Williamson RS, Zakai NA, Berry RJ, Judd SE, Qi YP, et al. Near-elimination of folate-deficiency anemia by mandatory folic acid fortification in older US adults: Reasons for Geographic and Racial Differences in Stroke study 2003-2007. The American Journal of Clinical Nutrition. 2013;**98**(4):1042-1047

[86] Azam A, Khan I, Mahmood A, Hameed A. Yield, chemical composition and nutritional quality responses of carrot, radish and turnip to elevated atmospheric carbon dioxide. Journal of the Science of Food and Agriculture. 2013;**93**(13):3237-3244

[87] Levine M, Wang Y, Padayatty SJ, Morrow J. A new recommended dietary allowance of vitamin C for healthy young women. Proceedings of the National Academy of Sciences. 2001;**98**(17):9842-9846

[88] Dattola A, Silvestri M, Bennardo L, Passante M, Scali E, Patruno C, et al. Role of vitamins in skin health: A systematic review. Current Nutrition Reports. 2020;**9**(3):226-235

[89] Liu M, Seo H-Y, Min S, Ku K-M. Phytonutrients and metabolism changes in topped radish root and its detached leaves during 1°C cold postharvest storage. Horticulturae. 2021;**8**(1):42

[90] Nicolle C, Gr S, Rock E, Amouroux P, Rémésy C. Genetic variability influences carotenoid, vitamin, phenolic, and mineral content in white, yellow, purple, orange, and dark-orange carrot cultivars. Journal of the American Society for Horticultural Science. 2004;**129**(4):523-529

[91] Lewis ED, Meydani SN, Wu D. Regulatory role of vitamin E in the immune system and inflammation. IUBMB Life. 2019;**71**(4):487-494

[92] Troesch B, Hoeft B, McBurney M, Eggersdorfer M, Weber P. Dietary surveys indicate vitamin intakes below recommendations are common in representative Western countries. British Journal of Nutrition. 2012;**108**(4):692-698

[93] Mahajan AS, Babbar R, Kansal N, Agarwal SK, Ray PC. Antihypertensive

and antioxidant action of amlodipine and vitamin C in patients of essential hypertension. Journal of Clinical Biochemistry and Nutrition. 2007;**40**(2):141-147

[94] Johnson JB, Budd C, Mani JS, Brown P, Walsh KB, Naiker M. Carotenoids, ascorbic acid and total phenolic content in the root tissue from five Australian-grown sweet potato cultivars. New Zealand Journal of Crop and Horticultural Science. 2022;**50**(1):32-47

[95] Matějková J, Petříková K. Variation in content of carotenoids and vitamin C in carrots. Notulae Scientia Biologicae. 2010;**2**(4):88-91

[96] Luby CH, Maeda HA, Goldman IL. Genetic and phenological variation of tocochromanol (vitamin E) content in wild (*Daucus carota* L. var. carota) and domesticated carrot (*D. carota* L. var. sativa). Horticulture Research. 2014;**1**:14015

[97] Bellani LM, Salvini L, Dell'Aquila A, Scialabba A. Reactive oxygen species release, vitamin E, fatty acid and phytosterol contents of artificially aged radish (*Raphanus sativus* L.) seeds during germination. Acta Physiologiae Plantarum. 2012;**34**(5):1789-1799

[98] Aremu M, Ajine PL, Omosebi MO, Baba NM, Onwuka JC, Audu SS, et al. Lipid profiles and health promoting uses of carrot (*Daucus carota* L.) and cucumber (*Cucumis sativus* L.). International Journal of Sciences. 2021;**10**(07):22-29

[99] Lee KB, Kim YJ, Kim HJ, Choi J, Kim JK. Phytochemical profiles of Brassicaceae vegetables and their multivariate characterization using chemometrics. Applied Biological Chemistry. 2018;**61**(2):131-144

[100] Han B, Jiang P, Jiang L, Li X, Ye X. Three phytosterols from sweet potato inhibit MCF7-xenograft-tumor growth through modulating gut microbiota homeostasis and SCFAs secretion. Food Research International. 2021;**141**:110147

[101] Lebot V, Faloye B, Okon E, Gueye B. Simultaneous quantification of allantoin and steroidal saponins in yam (Dioscorea spp.) powders. Journal of Applied Research on Medicinal and Aromatic Plants. 2019;**13**:100200

[102] Aumsuwan P, Khan SI, Khan IA, Ali Z, Avula B, Walker LA, et al. The anticancer potential of steroidal saponin, dioscin, isolated from wild yam (*Dioscorea villosa*) root extract in invasive human breast cancer cell line MDA-MB-231 in vitro. Archives of Biochemistry and Biophysics. 2016;**591**:98-110

[103] Son IS, Kim JH, Sohn HY, Son KH, Kim J-S, Kwon C-S. Antioxidative and hypolipidemic effects of diosgenin, a steroidal saponin of Yam (Dioscorea spp.), on high-cholesterol fed rats. Bioscience, Biotechnology, and Biochemistry. 2007;**71**(12):3063-3071

[104] Mikołajczyk-Bator K, Błaszczyk A, Czyżniejewski M, Kachlicki P. Characterisation and identification of triterpene saponins in the roots of red beets (*Beta vulgaris* L.) using two HPLC–MS systems. Food Chemistry. 2016;**192**:979-990

[105] Dini I, Tenore GC, Dini A. Saponins in Ipomoea batatas tubers: Isolation, characterization, quantification and antioxidant properties. Food Chemistry. 2009;**113**(2):411-419

[106] Muchlinski A, Ibdah M, Ellison S, Yahyaa M, Nawade B, Laliberte S, et al. Diversity and function of terpene synthases in the production of carrot aroma and flavor compounds. Scientific Reports. 2020;**10**(1):9989

[107] Lobo V, Patil A, Phatak A, Chandra N. Free radicals, antioxidants and functional foods: Impact on human health. Pharmacognosy Reviews. 2010;**4**(8):118-126

[108] Embugushiki RE, Mafulul SG, Okoye ZS. Protective effect of carrot juice pretreatment on cadmium-induced oxidative cytotoxic damage to some rat tissues. 2013.

[109] Potter AS, Foroudi S, Stamatikos A, Patil BS, Deyhim F. Drinking carrot juice increases total antioxidant status and decreases lipid peroxidation in adults. Nutrition Journal. 2011;**10**(1):1-6

[110] Klewicka E, Nowak A, Zduńczyk Z, Juśkiewicz J, Cukrowska B. Protective effect of lactofermented red beetroot juice against aberrant crypt foci formation, genotoxicity of fecal water and oxidative stress induced by 2-amino-1-methyl-6-phenylimidazo[4,5-b] pyridine in rats model. Environmental Toxicology and Pharmacology. 2012;**34**(3):895-904

[111] Sinaga FA, Hasibuan R, Risfandi M, Sinaga RN. Hepatoprotective effects of beetroot juice at maximum physical activity. Journal of Physics: Conference Series. 2020;**1462**(1):012007

[112] Salah-Abbès JB, Abbès S, Ouanes Z, Houas Z, Abdel-Wahhab MA, Bacha H, et al. Tunisian radish extract (*Raphanus sativus*) enhances the antioxidant status and protects against oxidative stress induced by zearalenone in Balb/c mice. Journal of Applied Toxicology: An International Journal. 2008;**28**(1):6-14

[113] Noviati TD, Tamtomo DG, Sugiarto S. The effect of purple sweet potato jelly on malondialdehyde and fasting blood sugar in experimental type 2 diabetic rat model. International Journal of Nutrition Sciences. 2019;**4**(4):181-185

[114] Chang S-J, Lee Y-C, Liu S-Y, Chang T-W. Chinese Yam (*Dioscorea alata* cv. Tainung No. 2) feeding exhibited antioxidative effects in hyperhomocysteinemia rats. Journal of Agricultural and Food Chemistry. 2004;**52**(6):1720-1725

[115] Tan D, Wang Y, Bai B, Yang X, Han J. Betanin attenuates oxidative stress and inflammatory reaction in kidney of paraquat-treated rat. Food and Chemical Toxicology. 2015;**78**:141-146

[116] Ahn M, Kim J, Hong S, Kim J, Ko H, Lee N-H, et al. Black radish (*Raphanus sativus* L. var. *niger*) extract mediates its hepatoprotective effect on carbon tetrachloride-induced hepatic injury by attenuating oxidative stress. Journal of Medicinal Food. 2018;**21**(9):866-875

[117] Zhang Z-F, Fan S-H, Zheng Y-L, Lu J, Wu D-M, Shan Q, et al. Purple sweet potato color attenuates oxidative stress and inflammatory response induced by d-galactose in mouse liver. Food and Chemical Toxicology. 2009;**47**(2):496-501

[118] Li X, Zhang Y, Yuan Y, Sun Y, Qin Y, Deng Z, et al. Protective effects of selenium, vitamin E, and purple carrot anthocyanins on d-galactose-induced oxidative damage in blood, liver, heart and kidney rats. Biological Trace Element Research. 2016;**173**(2):433-442

[119] Choi SJ, Kim JK, Kim HK, Harris K, Kim C-J, Park GG, et al. 2, 4-Di-tert-butylphenol from sweet potato protects against oxidative stress in PC12 cells and in mice. Journal of Medicinal Food. 2013;**16**(11):977-983

[120] El-din MME, editor The Protective Effect of Turnip Leaves

Against Oxidative Stress Induced by High Cholesterol Diet in Adult Rats. World Applied Sciences Journal. 2012;**20**(1):154-163

[121] Chang H-H, Lan Y-C, Chung S-D, Chien C-T. Sweet potato leaf feeding decreases cholesterol, oxidative stress and thrombosis formation in syrian hamsters with a high-cholesterol diet. Life. 2021;**11**(8):802

Chapter 4

# Root Vegetables Having Medicinal Properties: Their Possible Use in Pharmaceutical and Food Industries

*Saleem Siddiqui, Naseer Ahmed, Chongtham Allaylay Devi, Puthem Robindro Singh and Bawitlung Lalramhlimi*

## Abstract

Root, bulb, or tuber vegetables, which are borne underground, are reported to be dense in essential nutrients and come with several health benefits. Most of these root vegetables are the cultivated ones, but few are still underexploited. The root vegetables are consumed either wholly or partially and raw or after processing. They are high in fiber but low in fat and cholesterol. There are wide varieties of bioactive phytochemicals present in them that may contribute to their medicinal and nutraceutical properties. Although some research work has been conducted to uncover the pharmacological effects of root vegetables, their unlimited potential has yet to be fully exploited. The pharmaceutical industry can develop various health-promoting herbal formulations with medicinal properties. The food industry can employ novel processing technologies to preserve nutrition and prevent degradation of the phytochemicals during processing or for value addition of food products. The information presented in this chapter would be helpful for researchers, nutritional and medical professionals, pharmaceutical companies, and the food industry to design and develop effective medicines, drugs, and value-added food products by exploiting the specific as well as multiple modes of action of the various root vegetables.

**Keywords:** antioxidant, antimicrobial, bio-preservative, curing, food products, medicinal, nutrients, phytochemicals, root vegetables, value addition

## 1. Introduction

An increasing number of root vegetables have received attention due to the presence of bioactive compounds in them. Among them, root crops are important crops with swollen underground edible parts that are rich in carbohydrates, dietary fibers, protein, vitamins, fats, and minerals. The storage roots are enriched with specific bioactive substances that can control one or more metabolic pathways, hence promoting improved health conditions. Numerous bioactive elements found in root vegetables, including glucosinolates, isothiocyanates, phenolic compounds, flavonoids, organic acids, etc., have been discovered to contribute to a variety of health-beneficial effects, reduced risk of non-communicable diseases, and delayed onset of age-related disorders [1].

The importance of vegetables in our diet is well known. The root vegetables, unlike leaf or fruit vegetables, have a comparatively higher shelf life. They can be stored for a relatively longer time and have a wider consumption pattern. The nutritional value of different root vegetables varies with species, cultivar, cultivation method, maturity stage, and storage and processing conditions. The optimum utilization of the nutrients from root vegetables is, however, dependent on the method of consumption. The root vegetables may be consumed raw in the form of salad or after being cooked alone or in combination with other vegetables. Depending upon their nature and pattern of consumption, they can be freshly preserved, minimally processed, canned, frozen, or dried. The root vegetables are also processed into various novel value-added and shelf-stable food products. The by-products obtained from them or their unutilized peel and pomace can be used by both pharmaceutical and food industries. This chapter will highlight the beneficial medicinal and nutritional properties of some of the widely cultivated and underutilized root vegetable crops. To widen the scope of discussion, the vegetables having the edible part borne underground have been considered as root vegetables, though morphologically, they may not be representing root.

## 2. Different root crops and their health benefits

Different root vegetables are known for their distinctive nutritional and phytochemical constituents. The various phytochemicals present in them have different health benefits and great prospects in nutraceuticals and pharmaceutical industries. The various nutritional and bioactive compounds present along with the associated medicinal uses of different root crops are discussed below in alphabetical order.

### 2.1 Beetroot

Beetroot (*Beta vulgaris*), which belongs to the family Amaranthaceae, is also known as red beet, garden beet, table beet, or beet. Due to the high concentration of physiologically active ingredients including betalain, inorganic nitrates, polyphenols, vitamins, and folates as well as minerals (Na, K, Ca, and Mg) present in its tuberous root, it is consumed as a vegetable all over the world [2]. Its leaves are also edible. The plant also contains carbohydrates, proteins, fatty acids, vitamins, and fibers. The taproots have high sucrose content (15 to 20% of fresh weight), which makes them suitable for the industrial production of sugar [3, 4]. The roots contain phenolic acids (such as caffeic acid, ellagic acid, syringic acid, vanillic acid, and ferulic acid) and flavonoids (such as rutin, myricetin, kampferol, and quercetin) [2]. Beetroot also contains both water- and fat-soluble vitamins. In decreasing order of concentration, the vitamins are: vitamin B2 > > vitamin C > > vitamin B3 > vitamin E > vitamin B5 > vitamin B1 > vitamin B6 > vitamin K [4].

It has been reported by others [4, 5] that *B. vulgaris* has an abundance of a nitrogenous pigment betalain, which is water soluble. Betalain is a member group of secondary phytochemical and phenolic acids. Two major categories of betalains exist in the plant. The first is betacyanin, which is a red pigment, and the second is known as betaxanthin, which is a yellow pigment. The predominant betalains present in the beetroot are betaxanthin, isobetanin, and betanin (betacyanins). Betacyanins, mostly betanin and its isomers, are the major coloring pigments in red beetroot, whereas Vulgaxanthin I is responsible for the yellow pigments in beetroot.

Beetroot is used as a vegetable, and its juice and extracts are commonly used for medicinal and food purposes. However, the oxalic acid constitution of beetroot is relatively abundant (94.6–141.6 mg/100g and 300–525 mg/L). Oxalic acid encourages the development of nephroliths, and hence, beetroot is thought to be unhealthy, especially for people who are prone to kidney diseases [2].

The therapeutic properties of beetroot include antioxidant, antidepressant, antimicrobial, antifungal, anti-inflammatory, diuretic, aphrodisiac, expectorant, and carminative properties as well as anticarcinogenic and cardiovascular health protection [4]. Beetroot juice supplementation has been reported to be a cost-effective strategy in controlling diabetes and insulin hemostasis, blood pressure and vascular function, renal health, and the possible effect on microbiome abundance [6]. Betanin present in the root extract has been found to possess antioxidant activity that is 10 times higher than in tocopherol and three times higher than in catechin [2]. Chen et al. [7] reported that betalains minimize oxidative and nitrative stress by scavenging DPPH, preventing DNA deterioration, and lowering LDL. Nitrate in beetroots lowers blood lipids, glucose, and hypertension in various chronic conditions, enhances athletic performance, and reduces muscle soreness.

### 2.2 Black salsify

Black salsify (*Scorzonera hispanica* L.) is commonly known as Spanish salsify, black oyster plant, serpent root, viper's grass, or simply scorzonera. It belongs to the family Asteraceae. Lendzion *et al.* [8] reported that the scorzonera genus is the source of a wide range of bioactive chemicals that have the properties of wound healing, anti-inflammatory, pain-relieving, antioxidant, and cytotoxic agents against cancer cell lines. The primary benefits of inulin found in the roots of black salsify are its prebiotics and probiotic properties, regulation of lipid metabolism and diabetes, and immunomodulatory qualities. Numerous bioactive substances, such as triterpenoids, sesquiterpenoids, flavonoids, or derivatives of caffeic and quinic acids, have been found in extracts taken from the plant's aerial and subaerial portions. The anti-inflammatory, analgesic, and hepatoprotective effects of black salsify have also been identified, in addition to its antioxidant and cytotoxic capabilities [8].

Scorzonera species have activity against several bacteria and fungi strains. Their effectiveness in wound-healing therapy, treatment of microbial infections, viral infection-induced fever, and poisonous ulcers and as a lactation-inducing and diuretic agent has also been reported [8, 9]. Roots of *S. hispanica* L. have been used as mucolytic agents in pulmonary diseases; appetite stimulation; defeating a cold, and treating carbuncle, inflammation, and fever [10].

### 2.3 Carrot

Carrot (*Daucus carota* L.) is the most important crop of the *Apiaceae* family. Carrot root flesh can have a white, yellow, orange, red, purple, or a very dark purple hue. The majority of varieties of carrot has yellow and orange flesh. The color of the flesh is because of carotenoids. The common yellow carrot is a good source of pro-vitamin A and β- and α- carotene. The yellow color in carrots is due to the presence of lutein. The red color in red carrot is because of a high lycopene content, and the purple color in carrot is due to the presence of a higher concentration of anthocyanins. Meanwhile, carrots with a white flesh have relatively lesser pigments.

Besides being a rich source of vitamins (A, B, and C) and β-carotene, carrot also contains significant amounts of pantothenic acid, folic acid, and vitamins E and H. The pro-healthy nature of carrot is due to the presence of significant amounts of trace elements (K, Na, Ca, Mg, P, S, Mn, Fe, Cu, and Zn) along with vitamins and antioxidants, particularly carotene and phenolic compounds [11].

Carrots contain bioactive polyacetylenes including falcarindiol, falcarindiol-3-acetate, and falcarinol (a synonym for panaxynol) [12, 13]. The antioxidant, anti-carcinogenic, and immunity-boosting properties of carrots are due to the presence of carotenoids, polyphenols, and vitamins. In addition to these properties, carrots help in reducing the cholesterol level, prevent cardiovascular disease, and cure hypersensitivity and diabetes. Carrots also show anti-hypertensive, hepatoprotective, renoprotective, and wound-healing benefits. The carrot taproot and seed extracts are reported to have antibacterial, antifungal, anti-inflammatory, and analgesic properties [14, 15].

### 2.4 Cassava

Cassava, also known as manioc or yucca, belongs to the family Euphorbiaceae and is known for its nutty-flavored, starchy root vegetable or tuber. The varieties *Manihot esculenta* or *Manihot utilissima* (bitter cassava) are popular for their medicinal properties, and *Manihot dulcis* or *Manihot palmata* (sweet cassava) are cultivated for their tuberous roots, which yield important food products. Cassava root has moisture (5.85–7.30%), ashes (0.8–2.4%), proteins (0.25–1.25%), fat (2.01–3.70%), and fiber (0.98–2.31%). Cassava is also rich in calcium and manganese. The common anti-nutritional factors found in plants are cyanide, phytates, nitrates and nitrites, phenolic compounds, and oxalates [16, 17].

It contains carotenes, vitamin C, vitamin A, anthocyanins (flavonoids), saponins, steroids, and glycosides. Additionally, 10 antioxidant substances have been isolated and identified, including coniferaldehyde, isovanillin, 6-deoxyjacareubin, scopoletin, syringaldehyde, pinoresinol, p-coumaric acid, ficusol, balanophonin, and ethamivan [18]. Cassava has been used as a treatment for a variety of illnesses, including diabetes; celiac disease; bone and neurological health; cardiovascular disease; allergies; and issues with the prostate, gastro-intestinal tract, and blood pressure. It has high amounts of fibers and thereby eliminates constipation, bloating, and intestinal pain. However, if cassava is not prepared, processed, or cooked properly, it can be poisonous due to the presence of cyanide and other toxicants [16].

### 2.5 Garlic

Garlic (*Allium sativum* L.) belongs to the family *Amaryllidaceous*, and its bulb with cloves is mostly used for food, spice, and medicinal purposes. The composition of fresh, raw garlic bulbs includes 66% water, 27% carbohydrates, 2.5% protein, 1.3% amino acids, 1.6% fiber, fatty acids, phenols, trace minerals, and 2.4% sulfur-containing compounds [19].

Garlic contains various widely recognized types of phytochemicals. These bioactive compounds have the potential to treat a wide range of diseases and are essential for maintaining human health. Garlic has a unique nutritional profile with particular emphasis on its many bioactive components, which may be employed in various diet-based treatments of various ailments that are tied to a certain lifestyle. Polyphenols, amino acids, benzenoids, sulfur-containing substances, fatty acyls, glycerophospholipids, heteroaromatic substances, indoles, phenol lipids, pyrrolizines, quinolines, steroid derivatives, tetrahydrofurans, and other substances make up the majority of the phytochemicals. As shown in **Figure 1**, garlic consists of active compounds such as alliin, allicin, methiin, S-allylcysteine, diallyl sulfide, S-allymercapto cysteine, diallyl disulfide, diallyl trisulfide, and methyl allyl disulfide [19, 20]. The sulfur-containing compounds allicin, diallyl sulfide, diallyl disulfide, diallyl trisulfide, alliin, S-allylcysteine, and S-allylmercaptocysteine present in garlic are reported to cure cancer [21]. The beneficial effects of the consumption of garlic on health and the treatment of cancer have also been attributed to the seleno-compounds (Se-compounds) present in garlic [22].

Allicin is not produced *de novo* in fresh garlic; rather, an unstable compound known as alliin (S -3- (2-propenylsulfinyl)-l-alanine-) and a proteinous enzyme alliinase are *in situ*. On injury of the plant, alliinase converts alliin to allicin, which gives garlic its characteristic aroma. Allicin is also unstable and gets easily decomposed on heating to allyl and methyl sulfide derivatives (diallyl mono-, di-, and trisulfide). The allyl sulfur-based compounds, which are the product of the stream decomposition of allicin, are the major components of garlic oil [23]. Alliin, allicin, methiin, S-allylcysteine, diallyl sulfide, S-allymercapto cysteine, diallyl disulfide, diallyl trisulfide, and methyl allyl disulfide are the major bioactive compounds responsible for the oral health benefits of garlic bulbs [20]. In certain cases, allergies are caused by allyl methyl sulfide [19].

There are various commercially available garlic-based products in the market with certain health benefits. These products are divided into four main groups, that is, aged garlic extract, dried garlic powder, garlic oil, and garlic oil macerate. Garlic exhibits a range of anti-inflammatory, antibacterial, antifungal, immunomodulatory, hepatoprotective, digestive system protective, anti-cancer, anti-diabetic, anti-obesity, neuroprotective, renal protective, anti-Alzheimer, and antioxidant properties [23, 24]. The antioxidant and antibacterial effects of stored garlic are higher as compared to

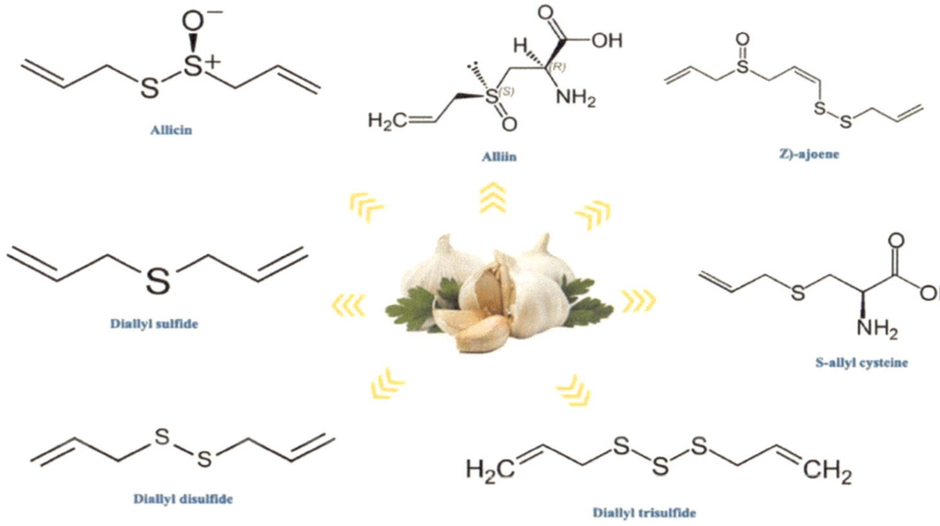


**Figure 1.**
*Structure of the important bioactive constituents present in garlic [20].*

those of freshly harvested garlic. Aged garlic shows more potent antioxidant and antibacterial effects than fresh garlic [25]. For use as a diuretic, diaphoretic, expectorant, and stimulant, garlic is prepared and offered in a variety of ways, including extract, decoct, infusion, tincture, and syrup. Easy-to-use allicin-based products such as chewing gum, garlic gel, mouth fresheners, and various confectionary products have also been developed [20].

### 2.6 Ginger

Ginger (*Zingiber offcinale* Roscoe) belongs to the Zingiberaceae family. The rhizomes of ginger are widely used for culinary and medicinal purposes. Fresh ginger rhizomes are mostly composed of water (80.7%), minerals (1.2%), protein (2.3%), fiber (2.4%), and fat (1.0%). Ginger contains the minerals calcium, magnesium, iron, phosphorus, sodium, and potassium [26]. Shoaib *et al.* [27] reviewed and reported that ginger contains 1–3% volatile oils, and these substances are also responsible for its flavor and scent. Dry ginger has a different scent and flavor from fresh ginger because the volatile aromatic components of ginger are lost during drying or heat processing. The three primary compounds in ginger's volatile oil are zingiberene, curcumene, and farnesene. A total of 40 distinct molecules are also found, with 1,8-cineole, linalool, borneol, neral, and geraniol being the most prevalent. The main ingredients in non-volatile oil include zingerone, paradols, shogaols, and gingerols, which give off a fiery taste or hot sensation in the tongue. Abdullahi *et al.* [28] reported that the volatile phytochemicals of domestic ginger are α-zingiberene (18.56%), geranial (13.88%), neral (10.75%), trans-caryophyllene (9.64%), eucalyptol (5.05%), β-phellandrene (5.51%), camphene (5.34%), α-pinene (2.05%), and heptan-2-ol (1.05%).

The primary polyphenols in fresh ginger are gingerols, including 6, 8, and 10-gingerol, which are the bioactive ingredients in ginger that exhibit a variety of health advantages. Gingerols can be changed into corresponding shogaols by heat treatment or extended storage. On hydrogenation, the shogaols can be converted into paradols. The other phenolic components found in ginger are quercetin, zingerone, gingerenone-A, and 6-dehydrogingerdione. In addition to these, ginger contains a number of terpene elements, including bisabolene, curcumene, zingiberene, farnesene, and sesquiphellandrene, which are thought to be the primary components of ginger essential oils. In addition to these, ginger also contains polysaccharides, lipids, organic acids, and raw fibers [29]. Along with hepatoprotective and antiallergic effects, antioxidant, anti-inflammatory, antibacterial, anticancer, neuroprotective, cardiovascular protective, respiratory protective, anti-obesity, antidiabetic, anti-nausea, and antiemetic are just a few of the biological effects of ginger.

The essential oils (EOs) of ginger have been reported to possess antimicrobial properties. Abdullahi *et al.* [28] reported that even at a lower concentration (1 ml/ml), EOs exhibited considerable inhibition of fungal pathogens. *Fusarium oxysporum* exhibited the highest inhibition, and the lowest was *Ganoderma boninense*. The order of the sensitivity (descending order) was *Fusarium oxysporum* > *Colletotrichum falcatum* > *Pyricularia oryzae* > *Rigidoporus microporus* > *Ganoderma boninense*. Similarly, ginger EOs showed significant antibacterial activity at a concentration range of 100 to 500 ml/ml. It was the most effective against *Xanthomonas oryzae* pv, oryzae strain A, *X. oryzae* pv. oryzae strain B, *Ralstonia solanacearum*, and *Klebsiella sp*. and least effective against *Bacillus* sp. Thus, ginger EOs are among the natural products that can serve as an alternative to natural antimicrobials because of their broad metabolite spectrum, which might open the door for new and more powerful compounds for controlling plant diseases. The nano-emulsion ginger essential oils have also been incorporated into gelatin-based films to produce activated films with improved physical properties and with antimicrobial and antioxidant activities [30].

### 2.7 Jerusalem artichoke

The Jerusalem artichoke (*Helianthus tuberosus*), also called sunchoke, belongs to the Asteraceae family. It is known for its carbohydrate-rich tubers, which can vary significantly in their size, shape, and color. The dried Jerusalem artichoke tubers contain crude protein (8.26%), crude fiber (5.92%), ash (6.82%), and inulin (73.50%) [31, 32].

Jerusalem artichoke tubers are used to obtain inulin, which is a chain of 1,2-D-fructose with a glucose terminal. When ingested or used in food products, inulin has a number of positive health effects. Inulin is not digestible by the human digestive system due to its distinct structure of fructose and glucose molecules. When it enters the large intestine and is fermented by microbes, its advantages become apparent. Prebiotic and probiotic benefits are stimulated by this process, which boosts the development of helpful bacteria and supports better digestive health. In addition, inulin may replace sugar or fat in diet and even makes it easier for the body to absorb minerals in the large intestine. Dietary inulins are hard to digest for humans, and hence, they serve as an ideal bulking agent, increase stool frequency, and show hypolipidemic functions [31].

A number of bioactive compounds have been isolated from the aerial parts of Jerusalem artichoke, demonstrating antifungal, antioxidant, anticancer activities, and other medicinal properties [32]. The various food, pharmaceutical, and chemical industries' applications of Jerusalem artichoke are presented in **Figure 2**.

### 2.8 Onion

Onion (*Allium cepa L.*), also known as bulb onion, belongs to the family of Amaryllidaceae. Based on color, there are 3 different types of onions: red, yellow, and white. Each has a unique flavor and level of pungency, ranging from moderate to extremely strong. Its underground bulbs are widely used for food and pharmaceutical purposes; however, young green leaf tops are also used for culinary purposes. Sami *et al.* [33] reported that onion bulbs were rich in proteins (9.22–13.21 g/100 g FW), with low fiber (1.7 g/100 g FW) and sugar (4.2 g/100 g FW) contents. Red and yellow

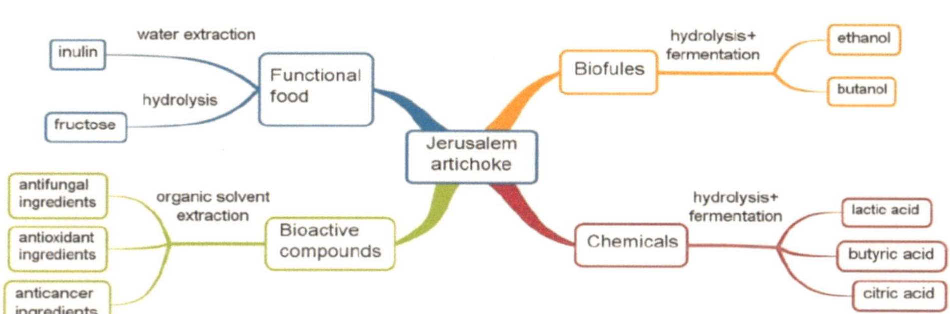


**Figure 2.**
*Food and non-food applications of Jerusalem artichoke [31].*

varieties showed high vitamin C (45.07 mg/100 g FW) and carotenoid (1.44 μg/mL FW) contents, respectively. The major amino acid was arginine, followed by glutamic and aspartic acids. The major elements present were calcium, iron, and sulfur.

Onions provide not only flavor but also health-promoting phytochemicals. The amounts of phytonutrients have been reported to be more in brown as compared to white and red onions. The four different diallyl sulfides, viz., diallylmonosulfide, diallyldisulfide, diallyltrisulfide, and diallyltetrasulfide, are the major organosulfur compounds that are present in onions [34]. Flavonoids, a subgroup of the polyphenol family, are abundant in onions. A major and important dietary flavonoid found in onions, quercetin, is a member of the flavonol subclass of flavonoids. In addition to quercetin, onions have also been shown to contain additional flavonols such as kaempferol and isorhamnetin [35]. The organo-sulfur compounds are attributed to the antibacterial, antiallergenic, anti-inflammatory, and antithrombotic properties of onion. In addition to having important biological functions for maintaining health, flavonols found in onions, such as quercetin and kaempferol, also protect the brain and have antiviral, antibacterial, anti-inflammatory, and anticancer properties [35, 36]. Onion contains sufficient amounts of ascorbic acid as antioxidant and fructo-oligosaccharides and prebiotics. These fructo-oligosaccharides retard the growth of potentially harmful bacteria, thus reducing the risk of emerging tumors in the colon, and also as work as prebiotics to promote the growth of healthy *Bifidobacterium* [37]. According to Zhao *et al.* [38], onion powder, juice, and extracts are beneficial for managing and preventing a number of illnesses, including obesity, hypertension, leukemia, heart disease, nephritis, respiratory issues, colitis, and sterility. They also showed that the ethanolic extract from onion peel might prevent some harmful bacteria from growing and retard oxidation in cooked beef [35]. Onion peels' antibacterial and antioxidant properties help in the inhibition of cancer cell growth [36].

The detailed effects of postharvest primary processing treatments on the bioactive compounds of onion have been reviewed by other workers [37, 39]. The processes involving damage of the tissue, viz., minimal processing, and unit operations including peeling, slicing, dicing, and chopping have been found to decrease the bioactive components of the onion. Similarly, drying and other secondary processing treatments also decreased the content of bioactive components [37], which has been summarized in **Table 1**.

### 2.9 Potato

Potato is a member of the Solanaceae family and is a major food crop grown all over the world for its starchy tubers. Raw potato comprises 2% protein, 17% carbs (88% of which is starch), 79% water, and hardly any fat. The vitamins C and B6 as well as the minerals potassium, magnesium, and iron are among the many vital nutrients found in potatoes [40]. Amylopectin, a branched chain glucose polymer, and amylose, a straight chain glucose polymer, make up the majority of potato starch in a comparatively constant 3:1 ratio. Potato starch, which makes up a minor part of total starch, is resistant starch (RS) and acts as a prebiotic by promoting the growth of beneficial colonic bacteria [41]. Two of the five types of resistant starch (RS) categories are found in potatoes: RS2, which is mostly present in raw potatoes, and RS3, which is generated when potatoes are processed and chilled enough for the starch to gelatinize and retrograde [42]. The peel or the thick periderm skin of the tubers is also rich in nutrients such as potassium, iron, riboflavin, folate, and vitamins [43].

| Postharvest/ processing practices | Parameters/conditions | Effect on bioactive compounds |
|---|---|---|
| Curing | The onions were cured under different conditions such as drying at 24°C, 20°C, and 28°C for three and six days, respectively, exposing to fluorescent light for three days. | Different curing treatments resulted in an increase in quercetin, quercetin glucosides, and anthocyanin levels in onion and decrease in flavanol content. |
| Minimal processing | The minimal processing treatments such as onion maceration for 5 h, peeling, chopping, trimming, and cutting were applied | The application of various minimal processing treatments resulted in decrease in flavanols and quercetin glucosides level. |
| Freezing | Diced onion frozen at −18°C and stored for 3, 4, and 5 months | The freezing temperature resulted in an increase in total flavanols and total anthocyanin content. |
| Frying | The different frying conditions were applied for frying of onions, i.e., in olive oil for 4–8 min at 180°C, in oil for 4–8 min, 5–15 min at 180°C, and sauteing for 5 minutes. | The frying in olive oil will not affect the bioactive compounds and flavonoids, but frying in any other oil for different time durations resulted in a decrease in quercetin content, and sauteing will result in decrease in flavonoid content. |
| Boiling | Boiling of onions for 10–60 min at 90–100°C | The boiling of onions resulted in a decrease in flavonoids, bioactive compounds, and quercetin content. |
| Heating | Various heating treatments such as blanching onion for 60 or 70°C for 3 or 1 min, respectively; pressure processing 400 MPa pressure at 5 °C temperature for 5 min; sterilization at 100 °C for 11–17 min; microwave heating for 4 min.; and heating between 36 and 120°C in the time range of 30 to 96 hrs. | The various heating treatments resulted in a decrease in quercetin and its derivatives, quercetin glucosides, and flavonoids and an increase in the total phenol content. |
| Roasting | Roasting for 15–30 min between 180 and 270°C | The roasting of onions resulted in a decrease in flavonoid content with no effect on the total quercetin. |
| Drying | Freeze-drying of onion slices at −70°C and 4.2 Pa pressure for 24 h and dehydration of onions at 70°C | The freeze-drying and dehydration resulted in an increase in flavonoid content and anthocyanin content and a decrease in ascorbic acid. |
| Irradiation | Peeled onion were treated with UV-irradiation at low (1.2 $KJ/m^2$) and medium (6.0 $KJ/m^2$) doses, gamma irradiation at 1.5 kGy and 1.0 kGy and 0.1% sodium benzoate treatment, and stored for 16 days. | The irradiation treatments resulted in a decrease in flavanol content and the total ascorbic acid and an increase in quercetin content, total phenols, polyphenols, and flavonoid content. |

**Table 1.**
*Effect of postharvest practices and minimal processing and conditions on bioactive compounds of onion [37].*

Several phytonutrients, including phenolic acids and carotenoids, are also present in potatoes. For every 100 g of fresh weight potatoes, the carotenoid content ranges from 35 μg to 795 μg. Compared to white flesh cultivars, dark yellow cultivars have 10 times more carotenoid content. Arylated petunidin glycosides (purple potatoes) and acylated pelargonidin glycosides (red and purple potatoes) are the anthocyanins found in potatoes in the highest concentrations. Up to 80% of the total phenolic content of potato tubers is made up of the colorless polyphenol chlorogenic acid [41]. The bioactive compounds present in potatoes have favorable impacts on human health. Potato anthocyanins, glycoalkaloids, and lectins are helpful as anti-tumor agents, while potato protein, resistant starches, and phosphorylated starches contribute cholesterol-lowering properties [44]. Antioxidants in particular have been linked to decreasing inflammation, a risk factor for diabetes, cardiovascular disease, and cancer [42]. Antioxidants present in potatoes are reported to reduce inflammation, cardiovascular disease, and cancer [45].

### 2.10 Radish

Radish (*Raphanus sativus L.* var. *niger*) belongs to the *Brassicaceae* family. The taproot is used for culinary and medicinal properties. It has significant amounts of minerals, vitamin C, and by-products. Radish is a low-caloric food but a good source of calcium, magnesium, copper, manganese, potassium, vitamin B6, vitamin C, folate, polyphenols, sinapine, raphanusanins, isoperoxidases, peroxidases, and alkaloids [46]. The anthocyanin pigments are responsible for the red color of roots, and the distinctive and pungent flavor is a result of their high potential to form isothiocyanates [47]. Pelargonidin and delphinidin are the predominant anthocyanidins in the red and pink radish cultivars, respectively, whereas cyaniding is the main anthocyanidin in the purple radish variant [48]. It contains glucosinolates and/or their derivatives (isothiocyanates, nitriles, and cyano-epithioalkanes), essential oils, flavonoids, and other polyphenolic compounds. Radish microgreens contain significantly greater concentrations of glucosinolates (3.8-fold) and isothiocyanates (8.2-fold) than mature radish taproots.

Radish is recommended as a hepatoprotectant, diuretic, antimicrobial, antioxidant, anti-inflammatory, anti-thrombotic, anti-scorbutic, expectorant, and astringent, while it is also used in urinary and syphilitic complaints, dysuria, calculus, and bronchial and chest troubles [49]. Radishes have also been found to possess anticancer and anxiety-reducing effects [46]. Lugasi et al. [50] reported that radish is a natural remedy for stomach bloating, inadequate digestion, gallstone prevention, and promotion of bile production and bile function.

### 2.11 Sweet potato

Sweet potato (*Ipomoea batatas* L.) belongs to the Convolvulaceae family. In addition to a variety of micronutrients including manganese, copper, potassium, iron, vitamin B complex, vitamin C, vitamin E, and provitamin A, its roots are also rich in macronutrients like starch and dietary fibers. The yellow- and orange-fleshed varieties are also rich in carotenoids. The amounts of protein and fat are quite low. The flesh may be white, cream, yellow, orange, or purple, whereas the skin is often brown, beige, red, or purple [51].

The sweet potato roots are a rich source of phytochemicals such as carotenoids, tocopherols, phenolic compounds, tannins, flavonoids, saponins, and anthocyanins; the concentration, however, varies with flesh color and varieties. Sweet potato is also rich in dietary fiber and resistant starch [52]. Grebla-Al-Zaben *et al.* [53] found that cianidine is a more common anthocyanin than peonidin. Polyphenols are composed of chlorogenic acid, caffeic acid, and their derivatives. From the coumarins family of chemicals, scopoletin, esculetin, and umbeliferon are also present in sweet potatoes. Triterpenes and calistegines are found in comparatively lesser concentrations [52].

Anthocyanin, polyphenolic compounds, coumarins, calystegines, and triterpenes in sweet potatoes stimulate immune function, act as antioxidants, reduce cardiovascular disease risk, suppress cancer cell growth, prevent and improve symptoms of diabetes and hypoglycemia, suppress HIV symptoms, and act as hepatoprotective agents [53]. These bioactive phytochemicals present in sweet potatoes, individually and together, also have bowel-regulating qualities and neuroprotective and anti-inflammatory capabilities [52].

## 2.12 Turmeric

Turmeric (*Curcuma longa* L., sometimes known as *Curcuma domestica* Valeton) belongs to the family Zingiberaceae and is widely cultivated in Asia's tropical regions. Its root (rhizome), in both processed and raw forms, is the part that is most frequently utilized for food and medicinal purposes (**Table 2**). Turmeric has a lot of fiber and carbohydrates. Additionally, it has certain proteins and lipids but no cholesterol and minerals, thus rendering it among the naturally occurring foods with a high nutritional value [54, 55]. Among the bioactive components, the major component is curcumin (1,7-bis(4-hydroxy-3-methoxyphenyl)-1,6-heptadiene-3,5-dione), followed by other significant components including curcuminoids atlantone, dimethoxycurcumin, diarylheptanoids, tumerone, and flavonoid curcumin. Curcumin can act as antioxidant, antimicrobial, analgesic, anti-inflammatory, antiseptic, anticarcinogenic, antiobesity, hypolipidemic, and cardioprotective [54]. Additionally, it protects against several forms of cellular injury. In addition to offering neuroprotection, turmeric and its ingredients are also helpful in controlling the pathology of neurological conditions like Parkinson's and Alzheimer's diseases and shield COVID-19 patients from developing lung damage brought on by cytokine storms [56].

Turmeric powder with milk is known for its good healing capability. As a remedy for dysentery, roasted turmeric powder is consumed. Both herbal toothpaste and powders have turmeric as one of the ingredients. The advantage of oral use of turmeric and curcumin is that it is safe even at high levels. However, in a few cases,

| Product name | Description | Uses |
|---|---|---|
| Whole rhizome | *Appearance*: orange-brown, red-yellow, or pale yellow<br>*Chemical composition*: it may contain 3–15% curcuminoids and 1.5 to 5% essential oils | Medicinal purposes |
| Turmeric powder | *Appearance*: yellow or red-yellow<br>*Chemical composition*: curcuminoids and essential oils | Spices, dyes, medicines, and dietary supplements |
| Turmeric oil | *Appearance*: yellow to brown oil<br>*Chemical composition*: essential oils mostly contain monoterpenes and rhizomes oil (sesquiterpenes) | Spices, dyes, medicines, and dietary supplements |
| Turmeric oleoresins | *Appearance*: Dark yellow, reddish brown<br>*Chemical composition*: 25% essential oil and 36–56 percent curcuminoids | Food colorant, medicine, and dietary supplement |
| Curcumin | *Appearance*: yellow to orange-red crystalline powder<br>*Chemical composition*: bisdemethoxy- and demethoxy-derivatives | Medicines and dietary supplements |

**Table 2.**
*The main products of turmeric, their chemical composition, and uses [54].*

itching, redness of the tongue, tachycardia, and gastrointestinal problems (such as flatulence, diarrhea, nausea, and constipation) have been reported [57]. Besides curcumin, various volatile oils specifically atlantone, turmerone, and zingiberone are found to be the other active constituents of turmeric. The fresh juice and essential oils from turmeric can be used as biopesticides, as it is reported to possess pesticidal properties and also repellent properties against mosquitos [58].

Wada *et al.* [59] reported the antimicrobial action of extracts of turmeric against *Staphylococcus aureus*, *Escherichia coli*, and *Klebsiella pneumoniae*. Earlier, Kurhekar [60] had reported that the aqueous extract of turmeric was effective in inhibiting the growth of Gram-positive *Bacillus subtilis, S. aureus, Enterococcus fecalis*, and fungal pathogen *Candida albicans* and Gram-negative isolates, *E. coli, Salmonella typhi, Shigella flexneri, Pseudomonas aeruginosa*, and *Proteus vulgaris*. Lakmina *et al.* [61] reported that both turmeric methanolic extract and turmeric powder samples were effective against *Salmonella* spp. and *E. coli* spp.

The low solubility of curcumin in water has restricted its systemic bioavailability and therapeutic potential. However, Yong *et al.* [62] reported that fermentation with *Lactobacillus fermentum* significantly increased the curcumin content by 9.76% and showed no cytotoxicity. In comparison to unfermented turmeric, higher concentrations of curcumin, demethoxycurcumin, bisdemethoxycurcumin, phenolic compounds, and total flavonoid-curcuminoid were observed after solid-state fermentation of wild turmeric (*Curcuma aromatica*) with *Rhizopus oligosporus* for 5 days.

### 2.13 Turnip

*Brassica rapa* L., commonly known as turnip, belongs to the family of Brassicaceae and is known for its white fleshy taproot. Turnip root has low calorie but is a good storehouse of minerals (Cu, Fe, Ca, and Mn), vitamins, dietary fiber, and antioxidants. The major organic acids present in turnip in higher concentrations are malic, sinapic, ferulic, and their derivative acids. Turnip greens and tops have been shown to contain flavonoids, mostly in the form of derivatives of quercetin, kaempferol, and isorhamnetin, but not the roots. In addition to being a great source of vitamin K, turnip greens are also rich in antioxidants such as carotenoids, xanthins, lutein, vitamin A, and vitamin C. The turnip's top greens are rich in vitamin B complex, which includes riboflavin, pantothenic acid, and thiamine [63].

Turnip is mostly composed of glucosinolates and isothiocyanates, particularly 2-phenylethyl, 4-pentenyl, and 3-butenyl derivatives, which have a variety of bioactivities, particularly for the prevention of cancer. In addition, bioactive volatiles, indoles, phenolics, and flavonoids are found in turnip roots. **Table 3** lists the numerous phenolic substances that may be found in turnips. In traditional Chinese therapy, turnips are used to treat a variety of ailments, including gonorrhea, syphilis, rheumatism, oedemas, and rabies. It has antitumor, antihypertensive, antidiabetic, antioxidant, antimicrobial, anti-inflammatory, hepatoprotective, and nephroprotective effects. Hexahydrofarnesylacetone present in it is known for bactericidal effects. The anticancer property of turnip is associated with its 2-phenylethyl isothiocyanate, phenylpropionitrile, brassicaphenanthrene A, 6-paradol, and *trans*-6-shogaol [64, 65].

### 2.14 Yam

*Dioscorea* species, popularly known as yam, belong to the family of Dioscoreaceae. Yams are cultivated for the consumption of their starchy tubers. The species are

| Phenolic compounds |
|---|
| Kaempferol 3-O-sophoroside-7-O-glucoside |
| Kaempferol 3-O-(feruloyl/caffeoyl)-sophoroside-7-O-glucoside |
| Kaempferol 3,7-di-O-glucoside |
| Isorhamnetin 3,7-O-diglucoside |
| Sinapic acid |
| Hydroxycinnamoyl gentiobiosides |
| Kaempferol-3-O-glucoside |
| Quercetin-3-O-(sinapoyl)-sophotrioside-7-O-glucoside |
| Hydroxycinnamoylmalic acids |
| Hydroxycinnamoylquinic acids |
| Kaempferol 3-O-sophoroside-7-O-glucoside |

**Table 3.**
*Turnip phenolic compounds [63].*

known for their food values, medicinal values, and low anti-nutritional factors. Yam has many medicinal uses and is used as a post-pregnancy tonic, in piles and dysentery, and as an antidiabetic agent [66].

The underground and/or aerial tubers represent valuable sources of proteins, fats, and vitamins. The presence of different bioactive compounds in various *Dioscorea* species has been reviewed by other workers [67, 68]. The secondary metabolites of yams include steroids, clerodane diterpenes, quinones, cyanidins, phenolics, diarylheptanoids, and nitrogen-containing compounds. They are low in phytates. The most common secondary metabolites are saponins, and there are more than 100 steroidal saponins (based on aglycon part as stigmastanol, furostanol, spirostanol, cholestanol, ergostanol, and pregnanol glycosides). Tubers and roots contain steroidal sapogenins, mostly diosgenin as well as volatile compounds [68].

*Dioscorea* species are used as a cure for different diseases and ailments (cough, cold, stomach ache, leprosy, burns, fungal infections, dysentery, skin diseases, rheumatism, arthritis, etc.), stomach pain (colic), menstrual disorders, birth control, and a disease caused by parasitic worms known as schistosomiasis [66, 69]. Yam is also known for its hypolipidemic, antioxidant, anti-inflammatory, antimicrobial, hypoglycemic, androgenic, estrogenic, contraceptive, and antiproliferative activities [67].

## 3. Uses in food and food industry

Depending upon the availability and nutritional profile, several traditional food dishes of root vegetables, alone or in combination with other vegetables in raw or cooked form, are prepared and consumed. However, there are some novel value-added foods products that are also prepared from these vegetables. The advances and value addition for the routine culinary usages have been briefly discussed in the following sections for the individual root vegetables.

### 3.1 Beet root

Chauhan *et al.* [70] have enumerated food uses of beetroot. The deep-red roots of beetroot can be eaten raw, boiled, steamed, and roasted. It is a popular salad in raw and shredded forms, alone or in combination with other salad vegetables. Fermented and non-fermented pickles are made from boiled and spiced beets. Betanin is a natural pigment in beet juice and is used as a red food colorant for processed products like tomato paste and sauces, desserts, ice cream, sweets, jam and jellies, and spices. Beetroot is also used in manufacturing a food coloring agent known as E162 [2, 71]. It is used to impart a rich red color to dairy products (e.g., milk, ice cream, and yogurt), beverages (juices, wines, and vinegar), fruit candies, and bakery products (cookies, desserts, and cakes). Red beet itself can be used to prepare rich red, Burgundy-style wines.

Dhawan *et al.* [72] developed phytochemical rich and organoleptically acceptable beetroot *Barfi* and *Kanjhi* by incorporating beetroot flour at 1% and 4%, respectively. Similarly, Ashraf *et al.* [73] developed a beetroot-enriched energy booster drink and flavored milk that contained high antioxidant activity, high proteins and fats, and low carbohydrates but adequate total energy content. Beetroot peels' (waste left out during the processing of beetroots) powder, which is rich in antioxidant content, has been used as a natural ingredient in several value-added emulsions, including mayonnaise, dressings, sauces, and creams [74]. In meat products, red beet can be an efficient replacer of nitrate to preserve the red color [2].

### 3.2 Black salsify

The roots and aerial parts of black salsify are used for food purposes. Petkova [75] reported that the leaves are eaten in fresh or blanched forms in salads. The leaves are rich in vitamin C but lack inulin. The roots are long, blackish from the outside and white, milky on the inside. The roots are bitter in taste, but boiling can remove the bitterness and develop an oyster-like taste as salsify. The fresh (raw, seasoned, or cooked) roots can be consumed as such or processed and canned. It can also be dried and used later to be cooked as a vegetable. Roots contain about 17% polysaccharides, of which 13–16% is inulin, 3–5% is pectin, and 1.5–2.5% is fiber. Powder of black salsify roots can be added at 3–4% in ice cream for fiber fortification. Because of high levels of inulin in roots, they can also be roasted as a coffee substitute for diabetic patients. The inulin can be also used for encapsulation purposes, especially in oregano oil, catechins, or thyme extracts as an active ingredient [76].

### 3.3 Carrots

The presence of the coloring pigments anthocyanins and carotenoids and the sufficient quantity of bioactive compounds in carrots have resulted in an increased interest in the food use of carrots. Carrot roots can be consumed fresh, steamed, blanched, or cooked in stews and soups and stuffed in baked products like cakes and pies. Different carrot-based products like dehydrated carrots are used as an ingredient in making instant soups and healthy snacks without oil. *Kanji*, a naturally fermented probiotic appetizer drink popular in North India, is typically prepared using black carrots. The high anthocyanin content and associated high antioxidant activity make the black carrot a suitable raw material for the preparation of functional foods like jams and marmalades. Carrot powder when incorporated in *chapatti*, cake, and *halwa* increases their nutritional value and fiber content [77]. *Gazrella* is a traditional and popular Indian sweet that is made by cooking carrot shreds with condensed milk and cane sugar. Carrot oil, canned baby carrots, pickle, dessert mix, juice, and candy are some of the other promising high-value products made from carrots.

The high nutritional content, sweet taste, and bright color of carrot juice have resulted in an increased popularity of carrots. It is also mixed with other fruit beverages. From carrots, baby foods are also prepared. Riaz *et al.* [78] prepared the value-added products from carrots and reported that β-carotene content and antioxidant activity were higher in carrot jam, followed by carrot candies and carrot–orange juice. Varshney and Mishra [15] reported that carrot pomace, a waste of the carrot food industry which contains approximately 50% carotene and is rich in dietary fibers, may be used to enhance the quality of the cake, bread, and biscuits as well as to make a diversity of other useful ready-to-eat snacks, carrot crispy chips, etc.

### 3.4 Cassava

The roots of cassava have abundant starch content. Zekarias et al. [16] reported that its tubers contain 32–35% carbohydrate on fresh weight (FW) and 80–90% on dry matter (DM) basis, of which 80% is starch and about 17% sucrose. In its starch, 83% is amylopectin and 17% is amylose. Though cassava roots are energy rich, they have limitations in their usage as food. They can be highly toxic if not properly prepared, processed, or cooked. Cassava contains cyanide and other toxic substances that are harmful to humans. Fermentation followed by boiling and drying can, however, significantly reduce the anti-nutritional factors and ensure its nutritional quality [79].

Both fermented and unfermented products are traditionally prepared from cassava. Fermented cassava flour and starch, cassava bread, *fufu, lafun, akyeke* (or attieke), *agbelima*, and *gari* are some of the popular fermented products, whereas unfermented cassava flour and starch, cassava chips, and pellets are the unfermented products [80]. Ukwuru *et al.* [81] discussed the production of biofuel and ethanol, iodine-supplemented and protein-enriched high-quality cassava flour (HQCF), and processed products. The use of HQCF as a raw material for the production of modified starch, monosodium glutamate, cassava bread, pies, chips, cookies, biscuits, noodles, etc. has also been discussed. The functional properties of cassava starch have been extensively reviewed by Lambebo *et al.* [17]. Since the residue obtained from cassava tubers have high quantity of organic matter, it serves as a suitable substrate for microorganisms for the production of organic acids and flavor and aromatic compounds. According to Airaodion *et al.* [79], the microorganisms primarily responsible for fermentation are *Neurospora sitophila, Geotrichum candidum*, and *Rhizopus oryzae.* They observed a 95% reduction in cyanogen levels by heap fermentation of cassava roots followed by sun drying.

### 3.5 Garlic

Garlic is a vegetable that is widely used as a seasoning, flavoring, food dish, and functional food. The use of garlic is popular due to its excellent effectiveness, less side effects, low cost, and easy availability. Bioactive compounds present in it are effective in promoting health and developing functional foods or nutritional supplements. However, highly unstable thiosulfinates, such as allicin, are destroyed during processing and are quickly converted into various organosulfur components [24, 82].

Therefore, the efficiency and safety of garlic products and food supplements are influenced by the processing methods used. Different value-added processed products like minimally processed, osmotically dehydrated flakes, freeze dried powder, paste, pickle, oil, etc. have been prepared from garlic.

The distinctive flavor of fresh garlic is attributed to a variety of thiosulfinates and volatile compounds like S-alkyl substituted cysteine sulfoxide derivatives, alkyl canthiosulfinates, pyruvate, and ammonia produced by the action of allinases (EC4.4.1.4). The enzyme action starts as soon as garlic tissues are disrupted. Allinase enzyme, which is involved in thiosulfate conversion, gets inactivated by pH below 3.5 or by heat [23, 83]. Microwave radiation for even 1 minute inactivates allinase. The diallyl thiosulfinate (allicin) formed by enzyme action accounts for approximately 60–80% of the total thiosulfinates in garlic. The half-life of allicin is up to 16 hours at room temperature and 2.5 days when stored in a juice or crushed form. Fresh garlic and garlic powder, garlic oil, and steam-distilled garlic do not have significant amounts of alliin or allicin but instead contain various products of allicin transformation [83].

Black garlic, which is fermented white garlic, is prepared by heat treatment of fresh garlic without additives to reduce the pungent odour and taste. Fresh garlic is exposed to high temperature (60–90°C) and high humidity (60–80%) for 60–90 days. The black garlic that is produced is richer in various bioactive components; however, it has a reduced pungent smell and taste of garlic. Ma *et al.* [84] studied the potential of probiotic fermentation to further improve the quality of black garlic. It was reported that the pH was significantly lowered, and total acids, amino nitrogen, total polyphenols, and total flavonoids increased by *Lactobacillus* fermentation. The content of 5-hydroxymethylfurfural (a carcinogenic component) was reduced by 25.10–40.81%. The contents of furfural, 2-acetylfuran, 5-methylfurfural, etc. responsible for unpleasant baking flavor were decreased, while an increase was observed for green grass, floral, and fruit aromas. The *Lactobacillus* fermentation resulted in increased contents of functional components like Gly-Pro-Glu, sorbose, lactic acid, and α-CEHC (3,4-dihydro-6-hydroxy-2,5,7,8-tetramethyl-2H-1-benzopyran-2-propanoic acid).

The exogenous addition of garlic-derived organosulfur compounds in ground beef was found to be a more effective antioxidant than the α-tocopherol. It significantly reduced total aerobes and inhibited the growth of five inoculated pathogenic bacteria: *S. typhimurium, E. coli O157:H7, Listeria monocytogenes, S. aureus*, and *Campylobacter jejuni* [85]. Similarly, Kanza Aziz et al. [86] reported that garlic fortification improved the stability and sensory attributes of chicken bites.

### 3.6 Ginger

Ginger is used either fresh or after drying the whole or chopped ginger. As the moisture content in fresh ginger is high, it causes difficulty in its drying and converting it into dry spice. Dried ginger is a highly wrinkled product with low volatile oil content. Various value-added products that can be prepared from ginger are ginger candies and preserve, ginger puree and paste, ginger powder and sticks, ginger beer and wine, and ginger oil and oleoresin [87]. It is also used as a flavoring substance in foods, curries, and beverages (such as ginger ales); in the confectionery industry in products such as pickles, chutneys, vinegar, and marmalades; and in bakery products. Other new products that can be made from ginger are skin/stick, osmotically dried appetizing ginger flakes, ginger drinks, and ginger starch [88].

Amer *et al.* [89] developed value-added extruded maize enriched with ginger extract at 3%. The product had better functional and sensory properties. Kaushal *et al.* [71] developed ginger fruit bars and ginger appetizing tablets, which have significantly higher antioxidant activity, total phenolics, and crude fiber. Tanweer *et al.* [90] prepared value-added meatballs by adding 10% dried ginger powder. The product was yellowish due to the presence of shogaol, had higher total phenolics and antioxidant potential, and had acceptable organoleptic quality.

### 3.7 Jerusalem artichoke

Due to the high nutritive value and proportion of fructans and low content of nitrates, the flour of Jerusalem artichoke tubers may be fully utilized as functional food [31]. Nadir *et al.* [91] obtained concentrated gluten by washing wheat flour and replacing the removed starch with flour and inulin extract obtained from Jerusalem artichoke. Compound flour was used to prepare pasta. It was observed that the volume of cooked pasta increased with increasing levels of flour and inulin extract. The appearance, color, taste, fragility, and stickiness of the pasta improved with increasing levels of Jerusalem artichoke flour up to 30%. Previously, Shin *et al.* [92] developed quality noodles by replacing 25% of wheat flour with Jerusalem artichoke flour.

Inulin extracted from Jerusalem artichoke tuber powder (JATP) in the form of oligosaccharides is used as a sugar substitute and in prebiotics. The conventional enzymatic method of inulinases for the extraction of inulin from JATP is limited by the narrow temperature range of enzyme activity, complicated processes, low substrate solubility, longer reaction times, and high operating costs. Bui *et al.* [93] developed a method using a combination of microwave heating and HCl as a catalyst to extract most of the carbohydrate content of JATP and selectively convert it to fructo-oligosaccharides. The method required a low reaction temperature and a relatively short reaction time. Bakr *et al.* [94] produced Bio-Labneh probiotic from cow's milk using Jerusalem artichoke tuber powder inoculated with *Lactobacillus acidophilus* LA-5. The most organoleptically acceptable Bio-Labneh was produced with 1% JATP, while 3% JATP obtained the highest overall score and *L. acidophilus* growth.

### 3.8 Onion

Onion, also referred to as the "queen of the kitchen," is known for its extremely valuable flavor, aroma, and distinctive style. Onion is eaten raw as salad or used as an ingredient in a diverse variety of foods by enhancing their flavor and taste. If consumed in raw form, it provides health benefits due to the direct intake of phytochemicals. The good fiber and higher flavonoids in bulbs and the various by-products obtained from it have tremendous scope for food application. Fresh onion and its dried powder when added to various food products have been reported to increase the shelf life of processed food products due to onion's antifungal, antibacterial, and antioxidant properties [95]. Sulfur containing volatile oils (allicin, ajoene, and alliin) present in onion can be extracted and used as a flavoring substance and preservative and also as a health-promoting bioactive compound. As onions are rich in sugars and other nutrients, they can also be processed into value-added products like onion vinegar, wine, paste, and sauce [96].

### 3.9 Potato

Potato is consumed mainly as a carbohydrate source. The starch from raw potato is nearly indigestible but is more easily digestible from cooked potato. Raw, peeled potatoes after cooking are one of the common most ingredients in food dishes. The potatoes are also processed into several groups of food products like potato flour and starch, potato flakes, granules and dried products, potato chips and french fries, frozen potato products, canned potatoes, and fabricated french fries and chips [97]. It may be noted that the normal potato tubers that have been properly grown and stored have small quantities of glycoalkaloids, safe enough for human health. However, if exposed to light, the sprouts and skin of the tuber accumulate high concentrations of glycoalkaloids, which have harmful effects on human health. The tubers exposed to light also produce solanine, a toxin harmful to human consumption. Anti-nutrients in raw and baked potatoes include glycoalkaloids and acrylamides [98].

The potato-processing industry, depending on the product produced and the employed peeling method, generates potato peel or skin at 15–40% of the tubers' fresh weight. The potato peel/skin can be a potential source of fiber and other phytochemicals (**Figure 3**). Arora and Camire [99] prepared cinnamon muffins and cookies using extruded and non-extruded potato skins at 25% as a substitute for wheat flour. Baked products with skin had reduced contents of glycoalkaloids and peroxides; they were darker, more resistant to compression, and lower in height and possessed increased fiber content.

### 3.10 Radish

The taproot of radishes is consumed worldwide in the form of pickled vegetables, salads, and curries. The leaves and sprouts are usually eaten raw as part of salads. Radish juice is not palatable because of its pungent taste. Kaur *et al.* [100] developed a

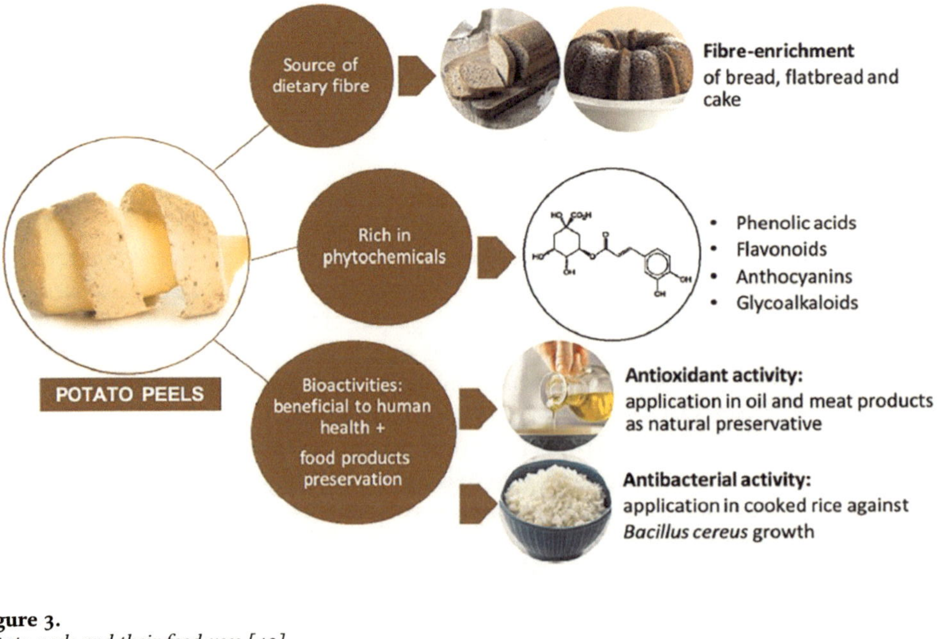


**Figure 3.**
*Potato peels and their food uses [43].*

high-acceptability nutritional drink by mixing radish juice with 30% sugarcane juice, 1 percent herbal extract (mint juice/coriander juice/citric acid in a ratio of 1:1:1 v:v:w), and 1.5% salt concentration (black salt/white salt in a ratio of 1:1). Lugasi *et al.* [50] reported no glucosinolates in pressed black radish root juice, as it got completely destroyed and converted into polyphenolic compounds by the pressing process and storage. Tanaka and Ohmiya [101] reported that anthocyanin derivatives from red radish can be used as a stable natural food colorant. The brick red/scarlet, red/magenta, and violet/blue color of radish are due to the presence of pelargonidin-, cyanidin-, and delphinidin-based anthocyanins, respectively.

### 3.11 Sweet potato

Sweet potatoes can be converted to several functional ingredients, food, and industrial products that have been summarized by Truong *et al.* [102] and presented in **Figure 4**.

Mitiku et al. [103] developed bread by blending of sweet potato flour with wheat flour. The developed bread was lower in protein but richer in ash, crude fiber, carbohydrates, iron, zinc, phosphorus, and vitamin A contents. The tannin and phytate contents of the composite bread were low. Similarly, Gracia et al. [104] developed a healthy cake by replacing 20% of maida (refined wheat flour) base with oven-dried *Ipomea batatas* powder. As compared to pure maida cake, the developed cake was high in fiber content and scored better organoleptically for color and taste.

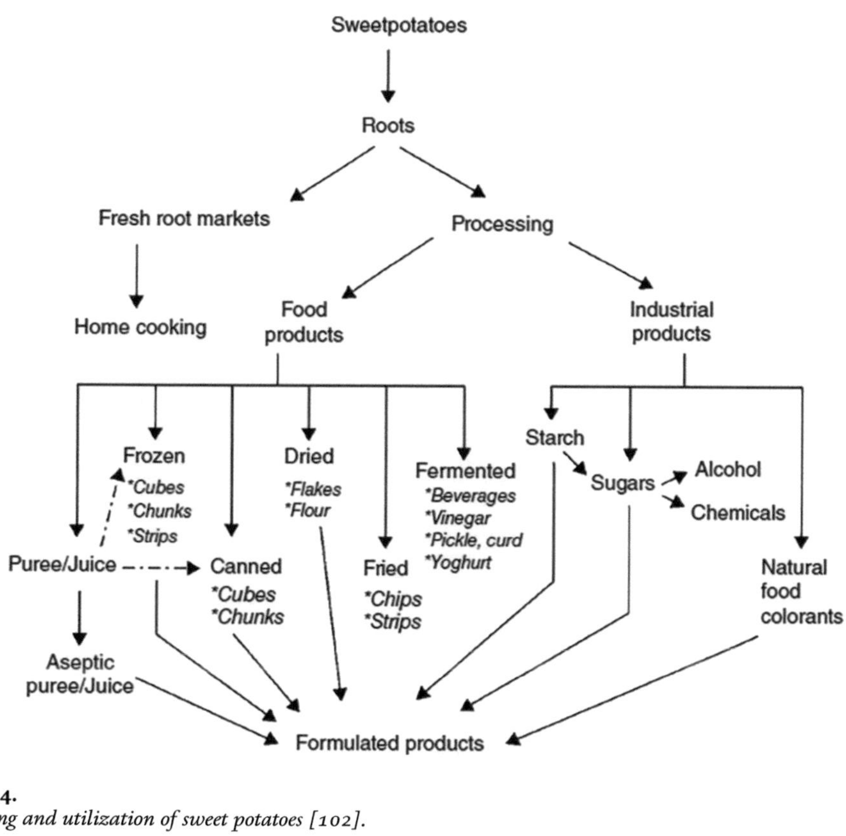


**Figure 4.**
*Processing and utilization of sweet potatoes [102].*

Supplemental foods based on root or tuber crops have an advantage of having significantly lower phytate content (by 3–20%) than the foods based on grains and legumes. Amagloh *et al.* [52] developed a complementary food for infants using sweet potato, soybean meal, soybean oil, and fish powder or skimmed milk powder. It was reported that sweet potato-based formulations had phytate and fructose levels superior to commercially available Weanimix.

### 3.12 Turmeric

The dried root powder of turmeric is used as a spice, food preservative, and flavoring and food-coloring agent. Turmeric is used in smoked meats, pickles, seafood, soups, rice, and various vegetable dishes. Turmeric extends the shelf life of seafood products by maintaining nutritional quality and attractiveness due to its antioxidant, antimicrobial, coloring, and flavoring properties [105]. Turmeric tea can be made from grated, dried, or powdered turmeric. The brewing methods such as hard, soft, and ambient infusion were tested to characterize the antioxidant potential of the turmeric tea. Among them, hard infusion shows the highest antioxidant potential as compared to soft and ambient infusion [106]. Ipar *et al.* [107] developed a ready-to-use turmeric latte by an innovative approach using microencapsulated turmeric oleoresin with a blend of gum acacia, maltodextrin, and dairy whitener with bio-enhancers. The microencapsulated powder obtained exhibited high encapsulation efficiency, wettability, dispersibility, superior antioxidant activity, and oral bioavailability. It showed the release of >95% of curcumin at pH 1.2. Choi *et al.* [108] observed that a novel processing technique of high-pressure gun puffing increased the brown color and porous structures of turmeric. Puffing also enhanced its antioxidant activity.

### 3.13 Turnip

Turnips have a shelf life longer than that of potatoes. Turnips can be stored for 4–5 months at a temperature of 0°C and relative humidity of 95–98%. Around the world, many types of dishes are prepared with turnips, especially from raw, boiled, or steamed roots, alone or with other vegetables or meat. Turnip roots are diced and frozen or dried and stored in a powder form. The traditional turnip dishes belong to the food groups of soups, stuffed turnips, baked goods, rice pilaf, salads, and juices. It can also be used as salad and garnishes. The greens of this vegetable are used in soups, stews, and various value-added food products [109].

Some new turnip root products have also been developed. Xue *et al.* [110] evaluated different methods such as hot air drying, explosion drying, infrared drying, and freeze drying (FD) to produce turnip chips from roots. It was observed that FD chips retained the most starch, total sugar, vitamin C, and volume ratio by maintaining better brittle values and rehydration rate. Tripathi and Yadav [111] reported that replacing wheat flour and refined flour with 15% turnip powder to make traditional Indian *Chappati, Namakpara,* noodles, etc. resulted in their improved Ca and Fe contents.

### 3.14 Yam

The white and yellow yams in West Africa are consumed by boiling the peeled yam tubers and then mashing them (sometimes together with cooked bananas) into a doughy paste to make a traditional dish of "pounded yam", locally called *Fufu*, *Foutou*, or *Iyan* [69]. Leng *et al.* [112] studied the nutritional composition and antioxidant activity of a yam-based formulated weaning feed. Dried yam slice flours purchased from the local market were enriched with soybean paste and groundnut to create a balanced diet. Carrots and eggshells were added as fortifying sources of micronutrients. The nutritional content and antioxidant properties of the flour mixtures indicated that the developed weaning feed met the recommended energy and macronutrient requirements according to the established standards. Dos Santos *et al.* [113] prepared Greek-type goat yogurt with added yam aqueous extract and goat milk casein powder. It was reported that the yam-fortified yogurt has increased yield, water-holding capacity, viscosity, sensory acceptance, and purchase intention score.

## 4. Conclusion and future prospects

Root vegetables have a distinct nutritional profile with special reference to their various bioactive components having innumerable medicinal properties. These bioactive components can be used in different pharmaceutical formulations or diet-based therapies to cure various ailments and disorders. Their advantages over traditional medicines are lower cost and fewer side effects. The endogenous synthesis of the nutrients and phytochemicals in these root vegetables, however, is also affected by pre-harvest farm practices adopted and biotic and abiotic stresses faced by the crop. The effect(s) of these factors on the quality and quantity of various bioactive compounds need to be studied in detail. Besides, further identification and characterization of newer bioactive compounds; the selection of proper cultivar or variety rich in bioactive component(s); development of effective methods of extraction; improvement in the technologies for stabilizing the formulations; use of encapsulation, microsome, liposomes, and nano-formulation strategies, etc. are some of the approaches that can be helpful to pharmaceutical industries in producing effective herbal drugs and herbal antibiotics with improved efficiencies. The phytoconstituents' bioavailability and compatibility with other phytonutrients; the safe dosage with respect to a person's age; and the potential risk of the formulations for persons suffering from ailments like bile stone, hypertension, diabetics, allergies, etc. are some of the other important aspects to be studied further to develop effective drugs from root vegetables.

Besides the routine culinary usage of root vegetables, there is tremendous scope for their utilization by food industries. The quality of traditional foods prepared from the local root vegetables can be improved and maintained to make them available and acceptable globally. The essential oils extracted from root vegetables like garlic, turmeric, ginger, etc. can serve as potential food biopreservatives because of their antimicrobial properties. The pigments from beetroot, turmeric, carrot, etc. can serve as food-grade natural coloring agents. The food industry can utilize modern non-thermal processing technologies to preserve the bioactive components during processing to develop pharma foods, possessing higher quantities of essential bioactive nutrients and phytochemicals than naturally existing processed food products. Further technologies need to be developed to utilize left-out peel and pomace of the root vegetable-processing industries to develop by-products and then utilize those by-products in the development of functional foods. The reduction of pesticide residue, heavy metals and microbial contamination, nitrate, and anti-nutrient contents are some of the important aspects that should be given due consideration by the food industries utilizing root vegetables. The development of newer varieties of root vegetables suitable for processing, studies of changes in nutrients during storage, and development of non-thermal technologies to preserve the nutrients' loss during processing can be the future line of work for the researchers engaged in root vegetables' production, utilization, and related aspects.

## Author details

Saleem Siddiqui[1*], Naseer Ahmed[2], Chongtham Allaylay Devi[1], Puthem Robindro Singh[3] and Bawitlung Lalramhlimi[4]

1 School of Agricultural Sciences, Sharda University, Greater Noida, India

2 Department of Food Technology, DKSG Akal College of Agriculture, Eternal University, Baru Sahib, Himachal Pradesh, India

3 ICAR Directorate of Onion and Garlic Research, Pune, India

4 ICAR RC for North Eastern Hilly Region, Mizoram Centre, Kolasib, Mizoram, India

*Address all correspondence to: saleem.siddiqui@sharda.ac.in

## References

[1] Ülger TG, Songur AN, Çırak O, Çakıroğlu FP. Role of vegetables in human nutrition and disease prevention. In: Asaduzzaman M, Asao T, editors. Vegetables-Importance of Quality Vegetables to Human Health. London, UK: IntechOpen; 2018. pp. 7-32. DOI: 10.5772/intechopen.77038

[2] Ceclu L, Nistor O. Red beetroot: Composition and health effects—A review. Journal of Nutritional Medicine and Diet Care. 2020;**6**(1):1-9

[3] Getz H. Sucrose accumulation and synthesis in sugar beet. In: Developments in Crop Science. Vol. 26. Amsterdam, Netherlands: Elsevier; 2000. pp. 55-77

[4] Afiomah CS, Iwuozor K. Nutritional and phytochemical properties of *Beta vulgaris* Linnaeus (Chenopodiaceae)–a review. Nigerian Journal of Pharmaceutical and Applied Science Research. 2020;**9**(4):38-44

[5] Masih D, Singh N, Singh A. Red beetroot: A source of natural colourant and antioxidants: A review. Journal of Pharmacognosy and Phytochemistry. 2019;**8**(4):162-166

[6] Mirmiran P, Houshialsadat Z, Gaeini Z, Bahadoran Z, Azizi F. Functional properties of beetroot (*Beta vulgaris*) in management of cardio-metabolic diseases. Nutrition & Metabolism. 2020;**17**(1):1-15

[7] Chen L, Zhu Y, Hu Z, Wu S, Jin C. Beetroot as a functional food with huge health benefits: Antioxidant, antitumor, physical function, and chronic metabolomics activity. Food Science & Nutrition. 2021;**9**(11):6406-6420

[8] Lendzion K, Gornowicz A, Bielawski K, Bielawska A. Phytochemical composition and biological activities of Scorzonera species. International Journal of Molecular Sciences. 2021;**22**(10):5128

[9] Wu QX, Su YB, Zhu Y. Triterpenes and steroids from the roots of *Scorzonera austriaca*. Fitoterapia. 2011;**82**:493-496

[10] Zhu Y, Wu QX, Hu PZ, Wu WS. Biguaiascorzolides a and B: Two novel dimeric guaianolides with a rare skeleton, from *Scorzonera austriaca*. Food Chemistry. 2009;**114**:1316-1320

[11] Bystrická J, Kavalcová P, Musilová J, Vollmannová A, Tomáš T, Lenková M. Carrot (Daucus carota L. ssp. sativus (Hoffm.) Arcang.) as source of antioxidants. Acta agriculturae Slovenica. 2015;**105**(2):303-311

[12] Ergun M, Süslüoğlu Z. Evaluating carrot as a functional food. Middle East Journal of Science. 2018;**4**(2):113-119

[13] Ahmad T, Cawood M, Iqbal Q, Ariño A, Batool A, Tariq RM, et al. Phytochemicals in *Daucus carota* and their health benefits. Food. 2019;**8**(9): 424

[14] Silva Dias JC. Nutritional and health benefits of carrots and their seed extracts. Food and Nutrition Sciences. 2014;**5**(22):2147-2156

[15] Varshney K, Mishra K. An analysis of health benefits of carrot. International Journal of Innovative Research in Engineering & Management. 2022;**9**(1): 211-214

[16] Zekarias T, Basa B, Herago T. Medicinal, nutritional and anti-nutritional properties of cassava (*Manihot esculenta*): A review. Academic Journal of Nutrition. 2019;**8**(3):34-46

[17] Lambebo T, Deme T. Evaluation of nutritional potential and effect of processing on improving nutrient content of cassava (*Mannihot esculenta* crantz) root and leaves. bioRxiv. 2022;**2**(4):479097. DOI: 10.1101/2022.02.04.479097

[18] Bahekar S, Kale R. Phytopharmacological aspects of *Manihot esculenta* Crantz (cassava)—A review. Mintage Journal of Pharmacy and Medical Sciences. 2013;**2**:4-5

[19] Suleria HAR, Butt MS, Khalid N, Sultan S, Raza A, Aleem M, et al. Garlic (*Allium sativum*): Diet based therapy of 21st century–a review. Asian Pacific Journal of Tropical Disease. 2015;**5**(4):271-278

[20] Sasi M, Kumar S, Kumar M, Thapa S, Prajapati U, Tak Y, et al. Garlic (*Allium sativum* L.) bioactives and its role in alleviating oral pathologies. Antioxidants. 2021;**10**(11):1847

[21] De Greef D, Barton EM, Sandberg EN, Croley CR, Pumarol J, Wong TL, et al. Anticancer potential of garlic and its bioactive constituents: A systematic and comprehensive review. In: Seminars in Cancer Biology. Vol. 73. Amsterdam, Netherlands: Academic Press; 2021. pp. 219-264

[22] Escudero LB, Monasterio RP, Lipinski VM, Filippini MF, Wuilloud RG. Selenized garlic: A future prospect or already a current functional food? Revista de la Facultad de Ciencias Agrarias. 2012;**44**(2):301-318

[23] Ezeorba TPC, Chukwudozie KI, Ezema CA, Anaduaka EG, Nweze EJ, Okeke ES. Potentials for health and therapeutic benefits of garlic essential oils: Recent findings and future prospects. Pharmacological Research-Modern Chinese Medicine. 2022;**5**:100075

[24] Shinde A, Dond S, Diwate V, Pawar S, Katkar R, Darandale S. Historical approach of garlic (*Allium sativum*) in food, spices and it's phytochemicals and therapeutic uses in medicine: A review. World Journal of Pharmaceutical Research. 2021;**10**(1):737-744

[25] Jang H-J, Lee H-J, Yoon D-K, Ji D-S, Kim J-H, Lee C-H. Antioxidant and antimicrobial activities of fresh garlic and aged garlic by-products extracted with different solvents. Food Science and Biotechnology. 2018;**27**(1):219-225

[26] Singletary K. Ginger: An overview of health benefits. Nutrition Today. 2010;**45**(4):171-183

[27] Shoaib M, Shehzad A, Butt MS, Saeed M, Raza H, Niazi S, et al. An overview: Ginger, a tremendous herb. Journal of Global Innovations in Agricultural and Social Sciences. 2016;**4**(4):172-187

[28] Abdullahi A, Khairulmazmi A, Yasmeen S, Ismail I, Norhayu A, Sulaiman M, et al. Phytochemical profiling and antimicrobial activity of ginger (*Zingiber officinale*) essential oils against important phytopathogens. Arabian Journal of Chemistry. 2020;**13**(11):8012-8025

[29] Mao Q-Q, Xu X-Y, Cao S-Y, Gan R-Y, Corke H, Beta T, et al. Bioactive compounds and bioactivities of ginger (*Zingiber officinale* roscoe). Food. 2019;**8**(6):185

[30] Alexandre EMC, Lourenço RV, Bittante AMQB, Moraes ICF, do Amaral Sobral PJ. Gelatin-based films reinforced with montmorillonite and activated with nanoemulsion of ginger essential oil for

food packaging applications. Food Packaging and Shelf Life. 2016;**10**:87-96

[31] Munim A, Rod MR, Tavakoli H, Hosseinian F. An analysis of the composition, health benefits, and future market potential of the *Jerusalem artichoke* in Canada. Journal of Food Research. 2017;**6**(5):69

[32] Yang L, He QS, Corscadden K, Udenigwe CC. The prospects of *Jerusalem artichoke* in functional food ingredients and bioenergy production. Biotechnology reports. 2015;**5**:77-88

[33] Sami R, Elhakem A, Alharbi M, Benajiba N, Almatrafi M, Helal M. Nutritional values of onion bulbs with some essential structural parameters for packaging process. Applied Sciences. 2021;**11**(5):2317

[34] Esienanwan EE, Luavese PA, Ezinne CC, Adepoju IO, Godwin O. Comparative qualitative phytochemical analysis of oil, juice and dry forms of garlic (*Allium sativum*) and different varieties of onions (*Allium cepa*) consumed in Makurdi metropolis. International Journal of Plant Physiology and Biochemistry. 2020;**12**(1):9-16

[35] Greeshma K, Muthulingam S, Thamizselvi R, Venkatamani GP. Phytochemical analysis and a review on biological importance of *Allium cepa*. L. GSC Advanced Research and Reviews. 2020;**2**(2):018-024

[36] Fredotović Ž, Puizina J, Nazlić M, Maravić A, Ljubenkov I, Soldo B, et al. Phytochemical characterization and screening of antioxidant, antimicrobial and antiproliferative properties of allium× cornutum Clementi and two varieties of *Allium cepa* L. peel extracts. Plants. 2021;**10**(5):832

[37] Sagar NA, Pareek S, Benkeblia N, Xiao J. Onion (*Allium cepa* L.) bioactives: Chemistry, pharmacotherapeutic functions, and industrial applications. Food Frontiers. 2022;**3**:135. DOI: 10.1002/fft2.135

[38] Zhao X-X, Lin F-J, Li H, Li H-B, Wu D-T, Geng F, et al. Recent advances in bioactive compounds, health functions, and safety concerns of onion (*Allium cepa* L.). Frontiers in Nutrition. 2021;**8**: 669805

[39] Ren F, Nian Y, Perussello CA. Effect of storage, food processing and novel extraction technologies on onions flavonoid content: A review. Food Research International. 2020;**132**:108953

[40] Khalid MZ, Aziz A, Tariq A, Ikram A, Rehan M, Younas S, et al. Nutritional composition and health benefits of potato: A review. Advanced Food and Nutritional Sciences. 2020;**5**: 7-16

[41] Beals KA. Potatoes, nutrition and health. American Journal of Potato Research. 2019;**96**(2):102-110

[42] McGill CR, Kurilich AC, Davignon J. The role of potatoes and potato components in cardiometabolic health: A review. Annals of Medicine. 2013;**45**(7): 467-473

[43] Sampaio SL, Petropoulos SA, Alexopoulos A, Heleno SA, Santos-Buelga C, Barros L, et al. Potato peels as sources of functional compounds for the food industry: A review. Trends in Food Science & Technology. 2020;**103**:118-129

[44] Camire ME, Kubow S, Donnelly DJ. Potatoes and human health. Critical Reviews in Food Science and Nutrition. 2009;**49**(10):823-840

[45] Jansky S, Navarre R, Bamberg J. Introduction to the Special Issue on the

Nutritional Value of Potato. American Journal of Potato Research. 2019;**96**(2): 95-97

[46] Ghayur MN, Gilani AH. Gastrointestinal stimulatory and uterotonic activities of dietary radish leaves extract are mediated through multiple pathways. Phytotherapy Research: An International Journal Devoted to Pharmacological and Toxicological Evaluation of Natural Product Derivatives. 2005;**19**(9):750-755

[47] Gamba M, Asllanaj E, Raguindin PF, Glisic M, Franco OH, Minder B, et al. Nutritional and phytochemical characterization of radish (*Raphanus sativus*): A systematic review. Trends in Food Science & Technology. 2021;**113**: 205-218

[48] Hanlon PR, Barnes DM. Phytochemical composition and biological activity of 8 varieties of radish (*Raphanus sativus* L.) sprouts and mature taproots. Journal of Food Science. 2011; **76**(1):C185-CC92

[49] Manivannan A, Kim JH, Kim DS, Lee ES, Lee HE. Deciphering the nutraceutical potential of *Raphanus sativus* - a comprehensive overview. Nutrients. 2019;**11**(2):402. DOI: 10.3390/nu11020402

[50] Lugasi A, Dworschák E, Blazovics A, Kery A. Antioxidant and free radical scavenging properties of squeezed juice from black radish (*Raphanus sativus* L. var Niger) root. Phytotherapy Research: An International Journal Devoted to Pharmacological and Toxicological Evaluation of Natural Product Derivatives. 1998;**12**(7):502-506

[51] Neela S, Fanta SW. Review on nutritional composition of orange-fleshed sweet potato and its role in management of vitamin a deficiency. Food Science & Nutrition. 2019;7(6):1920-1945

[52] Amagloh FC, Yada B, Tumuhimbise GA, Amagloh FK, Kaaya AN. The potential of sweetpotato as a functional food in sub-saharan Africa and its implications for health: A review. Molecules. 2021;**26**(10):2971

[53] Grebla-Al-Zaben B, Babalau-Fuss V, Biris-Dorhoi SE, Talos I, Tofana M. A review of the composition and health benefits of sweet potato. Bulletin of University of Agricultural Sciences and Veterinary Medicine Cluj-Napoca Food Science and Technology. 2021; **78**(1):1-10

[54] Ahmad RS, Hussain MB, Sultan MT, Arshad MS, Waheed M, Shariati MA, et al. Biochemistry, safety, pharmacological activities, and clinical applications of turmeric: A mechanistic review. Evidence-based complementary and alternative Medicine. 2020;**14**: 7656919. DOI: 10.1155/2020/7656919

[55] Enemor VHA, Ogbodo UC, Nworji OF, Ezeigwe OC, Okpala CO, Iheonunekwu GC. Evaluation of the nutritional status and phytomedicinal properties of dried rhizomes of turmeric (*Curcuma longa*). Journal of Biosciences and Medicines. 2020;**8**(08):163

[56] Ismail Z, Mazuki NAN. An overview of Golden Spice's (turmeric) medicinal properties for future development potential. Malaysian Journal of Science Health & Technology. 2020;**6**:53-57

[57] Soleimani V, Sahebkar A, Hosseinzadeh H. Turmeric (*Curcuma longa*) and its major constituent (curcumin) as nontoxic and safe substances. Phytotherapy Research. 2018;**32**(6):985-995

[58] Damalas CA. Potential uses of turmeric (*Curcuma longa*) products as alternative means of pest management in

crop production. Plant omics. 2011;**4**(3): 136-141

[59] Wada NM, Ambİ AA, Ibrahİm AA, Bello SK, Umar A, James DT. Mediterr J Infect Microb Antimicrob. 2021;**10**:6

[60] Kurhekar JV. *Curcuma longa* and *Allium sativum* as prebiotics. Bionano Frontier. 2013;**6**(2):327-329

[61] Lakmina D, Ranasinghe M, Abeyrathne E, Jayasena D. Antimicrobial activity of turmeric (*Curcurna longa*) against *salmonella* spp and *E. coli* Spp. In: Proceeding of the 2nd International Research Symposium. Badulla, Sri Lanka: Uva Wellassa University; 2018. p. 313

[62] Yong C, Yoon Y, Yoo H, Oh S. Effect of lactobacillus fermentation on the anti-inflammatory potential of turmeric. Journal of Microbiology and Biotechnology. 2019;**29**(10): 1561-1569

[63] Javed A, Ahmad A, Nouman M, Hameed A, Tahir A, Shabbir U. Turnip (*Brassica rapa* L.): A natural health tonic. Brazilian Journal of Food Technology. 2019;**22**:e2018253. DOI: 10.1590/ 1981-6723.25318

[64] Paul S, Geng CA, Yang TH, Yang YP, Chen JJ. Phytochemical and health-beneficial progress of turnip (*Brassica rapa*). Journal of Food Science. 2019;**84**(1):19-30

[65] Cao Q, Wang G, Peng Y. A critical review on phytochemical profile and biological effects of turnip (*Brassica rapa* L.). Frontiers in Nutrition. 2021;**8**: 721733

[66] Guchhait KC, Singh G, Sardar S, Kumbhojkar V, Marndi S, Kumar S. Yams of India and their medicinal values. Medico-Biowealth of India. 2022;**Vol V**:59

[67] Salehi B, Sener B, Kilic M, Sharifi-Rad J, Naz R, Yousaf Z, et al. Dioscorea plants: A genus rich in vital nutra-pharmaceuticals-a review. Iranian Journal of Pharmaceutical Research: IJPR. 2019;**18**(Suppl. 1):68

[68] Obidiegwu JE, Lyons JB, Chilaka CA. The Dioscorea genus (yam)—An appraisal of nutritional and therapeutic potentials. Food. 2020;**9**(9):1304

[69] Andres C, AdeOluwa OO, Bhullar GS. Yam (*Dioscorea* spp.)-a rich staple crop neglected by research. In: Encyclopedia of Applied Plant Sciences. Vol. 2. San Diego, USA: Academic Press; 2017. pp. 435-441

[70] Chauhan S, Chamoli K, Sharma S. Beetroot-a review paper. Journal of Pharmacognosy and Phytochemistry. 2020;**9**(2S):424-427

[71] Sawicki T, Wiczkowski W. The effects of boiling and fermentation on betalain profiles and antioxidant capacities of red beetroot products. Food Chemistry. 2018;**259**:292-303

[72] Dhawan D, Sharma S, Scholar D. Exploration of the nourishing, antioxidant and product development potential of beetroot (*Beta vulgaris*) flour. International Journal of Health Sciences & Research. 2019;**9**(6):280

[73] Ashraf S, Sayeed SA, Ali R, Vohra F, Ahmed N, Alam MK. Assessment of potential benefits of functional food characteristics of beetroot energy drink and flavored milk. BioMed Research International. 2022;**2022**:1971018. DOI: 10.1155/2022/1971018

[74] Lazăr S, Constantin OE, Horincar G, Andronoiu DG, Stănciuc N, Muresan C, et al. Beetroot by-product as a functional ingredient for obtaining value-added mayonnaise. PRO. 2022;**10**(2):227

[75] Petkova N. Characterization of inulin from black salsify (*Scorzonera hispanica* L.) for food and pharmaceutical purposes. Asian Journal of Pharmacy and Clinical Research. 2018;**11**:221-225

[76] Beirão-da-Costa S, Cl D, Bourbon I, Pinheiro A, Januário M, Vicente A, et al. Inulin potential for encapsulation and controlled delivery of oregano essential oil. Food Hydrocolloids. 2013; **33**:199-206

[77] Akubor PI, Eze JI. Quality evaluation and cake making potential of sun and oven dried carrot fruit. International Journal of Biosciences. 2012;**2**(10):19-27

[78] Riaz N, Yousaf Z, Yasmin Z, Munawar M, Younas A, Rashid M, et al. Development of carrot nutraceutical products as an alternative supplement for the prevention of nutritional diseases. Frontiers in Nutrition. 2021;**8**: 787351. DOI: 10.3389/fnut.2021.787351

[79] Airaodion AI, Airaodion EO, Ewa O, Ogbuagu EO, Ogbuagu U. Nutritional and anti–nutritional evaluation of garri processed by traditional and instant mechanical methods. Asian Food Science Journal. 2019;**9**(4):1-13

[80] Falade KO, Akingbala JO. Utilization of cassava for food. Food Reviews International. 2010;**27**(1):51-83

[81] Ukwuru M, Egbonu S. Recent development in cassava-based products research. Academia Journal of Food Research. 2013;**1**(1):1-13

[82] Amagase H, Petesch BL, Matsuura H, Kasuga S, Itakura Y. Recent advances on the nutritional effects associated with the use of garlic as a supplement. Journal of Nutrition. 2001; **131**(3):955S-962S

[83] Fesseha H, Goa E. Therapeutic value of garlic (*Allium sativum*): A review. Advances in Food Technology and Nutritional Sciences. 2019;**5**(3):107-117

[84] Ma L, Zhao C, Chen J, Zheng J. Effects of anaerobic fermentation on black garlic extract by lactobacillus: Changes in flavor and functional components. Frontiers in Nutrition. 2021;**8**:645416

[85] Yin MC, Cheng WS. Antioxidant and antimicrobial effects of four garlic-derived organosulfur compounds in ground beef. Meat Science. 2003;**63**(1): 23-28

[86] Kanza Aziz A, Masood Sadiq B, Faiza A, Rasul HA, S. Storage stability of garlic fortified chicken bites. Journal of Food Chemistry and Nanotechnology. 2017;**3**(2):80-85

[87] Kaushal M, Gupta A, Vaidya D, Gupta M. Postharvest management and value addition of ginger (*Zingiber officinale* Roscoe): A review. International Journal of Environment, Agriculture and Biotechnology. 2017; **2**(1):397-412

[88] Kaushal M, Vaidya D, Kaushik R, Verma A, Gupta A. Standardization of formulation for development of ginger supplemented confectionary products. International Journal of Bio-Resource and Stress Management. 2018;**9**(2): 237-242

[89] Amer SA, Rizk AE. Production and evaluation of novel functional extruded corn snacks fortified with ginger, bay leaves and turmeric powder. Food Production, Processing and Nutrition. 2022;**4**(1):1-17

[90] Tanweer S, Chughtai MFJ, Zainab S, Mehmood T, Khaliq A, Junaid-Ur-Rahman S, et al. Comparative study of

physicochemical and hedonic response of ginger rhizome and leaves enriched patties. Czech Journal of Food Sciences. 2021;**39**(5):402-409

[91] Nadir AES, Helmy I, Kamil MM. Effect of using Jerusalem artichoke and inulin flours on producing low carbohydrate high protein pasta. Australian Journal of Basic and Applied Sciences. 2011;**5**(12):2855-2864

[92] Shin J-Y, Byun M-W, Noh B-S, Choi E-H. Noodle characteristics of Jerusalem artichoke added wheat flour and improving effect of texture modifying agents. Korean Journal of Food Science and Technology. 1991; **23**(5):538-545

[93] Bui CV, Siriwatwechakul W, Tiyabhorn W, Wattanasiritham T, Limpraditthanont N, Boonyarattanakalin S. Conversion of *Jerusalem artichoke* tuber powder into fructooligosaccharides, fructose, and glucose by a combination of microwave heating and HCl as a catalyst. Thammasat International Journal of Science and Technology. 2016;**21**(3): 31-45

[94] Bakr ASHT, Elkot WF, Hussein AKHA. Impact of using *Jerusalem artichoke* tubers powder and probiotic strains on some properties of labneh. Journal of Food Technology & Nutrition Sciences. 2021;**3**(1):1-7

[95] Arshad MS, Sohaib M, Nadeem M, Saeed F, Imran A, Javed A, et al. Status and trends of nutraceuticals from onion and onion by-products: A critical review. Cogent Food & Agriculture. 2017;**3**(1): 1280254

[96] Gorrepati K, Khade YP. Mouth-watering value-added products of onion and garlic. Indian Horticulture. 2017;**62**(6):86-87

[97] Stearns LD, Petry TA, Krause MA. Potential Food and Nonfood Utilization of Potatoes and Related Byproducts in North Dakota. Fargo, North Dakota: North Dakota State University; 1994. 1189-2016-94269

[98] Wijesinha-Bettoni R, Mouillé B. The contribution of potatoes to global food security, nutrition and healthy diets. American Journal of Potato Research. 2019;**96**(2):139-149

[99] Arora A, Camire ME. Performance of potato peels in muffins and cookies. Food Research International. 1994;**27**(1): 15-22

[100] Kaur G, Kumar V, Goyal A, Tanwar B, Kaur J. Optimization of nutritional beverage developed from radish, sugarcane and herbal extract using response surface methodology. Nutrition & Food Science. 2018;**48**: 733-743. DOI: 10.1108/NFS-11-2017-0247

[101] Tanaka Y, Ohmiya A. Seeing is believing: Engineering anthocyanin and carotenoid biosynthetic pathways. Current Opinion in Biotechnology. 2008; **19**(2):190-197

[102] Truong V, Avula R, Pecota K, Yencho G. Sweetpotato production, processing, and nutritional quality. In: Saddiq M, Mark A, editors. Handbook of Vegetables and Vegetable Processing. Vol. 2. New Jersey, USA: John Wiley & Sons Ltd; 2018. pp. 811-838

[103] Mitiku DH, Abera S, Bussa N, Abera T. Physico-chemical characteristics and sensory evaluation of wheat bread partially substituted with sweet potato (*Ipomoea batatas* L.) flour. British Food Journal. 2018;**120**:1764-1775. DOI: 10.1108/BFJ-01-2018-0015

[104] Gracia J, Anitha D, Jayalakshmi M. Nutrient analysis of *Ipomoea batatas*

flour and sensory property of the formulated product. Journal of Emerging Technologies and Innovative Research. 2022;**9**(2):810-813

[105] Güneri N. A review on turmeric (*Curcuma longa* L.) and usage in seafood. Marine science and technology. Bulletin. 2021;**10**(1):71-84

[106] Mahjoob M, Stochaj U. Curcumin nanoformulations to combat aging-related diseases. Ageing Research Reviews. 2021;**69**:101364

[107] Ipar VS, Singhal RS, Devarajan PV. An innovative approach using microencapsulated turmeric oleoresin to develop ready-to-use turmeric milk powder with enhanced oral bioavailability. Food Chemistry. 2022; **373**:e131400

[108] Choi Y, Ban I, Lee H, Baik MY, Kim W. Puffing as a novel process to enhance the antioxidant and anti-inflammatory properties of *Curcuma longa* L. (turmeric). Antioxidants. 2019; **8**(11):506. DOI: 10.3390/antiox8110506

[109] Badem A. Traditional turnip meals of Konya. Journal of Tourism and Gastronomy Studies. 2021;**9**(2):725-743

[110] Xue Y-L, Chen J-N, Han H-T, Liu C-J, Gao Q, Li J-H, et al. Multivariate analyses of the physicochemical properties of turnip (*Brassica rapa* L.) chips dried using different methods. Drying Technology. 2020;**38**(4):411-419

[111] Tripathi D, Yadav L. Formulation and quality evaluation of turnip green (*Brassica rapa*) powder incorporated food products. The Pharma Innovation Journal. 2021;**10**(4):653-657

[112] Leng MS, Tobit P, Demasse AM, Wolf K, Gouado I, Ndjouenkeu R, et al. Nutritional and anti-oxidant properties of yam (*Dioscorea schimperiana*) based complementary food formulation. Scientific African. 2019;**5**:e00132

[113] dos Santos RA, de Lima RR, de Lima MBD, do Nascimento EB, de Carvalho AMB, de Almeida Gadelha CA, et al. Influence of aqueous yam extract and goat milk casein powder on the characteristics of goat Greek-style yogurt. International Journal of Gastronomy and Food Science. 2022;**27**: e100465

Chapter 5

# Phytochemical Changes in Root Vegetables during Postharvest Storage

*Elijah K. Lelmen and Jacqueline K. Makatiani*


## Abstract

Root vegetables contain phytochemicals that are essential for human nutrition, in addition to offering desirable health benefits such as anti-oxidative, anti-cancer, and immunomodulatory activities. The quantity and stability of these phytochemicals vary greatly among root vegetable cultivars and landraces. Besides, freshly harvested root vegetables deteriorate rapidly thus causing significant losses in their quality attributes. To minimize these losses, various postharvest technologies have been assessed and shown efficacy in prolonging the shelf-life of stored vegetables. However, postharvest technologies may contribute to deterioration of nutrients and/or accumulation of toxic compounds such as glycoalkaloids. Therefore, this chapter summarizes information that has been reported on the influence of varied pre-storage treatments and storage systems on the quality of root vegetables. Quality attributes that are highlighted include changes in: root vegetable morphology such as sprouting, dehydration, and greening; phytochemical content of phenolics, flavonoids, glycoalkaloids, alkaloids, glycosides, and terpenoids; and nutritional content of carbohydrates, protein, vitamins, and carotenoids.


**Keywords:** storage systems, pre-storage treatments, physico-chemical changes, phytonutrients

## 1. Introduction

Root vegetables are important components of a balanced diet for human health because of their richness in many biologically active compounds. These compounds, collectively termed phytochemicals or functional nutrients, provide numerous desirable health benefits such as anti-oxidative, anti-cancer hypoglycemic, antimicrobial, and immunomodulatory activities beyond basic nutritional benefits [1]. These compounds include terpenes, polyphenols, glucosinolates, phytoestrogens, and carotenoids [2–5]. They form part of a plant's secondary metabolite profile, and often accumulate as part of the plants' response mechanisms to abiotic and biotic stress factors [6–9]. Besides, some of the compounds are widely used as natural colorants in the textile, food, and drug industry. The quantity and stability of these phytochemicals vary greatly among root vegetable cultivars and landraces. Besides, root vegetables deteriorate rapidly after harvest with significant losses in their morphological (e.g., weight loss, sprouting, greening, and shriveling), phytochemical (e.g., deterioration of phytonutrients and/or accumulation of toxic compounds), physiological (e.g., softening of the tissues, color changes caused by the synthesis of new pigments or destruction of others) and biochemical (e.g., increased rate of respiration) attributes [10, 11]. To produce high quality processed vegetable products, majority of farmers are not only growing new, promising root vegetable varieties, but also minimizing these losses by preserving the root vegetable quality during storage. Various post-harvest technologies, including use of varied pre-storage treatments and storage systems, have been assessed and have shown efficacy in prolonging the shelf-life of fresh root vegetables [11–20]. Therefore, this chapter summarizes information reported on the effect of post-harvest technologies on the quality of the commonly utilized edible root vegetables that include beetroot (*Beta vulgaris* subsp. *vulgaris* L.), carrot (*Daucus carota* L.), cassava (*Manihot esculenta* Crantz), potato (*Solanum tuberosum* L.) and radish (*Raphanus sativus* L.).

## 2. Phytochemical and nutritional composition of root vegetables

### 2.1 Beetroot

Beetroot (*B. vulgaris* subsp. *vulgaris* L.) (Family: Chenopodiaceae) also known as red beetroot has fleshy root that is commonly consumed in form of supplemental juice, powder, bread, gel, oven-dried, pickled, pureed or jam-processed across different food cultures. Its bioactive phytochemicals include phenolics and carotenoids, and betalains hydrosoluble nitrogen containing pigments (e.g., betacyanin that are red-violet and betaxanthin that are yellow-orange) [21]. Betalains form 70–100% of the total phenolics and have antioxidant and anti-inflammatory properties and anti-carcinogenic potential [22]. Antiviral and antimicrobial effects of betalain pigments have also been reported [23]. Other antioxidant compounds include rutin, epicatechin, and caffeic acid. The sugar comprises mainly of sucrose (91.6%) [24], with a small and relatively similar proportion of glucose and fructose [25]. Red beetroot also contains dietary fiber, minerals (e.g., potassium, sodium, iron, copper, magnesium, calcium, phosphorus, and zinc), and vitamins (e.g., retinol, folate, ascorbic acid, and B-complex [26]. Moreover, beetroot contains a substantial amount of both non-essential and essential amino acids.

### 2.2 Cassava

Cassava (*M. esculenta* Crantz) (Family: Euphorbiaceae) is a starchy fibrous root crop used for traditional desserts, salad dressing, soup thickening, binding agent in sausages, high fructose syrup, and textile industries [27]. The sweet type of cassava root can be boiled or roasted and eaten as fresh root, or minimally processed into various products. Cassava contains alkaloids, as well as flavonoids that have antioxidant and hypolipidemic effects [28], and glycosides that are potent for heart disease [29]. It has low content of nutrients such as protein (<2%), fat (<0.2%), and fiber (<4%) [30]. Most of the carbohydrate is present as starch (>32% of fresh weight) with smaller amounts of free sugars (less than 1% of fresh weight). It has rich dry matter (>80%) thus making it a good source of energy [27]. The root also contains minimal amount of micronutrients (e.g., iron, potassium, magnesium, copper, zinc, and manganese. It also has some anti-nutrients and toxic substances (e.g., cyanogenic glucosides), which together with their breakdown products (cyanohydrins and free hydrocyanic acid [HCN]) that are formed during processing, can inhibit the digestibility and intake of major nutrients [31]. The yellow root varieties contain a significant concentration of β-carotene (up to 1 mg 100 $g^{-}$1,dry-weight) [32].

### 2.3 Carrot

Carrots (*D. carota* L.) (Family: Apiaceae [Umbelliferae]) play a major role in human nutrition, because of their high dietary value [33]. Phytochemicals present comprise mainly of phenolic compounds (e.g., phenolic acids, such as *p*-hydroxybenzoic, caffeic, and chlorogenic; and flavonoids such as anthocyanins), carotenoids (a precursor to vitamin A formation, which is involved in vision, cell differentiation, synthesis of glycoproteins, mucus secretion from the epithelial cells, and overall growth and development of bones) [34], polyacetylenes, and ascorbic acid. These chemicals aid in the risk reduction of cancer and cardiovascular diseases due to their antioxidant, anti-inflammatory, plasma lipid modification, and anti-tumor properties [35]. Phenolic acids are the potentially bitter compounds found in carrot peels. The moisture content of carrot varies from 86 to 89% [36]. They are good source of carbohydrates and minerals like calcium, phosphorus, iron, and magnesium, and also contain protein (0.9%), fat (0.2%), carbohydrate (10.6%), crude fiber (1.2%), and total ash (1.1%) [36].

### 2.4 Radish

Radish (*R. sativus* L.) (Family: Brassicaceae) has high nutritional value and is consumed in salads, or cooked or salted together with other vegetables. The roots can also be processed as dried or canned pickles. The extracts of radishes have been used to treat stomach disorders, urinary tract infections, hepatic inflammation, cardiac disorders, and ulcers [37]. Various reports have recorded the antimicrobial, anticancer, antioxidant, and anxiety reducing properties of radishes. Phytochemicals present in radish include alkaloids, reducing sugar, flavonoids, glycosides, cardiac glycosides, tannins, saponin, protein, amino acid, terpenoids, and steroids [38]. Anthocyanin pigments provide the red color of roots, while a high potential to form isothiocyanates contributes to the pungent flavor and distinctive taste [39]. Radish is low in calories but has a high content of vitamin C which helps to build tissue, blood vessels, bones and teeth. Other vitamins (e.g., B6 and folate) and minerals (e.g., potassium, magnesium, calcium, iron, zinc, copper, sodium, and phosphorous) are also found in radish roots. Also, radish has fiber and roughage that is effective in the treatment of constipation.

### 2.5 Potato

Potato (*S. tuberosum* L.) (Family: Solanaceae) is a staple food that is often cooked or processed into edible products such as fries and chips. It contains several phytochemicals such as phenolics, flavonoids, polyamines, and carotenoids. These phytochemicals have beneficial effects on human health hence they are highly desirable in diet. The phenolics, together with amino acids present in potatoes, confer anti-oxidant protection towards tissue damage, reactive oxygen species and diseases like atherosclerosis, diabetes mellitus, renal failure, and cancer [40]. Nutritionally, potato has complex carbohydrates (>20%) in the form of free sugars (with glucose and fructose as the principal monosaccharides and sucrose as the major disaccharide, crude proteins (>2.5%), crude fats (0.1%), crude fiber (0.6%), vitamins, and water (74%) and along with minerals (>0.8%), amino acids, and trace elements that include potassium, sodium, iodine, and magnesium, folic acid, pyridoxine, vitamin C, and iron. Additionally, potatoes contain glycoalkaloids, which are toxic steroidal glycosides and anti-nutritional substances. Although they play a role in plant resistance to bacterial and fungal diseases and pests, when in excessive amounts, they worsen taste, and a concentration above 200 $mg \cdot kg^{-1}$ of fresh weight has toxic effects on the human body [41].

## 3. Pre-storage treatment of root vegetables

Freshly harvested root vegetables are metabolically active, and therefore still undergoing the physiological and biological processes of senescence and maturation. The rates of these processes are influenced primarily by the produce temperature. To prolong postharvest nutritional and quality (e.g., appearance, texture, and flavor) attributes, freshly harvested produce is often exposed to one or several optimal pre-storage treatments that often work by; slowing down the senescence and maturation processes, reducing/inhibiting development of physiological disorders and growth of decay producing microorganisms, restricting enzymatic and respiratory activity, inhibiting water loss, and reducing ethylene production. Physical treatments include; heat [42], irradiation [19], coatings [15], pre-cooling [18] and curing [43].

Heat treatment methods that have been applied on carrot and potato include hot-water (sometimes accompanied by brushing), hot dry air, and steam [44–46]. These treatments activate or deactivate enzymatic activities that result in reduced effects on the phytochemical and nutrient content, besides reducing chilling injuries and controlling decay. The treatments can be of short (up to 1 h) or long-term (up to 4 days) duration, but they have high energy costs. Gamma irradiation and short wave ultraviolet radiation have been used to effectively inhibit growth and development of sprouts and microbial pathogens on potato [47, 48]. However, their use is still subject to strict legislation. Paraffin wax coat is often used in combination with the exclusion of oxygen or submerging roots in water or storing in an anaerobic environment that can inhibit the streaking of the cassava xylem tissue [20, 49].

Pre-cooling is the removal of field heat in the produce immediately following harvest by using methods that include; Hydro-cooling (i.e., submerging crops in cold water), forced-air cooling (i.e., cold air is directed directly through the crop at high velocity), room cooling (i.e., placing crops in a cold room where cold air passing through a fan) and package icing (i.e., placing crushed ice directly on top of the produce). The choice of the method depends on the product type and factors such as the airflow rate, air, and produce temperature, relative humidity, and the packing configuration [50]. Bunched beetroots (with tops) are pre-cooled to optimal level of below 4°C within 4–6 h of harvest, while mature beetroot are pre-cooled to below 5°C within 24 h after harvest. Forced-air cooling, prompt washing, and hydro-cooling in chlorinated water to under 5°C are essential to maintain carrot freshness. Bunched radish is often hydro-cooled in chlorinated water to an optimum level of 0–4.5°C [51]. Radish can also be pre-cooled using the package icing technique. Generally, room cooling has low or no cost involvement. Forced-air cooling has the risk of root desiccation. Hydro-cooling permits faster cooling but offers the moisture which some pathogens require to penetrate the skin of root vegetable [52]. Hence, it is recommended that washed root vegetables should be dried at room temperature before storage. Curing process, which is hardening the skin of potatoes and cassava under temperature and RH conditions that facilitate wound healing, depends on cultivar and on whether they are destined for industry or home consumption.

Surface treatment chemicals that include maleic hydroxide, α–naphthalene acetic acid, methylester, isopropyl N-(3chlorophenyl carbamate) chlorpropham) (CIPC) and 1, 2, 4, 5 tetra chloro-3nitrobenzene are applied on potatoes. These chemicals inhibit meristematic cell division and delay sprout development. Nevertheless, there are safety concerns about the potential toxic and carcinogenic properties of CIPC and its metabolites [53]. Safer alternatives include hydrogen peroxide plus (HPP) [54], 1,4-dimethylnaphthalene (1,4-DMN), essential oils, and ethylene [55]. Gaseous treatments include ozone that has been evaluated for postharvest disease control and other storage uses on potatoes and carrots [56, 57]. However, additional research is needed to define the potential and limits of effective use of ozone for postharvest treatment of whole and minimally-processed vegetables and fruits.

## 4. Storage systems of root vegetables

Most fresh cassava cultivars deteriorate within 2–3 days after harvest and therefore, processing the roots into storable forms (through sun drying and fermentation) at the farm level is a better option for extending the shelf life and eliminating some toxic compound like cyanide. Some of the commonly used traditional methods include; coating of root with a paste of mud or earth, in-ground storage, field clumps, storage in a box or trench with alternative layers of moist sawdust or wood shavings and cassava roots, and storage in plastic bags [49, 58–60]. Advanced methods that are mainly used on export produce include cold storage/refrigeration at lower temperature range of 0–4°C, freeze drying, and modified atmosphere packaging [20, 49]. Nonetheless, financial and technical constraints have limited the use of advanced methods in many developing countries. Although a number of researches have been conducted on beet, carrot, and radish roots focusing on the potential uses as minimally processed ready-to-use, fresh-cut produce, or as ingredients of processed foods, information on the maintenance of freshness and quality of whole roots during storage is very limited [61]. However, freshly harvested bunched (with tops) and topped beets, carrots, and radish are usually stored between layers of moist sand, leaves, or sawdust in a box in cool place with condition of 0–4°C and 90–95% relative humidity (RH) [62]. The success of Controlled Atmosphere (CA) storage technology requires that the precise levels of $CO_2$ and $O_2$ gases are achieved and maintained within the storage facility. Where $CO_2$ level is too high and $O_2$ level too low, then the root vegetables may be irrevocably damaged [63]. However, there is little or no benefit from controlled atmosphere storage of these vegetable roots.

Potatoes are used for seed, ware, and processing and hence, storage requirements vary depending upon the purpose for which potatoes are to be used. Seed potatoes are required only at the time of planting; therefore, they need to be stored for longer. Generally, seed potatoes are stored at low temperatures (2–4°C). However, most local farmers store seed potato in simple and low-cost diffused light stores (DLS) that use natural indirect light with good ventilation to control excessive sprouting and to produce sprouts which are short, stout, green colored and with higher vigor [64]. Ware and processing potatoes are in demand throughout the year and hence both short- and long-term storage are needed. Ware and processing potatoes are stored at higher temperatures (8–12°C at 85–90% RH) [65]. Majority of smallholder farmers use traditional storage methods of ware potato. These include piling on the floor or corner in houses, dark stores, DLS, covering potato deep in the soil, stacking tubers in sacks, and heaping potato tubers under the tree shades. Majority of these methods allow keeping potato in good quality for a short period of 2–5 weeks only, depending on the potato variety [66]. Ambient ware potato storage units are also in use and can maintain the marketability of ware potato up to 9 weeks. While advanced ware potato storage methods like evaporative cool storage and cold storage exist, they are not used in most developing countries due to high costs.

## 5. Effect of pre-storage treatment and storage conditions on quality of root vegetables

Freshly harvested root vegetables deteriorate over a short period of time if not handled appropriately. Several morphological, biochemical, and physiological changes that are essential to the root tissue occur. For example, increased respiration rate, moisture loss and desiccation, spoilage caused by fungi and bacteria, earthy odors and flavors, sprouting and development of white blush on damaged surfaces, softening of tissues, color changes resulting from the synthesis of new pigments and destruction of others, and changes in the phytochemical and nutritional composition. Most of these changes are temperature-dependent.

Postharvest quality changes in carrot includes weight loss, bitterness, bacterial deterioration, rooting, and sprouting. The carrot has low metabolic activity at low temperatures, and can be stored for 6–8 months without loss of quality under optimal storage conditions of 0°C temperature and 98% RH [45]. **Table 1** provides a summary of some quality changes that occur in root vegetables when exposed to varied pre-storage treatments. Post-harvest hot-water treatment (50°C for 1 min) can be used for preserving their β-carotene and vitamin C content, although for carrots not destined for storage (**Table 1**) [45]. Pre-storage treatment involving carrot exposure to Ozone atmosphere 50± 10 nL/L treatment at 0.5°C and ≥95% RH recorded reduced severity of watery soft rot and gray mold fungal diseases, and blotches of discolored brown periderm tissue after a storage period of 180 days [56]. Similarly, reduced severity of fungal watery soft rot disease was observed on carrots that were exposed to 5 s pre-storage treatment of steam under 0.2 MPa pressure and 70°C prior to storage at 0.5°C for 60 days, and an additional 14 days at 20°C [46]. A UV-C (0.88 kJ/m2) treatment of carrot at 10°C and 90% RH resulted in reduced severity of watery soft rot and gray mold fungal diseases on carrots stored for 15 days (**Table 1**) [67].

Generally, when plants are subjected to postharvest abiotic stresses they synthesize secondary metabolites, such as phenolic compounds. **Table 2** summarizes the effect that different storage systems on root vegetables.

In a study where carrots were stored for 48 h at 20°C, a significant increase in the phenolic content was found [79]. In [76], study results showed that when black carrots were stored at 4°C retained a high level (53.4–81.0%) of anthocyanins than samples stored at 25°C for 20 weeks (7.8–69.3%). Similarly, in a study that investigated the effect of controlled atmosphere on baby carrots revealed that that controlled atmosphere of 5 kPa O2 and 5 kPa CO2 significantly increased the phenolic content, particularly chlorogenic acid [74]. Slight variations in α- or β-carotene have been demonstrated when carrots were stored at 0°C for 6 months [93]. According to Imsic

| Root vegetable | Pre-storage treatment | Effect on root quality | References |
|---|---|---|---|
| Carrot | HW; 50°C for 1 min | • β-carotene and vitamin C levels remained unchanged | [45] |
| | Ozone atmosphere 50 ± 10 nL/L; 0.5°C; ≥95% RH; 180 d | • reduced severity of watery soft rot and gray mold diseases; blotches of discolored brown periderm tissue | [56] |
| | ST (0.2 MPa pressure); 70°C; 5 s application time; 0.5°C; 60 d + 14 d at 20°C | • reduced severity of fungal watery soft rot disease | [46] |
| | UV-C (0.88 kJ/m2); 10°C; 90% RH; 15 d | • reduced severity of watery soft rot and gray mold diseases | [67] |
| Cassava | PW coat;55–65°C; few seconds; 60 d | • prolonged shelf life | [49, 68] |
| Potato | Ethylene treatment; continuous application | • sprouting inhibited; increased reducing sugars content | [69, 70] |
| | HPP treatment for 10 h; 180 d at 10 ± 1°C | • complete sprout suppression | [54] |
| | UV-C light; 20°C; 80% RH; 40 d (in dark) | • reduction in sprout length and development up to 20 d | [48] |
| | GI at 10, 30, and 50 d after harvest; 8 and 16°C; 150 d | • early irradiation and higher irradiation levels had decreased sprouting, percent weight loss and specific gravity; delayed irradiation had decreased ascorbic acid and contents of reducing and non-reducing sugars; decrease in ascorbic acid at higher storage temperature | [47] |

*d: days of storage; EC: edible coating (carboxy methyl cellulous and cellophane); PW: paraffin wax; FT: fungicide treatment; HPP: hydrogen peroxide plus; GI: gamma irradiation (0, 50, 100, and 150 Gy); HWT: hot water treatment; ST: steam treatment; min: minute; RH: relative humidity; and UV-C: ultra violet light (intensities 0.0, 3.4, 7.1, 10.5, and 13.6 kJ $m^{-2}$).*

**Table 1.**
*Summary of studies on effects of pre-storage treatments on the quality of selected root vegetables.*

et al. [77], storing carrots at either 4°C or 20°C resulted in increases in (all-*E*)-β-carotene of 20.3% after 3 days at 4°C and 34.4% after 14 days at 20°C, respectively. In contrast, another study [80] reported that β-carotene contents were reduced after 8 days of storage at different temperatures, by 46% (7.5–8.5°C), 51% (17–21°C), and 70% (22–37.5°C). Significantly high concentrations of polyacetylenes (falcarinol, falcarindiol, and falcarindiol-3-acetate) were documented in whole carrots that were refrigerated for 4 months at 1°C [75]. This indicates that polyacetylenes were produced during postharvest storage or there was little degradation in intact carrots after cold storage [75]. In a separate study [76], the level of vitamin C in baby carrots reduced during cold storage in high and moderate O2 conditions but under a low O2 atmosphere, baby carrots retained the highest amount of vitamin C. Freezing also has a negative impact on vitamin C content of carrots as in [78] where a decrease of 4.1% was recorded. Also, prolonged storage duration has been shown to lower the concentration of vitamin C from 15 to 49% [94]. Similarly, Kjellenberg et al. [95] noted that during storage, there is a decrease of glucose and fructose and a development of polyacetylenes, which causes a reduction of soluble sugars. The decreased

| Root vegetable | Storage condition | Effect on root quality | References |
|---|---|---|---|
| Beetroot | CS; 5°C; 196 d | • decreased betanin and increased isobetanin content in first 140 and 98 d, respectively | [71] |
| | CS; 30 d | • increased total dry matter content after 30 d, and a decrease after 120 d; increased total phenolic content | [72] |
| | CS; 1 ± 1°C; 90–95% RH; 210 d | • retained total soluble sugar content after 30 d, and a decrease after 120 d | [61] |
| | CS; 120 d | • increased nitrate content after 30 d with no further change | [73] |
| Carrot | CA 5 kPa $O_2$; 5 kPa $CO_2$; 4°C; 8 d | • increased phenolic (chlorogenic acid) content; maintained overall visual quality | [74] |
| | CS; 1°C; 120 d | • increased polyacetylenes (falcarinol, falcarindiol and falcarindiol-3-acetate) content | [75] |
| | CS; low, moderate and high $O_2$ condition | • Decreased vitamin C in high and moderate $O_2$ conditions; vitamin C content retained at low $O_2$ | [76] |
| | CS;4°C; 25°C; 140 d | • increased β-carotene after 3 d at 4°C, and 14 d at 20°C | [77] |
| | Freezing | • decreased vitamin C content | [78] |
| | 20°C; 48 h | • increased phenolics content | [79] |
| | CS;4°C; 25°C; 140 d | • high anthocyanins content retained at 4°C than at 25°C | [76] |
| | Temperature range 7.5–37.5°C | • decreased β-carotene contents after 8 d | [80] |
| Cassava | In-ground storage; >12 months | • increased fibrous and woody content, reduced extractable starch; increased susceptibility to pathogen attack | [49, 58] |
| | Curing and clumps; >90 d | • no internal discoloration, reduction in hydrogen cyanide content, increased soluble sugars, softening of central root core | [75] |
| | Boxes with moist sawdust; 28 d | • roots remained healthy | [60] |

| Root vegetable | Storage condition | Effect on root quality | References |
|---|---|---|---|
| Potato | CS; 3°C; 90 d | • increased phenolic (chlorogenic, caffeic and sinapic acids) content | [81] |
| | Curing | | |
| | CS; 4 °C; 120 d | • decreased vitamin C | [82] |
| | Pit/ underground storage | • increased respiratory heat that promoted growth and development of potato sprouts | [83, 84] |
| | Store (ACS, GTHS, DLS); 9.7–19.4°C; 65–93 RH; 56 d | • increased weight loss, greening, shrinkage, and reducing sugar content in DLS; decreased starch content with increasing storage time | [85] |
| | 4 or 10°C; 42 and 92 d | • increased glycoalkaloids content at low temperatures within 14 d | [86] |
| | CS 4°C; also room temperature and 15°C; 90 d | • all phenolic acid content increased with storage time, except para-coumaric acid which decreased at 4°C; decreased vitamin C content | [87] |
| | CA; 5 or 10°C; gas mixture (0.5% $CO_2$ and 21.0% $O_2$ or 9, 6.4, 3.6, or 0.4% $CO_2$ all combined with 3.6% $O_2$); 100 d | • Complete sprout inhibition at 9.4% $CO_2$ with 3.6% $O_2$ at 5°C; decreased weight loss; healthy skin maintained | [63] |
| Radish | CS airtight; 1°C; 45 d | • reduction in total aliphatic glucosinolates after 5 d | [88] |
| | CS; 0°C; 120 d | • reduction in sulforaphene isothiocyanate concentration | [89] |
| | Curing and packed in micro perforated HDPE film in plastic crate | • decreased weight loss in cured and non-cured samples; maintained total soluble solids level, and flesh and skin firmness of cured; reduced black spot disease severity and shrinkage | [90] |
| | Temperature; 20 ± 2°C; without leaves | • increased weight loss after 3 d | [91] |
| | Temperature 1, 5, 10°C; 10 d | • increased weight loss at 10°C | [92] |

*d: days of storage; CS: cold storage; CA: controlled atmosphere; RH: relative humidity; h: hour; ACS: ambient charcoal store; GTHS: grass thatched hut store; DLS: diffused light store; and HDPE: high-density polyethylene.*

**Table 2.**
*Summary of studies on effects of storage system on the quality of selected root vegetables.*

sugar content is linked to the development of harsh, and oily flavor in carrots during prolonged storage [96].

Significant differences have been observed in the content of main beetroot betcyanins (betanin and isobetanin) during cold storage of red beetroot at 5°C for 196 days [71]. The content of betanin in red beetroot peel decreased in the first 140 days of cold storage and then slightly increased (**Table 2**). In terms of isobetanin, until 98 days, an increasing trend and afterward, up to 140 days of storage, a light decrease were noticed [71]. In [72] it was shown that after 1 month of cold storage, there was an increase in beetroot water loss that resulted into increased total dry matter content. The total phenolic content of beetroot in cold storage increased after 4 months of storage. In another study, the nitrate content in beetroot increased significantly after 1 month of cold storage [73], and no further significant increase was observed after a cold storage period of 4 months. In [61], there was no significant change in total sugar content of 11 beetroot genotypes stored under optimal cold storage conditions at 1 ± 1°C and 90–95% RH for 7 months. Additionally, beetroots retained their levels of total soluble sugar contents after 1 month of cold storage, and the levels decreased after 4 months of cold storage. Therefore, it is beneficial to use these beetroot genotypes freshly or within the first month of storage, when a high sugar content is desired.

Freshly harvested cassava roots transpire and loose moisture, which reduces their quality during storage. Cassava undergoes postharvest physiological deterioration (PPD) once the roots are separated from the main plant. As a result of this mechanical damage (wounding), the roots respond with a healing mechanism that initiates about 15 min after damage, and fails to switch off in harvested roots [97]. This is observed as a blue/black or brown discoloration of the vascular parenchyma (vascular streaking) within 24–72 h of harvest [20] rendering it unpalatable. As described by Beeching [98, 99], this mechanism involves an oxidative burst of the superoxide radical ($O_2-$), which is followed by the further production of reactive oxygen species, altered gene expression, and the accumulation of secondary metabolites. This physical and biochemical change is often followed by microbial deterioration and root tissue softening.

Traditional methods have been shown to extend cassava shelf life for few days during which, the cyanide, moisture, and starch content in the root decreases while the ash, sugar, crude fiber as well as the acidic content increases with the length of storage [49, 58, 59]. Generally, temperatures of about 20–30°C and low RH between 65 and 80% encourage deterioration [100]. Therefore, storage of cassava under high relative humidity and limited oxygen conditions, for example, in polyethylene bags, storage boxes, and coating with paraffin wax, can reduce water loss and oxidative stress [101]. Besides, Rickard [102] reported that at 80–90% RH, cassava roots showed a typical wound-healing response with periderm formation occurring in 7–9 days at 35°C and 10–14 days at 25°C.

Prolonged in-ground storage has been shown to increase the size of the root. However, the roots become more woody and fibrous, and with decreased palatability and amount of extractable starch, besides becoming more susceptible to attack by pathogenic microorganisms (**Table 2**) [49, 58]. Covering cassava roots with paraffin wax by dipping the root in paraffin wax (at a temperature of 55–65°C for a few seconds) after treatment with fungicide has been reported to prolong shelf-life of cassava roots up to 2 months (**Table 1**) [49, 68].

Cassava roots can also be stored for 2 weeks between 0 and 4°C without any internal deterioration, and after 6.5 months of storage, the part of the root without decay usually is in excellent condition for human consumption [49]. At temperatures above 4°C, roots develop the PPD symptoms more rapidly and have to be discarded after 2 weeks of storage [49]. A study on the effect of the total carotenoids content in cassava root on the reduction and delay of postharvest deterioration showed that cassava roots kept at 10°C and 80% RH can remain fresh till after 2 weeks [100]. Thus, higher carotenoid can reduce or delay the onsets of PPD and extend the shelf life of the root. Wijesinghe and Sarananda [103] reported that freezing of fresh cassava for up to 8 weeks resulted into loss in water and increase in dry matter, that was driven by the high vapor pressure deficit created between the product and the low RH in the refrigerator environment. Consequently, the water stress which remained of acceptable eating quality although none remained as good as the freshly harvested ones.

Physicochemical parameters that determine quality of radish are well maintained in lower storage temperature of 0°C hence it is recommended for extended storage period of radish. Studies by Chandra et al. [90] showed that weight losses of radish roots were remarkably lower (<3%) in radish packed with micro perforated HDPE film in plastic crate and that cured then packed in micro perforated HDPE film in PC while the cured sample maintained its total soluble solids, and flesh and skin firmness. Both samples also recorded lower scores of black spot, surface shrinkage, and fungal infection incidence (**Table 2**). About 1.3% weight loss of unpacked topped radish root was noted after 9 days of storage at either 5°C or 10°C [104]. However, severe weight loss of radishes (about 52%) was noted just after 3 days of storage when radishes were stored at room temperature (20 ± 2°C) without leaves [91]. In a related study, a weight loss of about 2.5% was noted when whole radishes were stored at 1°C or 5°C for 10 days, whereas this loss reached nearly double when they were stored at 10°C [92]. These results indicate that storage temperature greatly affects the fresh weight retention ability of radish. In another study, the concentration of isothiocyanate sulforaphene and myrosinase activity were measured in two radish cultivars, namely "Chungwoon plus" (CP) and "Taebaek" (TB), during storage at 0°C for 4 months. After the storage period, the sulforaphene concentrations in the CP and TB radish cultivars decreased by 81% and 40%, respectively [89]. Also, the myrosinase activity decreased in both cultivars which subsequently decreased the formation of sulforaphene [89]. Glucosinolates, lipid-soluble vitamins E and K contents, and primary metabolites were measured from topped radish root stored at 1°C for 90 days. The results indicated that the tested storage conditions had no effect on the concentration of aliphatic glucosinolates present in radish [88].

Use of low temperatures (i.e., 2–4°C) and potato sprout inhibitors are the widely used storage treatments on freshly harvested potatoes. Curing potatoes allows the formation of a protective layer (wound periderm) over areas of potato that could have been damaged during harvesting. Curing has been reported to limit weight loss and to prevent the penetration of pathogenic microorganisms. In a study in which the effects of various curing and storage conditions (i.e., duration, temperature, and RH) on the quality of two potato cultivars, "Moonlight" and "Nadine" were investigated, high curing RH (93%) led to significantly lower skin browning, shriveling, and weight loss in both cultivars, and significantly lower incidence of rot in "Nadine" than low curing RH (62%) [105]. A 7 days curing at >90% RH and at least 15°C was recommended.

Sprout inhibitors (e.g., ethylene) and treatments to inhibit microbial establishment on harvested potatoes are reported to be effective but with varied secondary effects on potatoes. Previous studies have reported that ethylene application can either shorten or delay potato dormancy period depending on both treatment duration and concentration [106]. Furthermore, a continuous application of ethylene [69] and/or early application (applied after appearance of first sprouting) [70] is reported to prevent potato sprouting. However, this can also increase the content of reducing sugars (primarily glucose and fructose) in potato, thus limiting its use for processing potatoes. High levels of reducing sugars in processing potatoes causes cold induced sweetness [107] that is responsible for the dark brown color on processed potato products that gives a bitter taste [108]. In a separate study, a single treatment with HPP or CIPC resulted, after 6 months of storage at 10 ± 1°C, in sprouting rates of 61% and 58%, respectively, vs. 87% in the untreated control [54]. From preliminary experiments in [48], potatoes exposed to UV-C light at five different intensities (0.0, 3.4, 7.1, 10.5, and 13.6 kJ $m^{-2}$) and stored in the dark at 20°C and 80% RH for 40 days showed reduction in sprout length and development up to 20 d. However, this effect diminished during storage. Also in [47], potatoes exposed to gamma Irradiation (0, 50, 100, and 150 Gy levels) on different dates (10, 30, and 50 days after harvest) were studied during prolonged storage at 8 and 16°C. Results indicated that indicated that early and higher irradiation levels significantly decreased sprouting, percent weight loss and specific gravity of potato. However, the loss of ascorbic acid, and reducing and non-reducing sugars significantly increased by delay in irradiation whereas the sugars and ascorbic acid content was decreased by irradiation. Higher storage temperature (16°C) caused greater loss of ascorbic acid. A delay in irradiation and storage at high temperature was not recommended [47]. A study in [83] showed that potatoes buried at deeper depths (overground/pit storage) accumulated a lot of respiratory heat. This heat has been reported to promote potato sprouts [84]. Sprouting results in remobilization of storage compounds mainly starch and proteins as sprout tissue is built from the potato reserves. This increases the rate of respiration as well as evaporation [109], and consequently weight loss. Also, vitamin C is adversely affected by sprouting [47]. Sprouting and sprout growth contributes to formation of toxic glycoalkaloids (TGA). This involves the buildup of chlorophyll beneath the peel, a process known as "greening" [110]. Greening in potatoes is associated with the TGA solanine accumulation. In a study where six potato cultivars were analyzed for TGA content after 6 and 14 weeks of storage at either 10 or 4°C, results indicated that the exposure of some cultivars, to low temperatures within 2 weeks of harvest resulted in a relatively rapid accumulation of TGA to levels close to or exceeding the recommended safe maximum level of 200 mg of TGA per kilogram of fresh weight [86].

Nevertheless, storage of potatoes at low temperatures has been shown to significantly decreases the content of vitamin C [87, 111, 112], although in [82] it was noted that even with the decrease in vitamin C, a significant amount was still retained. The decrease in vitamin C content is attributed to its use, in the potato, as an antioxidant compound in response to oxidative stress caused by low temperature storage. In [87], an evaluation for antioxidant parameters of potatoes stored at room temperature, 15°C and 4°C for 90 days showed that all phenolic acid content increased with storage time, except para-coumaric acid which decreased at 4°C. Similarly, a separate study showed an increase in chlorogenic acid, caffeic acid, and sinapic acid content during storage of potatoes for 90 days at 3°C [81].

High temperatures can also influence respiration rate, development of decay causing organisms, greening, and shrinkage in stored potatoes [113]. A study on the effect of three store types (ambient charcoal-cooled, traditional grass thatched hut and diffused light store [DLS]) on the quality of potatoes stored for 56 days at temperatures 9.7–19.4°C and RH 65–93%, results showed increased weight loss, greening ,and shrinkage in potatoes stored in DLS (temperature 16.15–19.35°C; RH 65–89%) [85]. Additionally, there was a significant increase in the reducing sugar content. The starch content in potato samples from the three stores decreased with increasing storage time.

## 6. Conclusion

The quality of root vegetables deteriorate gradually during storage in response to several endogenic factors and environmental conditions. Processes such as transpiration, respiration, senescence, and maturation, as well as attacks by pathogens lead not only to quantitative and quality losses. The range of pre-storage treatment methods and storage conditions that are commonly used have been reported to be safe and effective at mitigating several postharvest deterioration in root vegetable. Nevertheless, they can also influence the quality of freshly harvested root vegetables. Therefore, choice of appropriate treatments and storage conditions that do not significantly decrease the quality of stored produce should be considered.

## Acknowledgements

Authors thank the Kenya Climate Smart Agriculture Project (KCSAP) under Project Title "Climate-smart adaptive technologies for sustainable production and postharvest management and utilization of ware potato in Kenya" that provided the background and interest on storage of root vegetables. Dr. Viola Kosgei, Moi University (Kenya), Department of Chemistry and Biochemistry, for the discussions on carotenoids.

## Conflict of interest

The authors declare no conflict of interest.

## Author details

Elijah K. Lelmen[1] and Jacqueline K. Makatiani[2,3]*

1 Department of Biochemistry and Molecular Biology, Egerton University, Egerton, Nakuru, Kenya

2 Department of Biological Sciences, Moi University, Eldoret, Kenya

3 Africa Center of Excellence in Phytochemicals, Textile and Renewable Energy, Moi University, Eldoret, Kenya

*Address all correspondence to: jkmakatiani@mu.ac.ke

## References

[1] Baião DS, Silva FO, d'El-Rei J, Neves MF, Perrone D, del Aguila EM, et al. A new functional beetroot formulation enhances adherence to nitrate supplementation and health outcomes in clinical practice. SDRP Journal of Food Science & Technology. 2018;**3**:484-498. DOI: 10.25177/JFST.3.6.1

[2] Horáková L. Flavonoids in prevention of diseases with respect to modulation of Ca-pump function. Interdisciplinary Toxicology. 2011;**4**(3):114-124

[3] Barba FJ, Nikmaram N, Roohinejad S, Khelfa A, Zhu Z, Koubaa M. Bioavailability of glucosinolates and their breakdown products: Impact of processing. Frontiers in Nutrition. 2016;**3**:24

[4] Rietjens IMCM, Louisse J, Beekmann K. The potential health effects of dietary phytoestrogens. British Journal of Pharmacology. 2017;**174**(11):1263-1280

[5] Eggersdorfer M, Wyss A. Carotenoids in human nutrition and health. Archives of Biochemistry and Biophysics. 2018;**652**:18-26

[6] Jan R, Asaf S, Numan M, Lubna KK-M. Plant secondary metabolite biosynthesis and transcriptional regulation in response to biotic and abiotic stress conditions. Agronomy. 2021;**11**:968. DOI: 10.3390/agronomy11050968

[7] Forni C, Facchiano F, Bartoli M, Pieretti S, Facchiano A, D'Arcangelo D, et al. Beneficial role of phytochemicals on oxidative stress and age-related diseases. BioMed Research International. 2019 Apr;7(2019):8748253. DOI: 10.1155/2019/8748253

[8] Yang L, Wen K-S, Ruan X, Zhao Y-X, Wei F, Wang Q. Response of plant secondary metabolites to environmental factors. Molecules. 2018;**23**(4):762

[9] Machado RMA, Alves-Pereira I, Ferreira RMA. Plant growth, phytochemical accumulation and antioxidant activity of substrate-grown spinach. Heliyon. 2018;**4**(8):e00751-e

[10] Hossain MA, Miah MAM. Final Report (CF 2/08): Postharvest losses and technical efficiency of potato storage systems in Bangladesh. Funded by USAID and EC & jointly implemented by FAO and FPMU of MoFDM. 2009. p. 88

[11] Westby A. Ch 14. Cassava utilization, storage and small-scale processing. In: Hillocks RJ, Thresh JM, Bellotti AC, editors. Cassava: Biology, Production and Utilization. Wallingford, UK: CAB International; 2002. pp. 281-300

[12] Wang Q, Cao Y, Zhou L, Jiang C, Feng Y, Wei S. Effects of postharvest curing treatment on flesh colour and phenolic metabolism in fresh-cut potato products. Food Chemistry. 2015;**169**:246-254. DOI: 10.1016/j.foodchem.2014.08.011

[13] Fallick E, Ilić Z. Control of decay of fruits and vegetables by heat treatments; the risks and the benefits. In: Palou L, Smilanick JL, editors. Postharvest Pathology of Fresh Produce. Boca Raton: CRC Press; 2020. pp. 521-537. DOI: 10.1201/9781315209180

[14] Mani F, Bettaieb T, Doudech N, Hannachi C. Physiological mechanisms for potato dormancy release and sprouting: A review. African Crop Science Journal. 2014;**22**:155-174

[15] Gol NB, Patel PR, Rao TVR. Improvement of quality and shelf life of strawberries with edible coatings enriched with chitosan. Postharvest Biology and Technology. 2013;**85**:185-195. DOI: 10.1016/j.postharvbio.2013.06.008

[16] Fadeyibi A. Storage methods and some uses of cassava in Nigeria. Continental Journal of Agricultural Science. 2012;**5**:12-18

[17] Karim O, Fasasi O, Oyeyinka S. Gari yield and chemical composition of cassava roots stored using traditional methods. Pakistan Journal of Nutrition. 2009;**8**:1830-1833

[18] Fricke BA. Precooling fruits and vegetables using hydrocooling. ASHRAE Journal. 2006;**48**(2):20-28

[19] Farkas J. Food technologies: food irradiation. Encyclopedia of Food Safety. 2014;**3**:178-186

[20] Reilly K, Góomez-Váasquez R, Buschmann H, Tohme J, Beeching JR. Oxidative stress responses during cassava post-harvest physiological deterioration. Plant Molecular Biology. 2004;**56**:625-641

[21] Nemzer B, Pietrzkowski Z, Spórna A, Stalica P, Thresher W, Michałowski T, et al. Betalainic and nutritional profiles of pigment-enriched red beet root (*Beta vulgaris* L.) dried extracts. Food Chemistry. 2011;**127**:42-53. DOI: 10.1016/j.foodchem.2010.12.081

[22] Kapadia GJ, Rao GS, Ramachandran C, Lida A, Suzuki N, Tokuda H. Synergistic cytotoxicity of red beetroot (*Beta vulgaris* L.) extract with doxorubicin in human pancreatic, breast and prostate cancer cell lines. Journal of Complementary and Integrative Medicine. 2013;**10**(1):1-10. DOI: 10.1515/jcim-2013-0007

[23] Strack D, Vogt T, Schliemann W. Recent advances in betalain research. Phytochemistry. 2003;**62**:247-269

[24] Bavec M, Turinek M, Grobelnik-Mlakar S, Slatnar A, Bavec F. Influence of Industrial and alternative farming systems on contents of sugars, organic acids, total phenolic content, and the antioxidant activity of red beet (*Beta Vulgaris* L. Ssp. *Vulgaris* Rote Kugel). Journal of Agricultural and Food Chemistry. 2010;**58**:11825-11831

[25] Vasconcellos J, Conte-Junior C, Silva D, Pierucci AP, Paschoalin V, Alvares TS. Comparison of total antioxidant potential, and total phenolic, nitrate, sugar, and organic acid contents in beetroot juice, chips, powder, and cooked beetroot. Food Science and Biotechnology. 2016;**25**:79-84

[26] Sobhy ES, Abdo E, Shaltout O, Abdalla A, Zeitoun A. Nutritional evaluation of beetroots (*Beta vulgaris* L.) and its potential application in a functional beverage. Plants (Basel). 2020;**9**:1752. DOI: 10.3390/plants9121752

[27] Montagnac JA, Davis CR, Tanumihardjo SA. Nutritional value of cassava for use as a staple food and recent advances for improvement. Comprehensive Reviews in Food Science and Food Safety. 2009;**8**:181-194

[28] Chen J, Li X. Hypolipidemic effect of flavonoids from mulberry leaves in triton WR-1339 induced hyperlipidemic mice. Asia Pacific Journal of Clinical Nutrition. 2007;**16**(Suppl 1):290-294

[29] Feng Q, Leong WS, Liu L, Chan WI. Peruvoside, a cardiac glycoside, induces primitive myeloid leukemia cell death. Molecules. 2016;**21**:534

[30] Charles A, Sriroth K, Huang T. Proximate composition,mineral contents,

hydrogen cyanide and phytic acid of 5 cassava genotypes. Food Chemistry. 2005;**92**:615-620

[31] Amarachi DU, Oluwafemi JC, Umezuruike LO. Postharvest handling and storage of fresh cassava root and products: A review. Food and Bioprocess Technology. 2015;**8**:729-748. DOI: 10.1007/s11947-015-1478-z

[32] McDowell I, Oduro KA. Investigation of the β-carotene in yellow varieties of cassava (*Manihot esculenta*). Journal of Plant Foods. 1983;**5**:169-171

[33] Leja M, Kamińska I, Kramer M, Maksylewicz-Kaul A, Kammerer D, Carle R, et al. The content of phenolic compounds and radical scavenging activity varies with carrot origin and root color. Plant Foods for Human Nutrition. 2013;**68**:163-170. DOI: 10.1007/s11130-013-0351-3

[34] Krinsky NI, Johnson EJ. Carotenoid actions and their relation to health and disease. Molecular Aspects of Medicine. 2005;**26**:459-516

[35] Ahmad T, Cawood M, Iqbal Q, Ariño A, Batool A, Tariq R, et al. Phytochemicals in *Daucus carota* and their health benefits-Review article. Foods (Basel, Switzerland). 2019;**8**:424. DOI: 10.3390/foods8090424

[36] Gopalan C, Ramasastry BV, Balasubramanian SC. Nutritive value of Indian foods. Hyderabad: National Institute of Nutrition; 1991. p. 47

[37] Goyeneche R, Roura S, Ponce A, Vega-Galvez A, Quispe-Fuentes I, Uribe E, et al. Chemical characterization and antioxidant capacity of red radish (*Raphanus sativus* L.) leaves and roots. Journal of Functional Foods. 2015;**16**:256-264

[38] Ramesh CK, Raghu KL, Jamuna KS, Joyce GS, Mala RS, Avin BR. Comparative evaluation of antioxidant property in methanol extracts of some common vegetables of India. Annals of Biological Research. 2011;**2**:86-94

[39] Gupta K, Talwar G, Jain V, Dhawan K, Jain S. Salad crops: Root, bulb, and tuber crops. In: Caballero B, Trugo LC, Finglas PM, editors. Encyclopedia of Food Sciences and Nutrition. 2nd ed. San Diego, California: Academic Press; 2003. pp. 5060-5073

[40] Burgos G, Zum FT, Andre C, Kubow S. The potato and its contribution to the human diet and health. In: Campos H, Ortiz O, editors. The Potato Crop. Cham, Switzerland: Springer; 2020. pp. 37-74. DOI: 10.1007/978-3-030-28683-5_2

[41] Şengül M, Keleş F, Keleş MS. The effect of storage conditions (temperature, light, time) and variety on the content of potato tubers and sprouts. Food Control. 2004;**15**:181-186

[42] Fallik E. Pre-storage hot water treatments (immersion, rinsing and brushing). Postharvest Biology and Technology. 2004;**32**:125-134. DOI: 10.1016/j.postharvbio.2003.10.005

[43] Pinhero RG, Coffin R, Yada RY. Chapter 12 Post-harvest storage of potatoes. In: Jaspreet S, Lovedeep K, editors. Advances in Potato Chemistry and Technology. New York, USA: Academic Press; 2009. pp. 339-370. DOI: 10.1016/B978-0-12-374349-7.00012-X

[44] Lurie S. Prestorage heat stress to improve storability of fresh produce: A review. Israel Journal of Plant Sciences. 2006;**63**(1):17-21. DOI: 10.1080/07929978.2016.1159411

[45] Ilić SZ, Šunić L, Barać S, Stanojević L, Cvetković D. Effect of postharvest treatments and storage conditions on quality parameters of carrots (*Daucus carota* L.). The Journal of Agricultural Science. 2013;**5**:100-106

[46] Afek U, Orenstein J, Nuriel E. Steam treatment to prevent carrot decay during storage. Crop Protection. 1999;**18**:639-642. DOI: 10.1016/S0261-2194(99)00065-4

[47] Rezaee M, Almassi M, Farahani AM, Minaei S, Khodadadi M. Potato sprout inhibition and tuber quality after postharvest treatment with gamma irradiation on different dates. Journal of Agricultural Science and Technology. 2011;**13**:829-841

[48] Pristijono P, Bowyer MC, Scarlett CJ, Vuong QV, Stathopoulos CE, Golding JB. Effect of UV-C irradiation on sprouting of potatoes in storage. Acta Horticulturae. 2018;**1194**:475-478

[49] Ravi V, Aked J, Balagopalan C. Review on tropical root and tuber crops I. Storage methods and quality changes. Critical Reviews in Food Science and Nutrition. 1996;**36**:661-709

[50] Dehghannya J, Ngadi M, Vigneault C. Mathematical modeling procedures for airflow, heat and mass transfer during forced convection cooling of produce: Review. Food Engineering Reviews. 2010;**2**:227-243. DOI: 10.1007/s12393-010-9027-z

[51] Brooker J, Pearson JL, editors. Planning data for marketing selected fruits and vegetables in the south, Part III - Fresh Vegetable Packing Handbook. Southern Cooperative Series Bulletin 152. Raleigh, USA: University of North Carolina Agricultural Experiment Station; 1970. pp. 135

[52] Francis GA, O'Beirne D. Effects of vegetable type, package atmosphere and storage temperature on growth and survival of *Escherichia coli* O157:H7 and *Listeria monocytogenes*. Journal of Industrial Microbiology & Biotechnology. 2001;**27**:111-116

[53] El-Awady AA, Moghazy AM, Gouda AEA, Elshatoury RSA. Inhibition of sprout growth and increase storability of processing potato by antisprouting agent. Trends in Horticultural Research. 2014;**4**:31-40. DOI: 10.3923/thr.2014.31.40

[54] Afek U, Orenstein J, Nuriel E. Inhibition of potato sprouting during storage by HPP (hydrogen peroxide plus). American Journal of Potato Research. 2000;**77**:63-65

[55] Thoma J, Zheljazkov VD. Sprout suppressants in potato storage: Conventional options and promising essential oils—A review. Sustainability. 2022;**14**(11):6382. DOI: 10.3390/su14116382

[56] Hildebrand PD, Forney CF, Song J, Fan L, McRae KB. Effect of a continuous low ozone exposure (50nLL− 1) on decay and quality of stored carrots. Postharvest Biology and Technology. 2008;**49**(3):397-402

[57] Forney CF, Song J, Hildebrand PD, Fan L, McRae KB. Interactive effects of ozone and 1-methylcyclopropene on decay resistance and quality of stored carrots. Postharvest Biology and Technology. 2007;**45**(3):341-348

[58] Wenham JE. Post-harvest deterioration of cassava. A biotechnology perspective. FAO Plant Production and Protection Paper 130. 1995. NRI/FAO. Rome. p. 90.

[59] Booth RH, Coursey DG. Storage of cassava and related post-harvest problems. In: Cassava Processing and Storage. Ottawa, Canada: International

Development Research Centre Monographs IDRC-03le; 1974. pp. 43-49

[60] Nahdy SM, Odong M. Storage of fresh cassava tuber in plant based storage media. In: Proceeding of the Workshop on PostHarvest technology Experience in Africa; 4-8 July 1994. 1995. Accra, Ghana. Edited by FAO, Rome

[61] Viskelis J, Nevidomskis S, Bobinas C, Urbonaviciene D, Bobinaite R, Karkleliene R, et al. Evaluation of beetroot quality during various storage conditions. In: Proceedings of the FOODBALT 2019 13th Baltic Conference on Food Science and Technology. Food, Nutrition, Well-Being 2-3 May 2019. Jelgava. Latvia. pp. 170-175

[62] Gross KC, Wang CY, Saltveit M, editors. The commercial storage of fruits, vegetables, and florist and nursery stocks. In: Agriculture Handbook 66. Beltsville, Maryland, USA: Agricultural Research Service, United States Department of Agriculture; 2016. pp. 792

[63] Thompson AK, Prange RK, Bancroft RD, Puttongsiri T. Controlled Atmosphere Storage of Fruit and Vegetables. 3rd ed. Boston MA, USA: CABI; 2018. pp. 445

[64] Muthoni J, Kabira JN, Kipkoech D, Abong GO, Nderitu JH. Feasibility of low-cost seed potato storage in Kenya: The case of diffused light storage in Nyandarua county. The Journal of Agricultural Science. 2014;**6**(1):59-65. DOI: 10.5539/jas.v6n1p59

[65] Gottschalk K, Ezekiel R. Chapter 13 Storage. In: Gopal J, Khurana SMP, editors. Handbook of Potato Production, Improvement, and Postharvest Management. Boca Raton, Florida, USA: CRC Press; 2006. pp. 489-522

[66] Wauters P, Naziri D, Turinawe A, Akello R, Parker ML. Economic analysis of alternative ware potato storage technologies in Uganda. American Journal of Potato Research. 2022. pp. 12. Available from: https://link.springer.com/article/10.1007/s12230-022-09874-3. DOI: 10.1007/ s12230-022-09874-3. [Accessed: 2022-08-10]

[67] Ojaghian MR, Zhang J-Z, Xie G-L, Wang Q, Li X-L, Guo D-P. Efficacy of UV-C radiation in inducing systemic acquired resistance against storage carrot rot caused by *Sclerotinia sclerotiorum*. Postharvest Biology and Technology. 2017;**130**:94-102

[68] Aristizabal J, Sanchez T. Technical guide for the production and analysis of cassava starch. Bulletin of Agriculture Services of the FAO 130. Rome Italy; 2007. p. 134

[69] Suttle JC. Physiological regulation of potato tuber dormancy. American Journal of Potato Research. 2004;**81**:253-262

[70] Foukaraki SG, Cools K, Chope GA, Terry LA. Impact of ethylene and 1-MCP on sprouting and sugar accumulation in stored potatoes. Postharvest Biology and Technology. 2016;**114**:95-103

[71] Kujala TS, Loponen JM, Klika KD, Pihlaja K. Phenolics and betacyanins in red beetroot (*Beta Vulgaris*) root: distribution and effect of cold storage on the content of total phenolics and three individual compounds. Journal of Agricultural and Food Chemistry. 2000;**48**:5338-5342

[72] Yasaminshirazi K, Hartung J, Fleck M, Graeff-Hönninger S. Impact of cold storage on bioactive compounds and their stability of 36 organically grown beetroot genotypes. Food. 2021;**10**:1281. DOI: 10.3390/foods10061281

[73] Kosson R, Elkner K, Szafirowska A. Quality of fresh and processed red

beet from organic and conventionl cultivation. Vegetable Crops Research Bulletin. 2011;**75**:125-132

[74] Simoes ADN, Allende A, Tudela JA, Puschmann R, Gil MI. Optimum controlled atmospheres minimize respiration rate and quality losses while increase phenolic compounds of baby carrots. Lebensmittel-Wissenschaft und Technologie—Food Science and Technology. 2011;**44**(1): 277-283

[75] Kidmose U, Hansen SL, Christensen LP, Edelenbos M, Larsen E, Nørbaek R. Effects of genotype, root size, storage, and processing on bioactive compounds in organically grown carrots (*Daucus carota* L.). Journal of Food Science. 2004;**69**:S388-S394

[76] Kamiloglu S, Pasli AA, Ozcelik B, Van Camp J, Capanoglu E. Colour retention, anthocyanin stability and antioxidant capacity in black carrot (*Daucus carota*) jams and marmalades: Effect of processing, storage conditions and in vitro gastrointestinal digestion. Journal of Functional Foods. 2015;**13**:1-10

[77] Imsic M, Winkler S, Tomkins B, Jones R. Effect of storage and cooking on β-Carotene isomers in carrots (*Daucus carota* L. cv. 'Stefano'). Journal of Agricultural and Food Chemistry. 2010;**58**:5109-5113

[78] Cortés C, Esteve MJ, Frígola A, Torregrosa F. Changes in carotenoids including geometrical isomers and ascorbic acid content in orange–carrot juice during frozen storage. European Food Research and Technology. 2005;**221**:125-131

[79] Jacobo-Velázquez DA, Martinez-Hernandez GB, Rodriguez S, Cao CM, Cisneros-Zevallos L. Plants as biofactories: Physiological role of reactive oxygen species on the accumulation of phenolic antioxidants in carrot tissue under wounding and hyperoxia stress. Journal of Agricultural and Food Chemistry. 2011;**59**(12):6583-6593

[80] Negi PS, Roy SK. Effect of low-cost storage and packaging on quality and nutritive value of fresh and dehydrated carrots. Journal of the Science of Food and Agriculture. 2000;**80**:2169-2175

[81] Madiwale GP, Reddivari L, Holm DG, Vanamala J. Storage elevates phenolic content and antioxidant activity but suppresses antiproliferative and pro-apoptotic properties of colored-flesh potatoes against human colon cancer cell lines. Journal of Agricultural and Food Chemistry. 2011;**59**:8155-8166

[82] Külen O, Stushnoff C, Holm DG. Effect of cold storage on total phenolics content, antioxidant activity and vitamin C level of selected potato clones. Journal of the Science of Food and Agriculture. 2013;**93**:2437-2444

[83] Gichohi EG, Pritchard MK. Storage temperature and maleic hydrazide effects on sprouting, sugars, and fry color of shepody potatoes. American Potato Journal. 1995;**72**:737-747

[84] Forbush T, Brook R. Influence of air flow on chip potato storage management. American Potato Journal. 1993;**70**(12):869

[85] Lelmen EK, Makatiani JK. Performance of Different Storage Systems on Physico-Chemical Quality of Ware Potato in Elgeyo Marakwet County. Kenya: Egerton University International Conference; 2022

[86] Griffiths DW, Bain H, Finlay M, Dale B. Effect of storage temperature on potato (*solanum tuberosum* l.) tuber glycoalkaloid content and the subsequent

accumulation of glycoalkaloids and chlorophyll in response to light exposure. Journal of Agricultural and Food Chemistry. 1998;**46**(12):5262-5268. DOI: 10.1021/jf9800514

[87] Galani JHY, Mankad PM, Shah AK, Patel NJ, Acharya RR, Talati JG. Effect of storage temperature on vitamin C, total phenolics, UPLC phenolic acid profile and antioxidant capacity of eleven potato (*Solanum tuberosum*) varieties. Hort. The Plant Journal. 2017;**3**:73-89

[88] Liu M, Seo H-Y, Min S, Ku K-M. Phytonutrients and metabolism changes in topped radish root and its detached leaves during 1 °C Cold postharvest storage. Horticulturae. 2022;**8**:42. DOI: 10.3390/horticulturae8010042

[89] Lim S, Lee EJ, Kim J. Decreased sulforaphene concentration and reduced myrosinase activity of radish (*Raphanus sativus* L.) root during cold storage. Postharvest Biology and Technology. 2015;**100**:219-225. DOI: 10.1016/j.postharvbio.2014.10.007

[90] Chandra D, Lee J-S, Choi H, Kim J. Effects of packaging on shelf life and postharvest qualities of radish roots during storage at low temperature for an extended period. Journal of Food Quality. 2018;**1**:1-12. DOI: 10.1155/2018/3942071

[91] Ayub RA, Spinardi B, Gioppo M. Storage and fresh cut radish. Acta Scientiarum Agronomy. 2013;**35**(2):241-245

[92] del Aguila JS, Sasaki FF, Heiffig LS, Ortega EMM, Jacomino AP, Kluge RA. Fresh-cut radish using different cut types and storage temperatures. Postharvest Biology and Technology. 2006;**40**(2):149-154

[93] Koca N, Karadeniz F. Changes of bioactive compounds and anti-oxidant activity during cold storage of carrots. International Journal of Food Science and Technology. 2008;**43**:2019-2025

[94] Matejková J, Petříková K. Variation in content of carotenoids and vitamin C in carrots. Notulae Scientia Biologicae. 2010;**2**:88-91

[95] Kjellenberg L, Johansson E, Gustavsson KE, Olsson ME. Polyacetylenes in fresh and stored carrots (*Daucus carota*): relations to root morphology and sugar content. Journal of the Science of Food and Agriculture. 2012;**92**:1748-1754

[96] Kjeldsen F, Christensen LP, Edelenbos M. Changes in volatile compounds of carrots (*Daucus carota* L.) during refrigerated and frozen storage. Journal of Agricultural and Food Chemistry. 2003;**51**:5400-5407

[97] Sánchez H, Ceballos F, Calle JC, Pérez C, Egesi CE, Cuambe AF, et al. Tolerance to postharvest physiological deterioration in cassava roots. Crop Science. 2010;**50**(4):1333-1338

[98] Beeching J. Identifying Target Points for the Control of Post-Harvest Physiological Deterioration in Cassava. Final Technical Report, Project R6983 (ZB0098). University of Bath, UK: Crop Post Harvest Programme; 2001

[99] Reilly K, Bernal D, Cortes DF, Gomez-vasquez R, Tohme J, Beeching JR. Towards identifying the full set of genes expressed during cassava postharvest deterioration. Plant Molecular Biology. 2007;**64**:187-203

[100] Sánchez T, Chávez AL, Ceballos H, Rodriguez-Amaya DB, Nestel P, Ishitani M. Reduction or delay of post-harvest physiological deterioration in cassava roots with higher carotenoid content. Journal of the Science of Food

and Agriculture. 2006;**86**(4):634-639. DOI: 10.1002/jsfa.2371

[101] Marriott J, Been BO, Perkins C. The aetiology of vascular discoloration in cassava roots after harvesting: association with water loss from wounds. Plant Physiology. 1978;**44**:38-42

[102] Rickard IE. Physiological deterioration of cassava roots. Journal of the Science of Food and Agriculture. 1985;**36**:167-176

[103] Wijeslinghe WAJP, Sarananda KH. Utilization of cassava through freezing. Journal of Food Science and Technology. 2002;**50**(6):611-654

[104] Tsouvaltzis P, Brecht JK. Changes in quality and antioxidant enzyme activities of bunched and topped radish (*Raphanus sativus* L.) plants during storage at 5 or 10 °C. Journal of Food Quality. 2014;**37**(3):157-167

[105] Mahilum H, Li M, Heyes J. Postharvest curing and storage of New Zealand-grown potato cultivars ('Moonlight' and 'Nadine') for the fresh export market. New Zealand Journal of Crop and Horticultural Science. 2022;**50**(2-3):162-177. DOI: 10.1080/01140671.2021.2012488

[106] Shi Z, Halaly-Basha T, Zheng C, Weissberg M, Ophir R, Galbraith DW, et al. Transient induction of a subset of ethylene biosynthesis genes is potentially involved in regulation of grapevine bud dormancy release. Plant Molecular Biology. 2018;**98**:507-523

[107] Sonnewald U. Control of potato tuber sprouting. Trends in Plant Science. 2001;**6**:333-335

[108] Frazier MJ, Olsen N, Kleinkopf G. Organic and Alternative Methods for Potato Sprout Control in Storage. Moscow, ID: University of Idaho; 2004. Available from: http://www.extension.uidaho.edu/publishing/pdf/cis/cis1120.pd

[109] Wustman R, Struik PC. The canon of potato science: seed and ware potato storage. Potato Research. 2007;**50**(3):351-355

[110] Tanios S, Eyles A, Tegg R, Wilson C. Potato tuber greening: A review of predisposing factors, management and future challenges. American Journal of Potato Research. 2018;**95**:248-257

[111] Abong GO, Okoth MW, Imugi JK, Kabira JN. Losses in ascorbic acid during storage of fresh tubers, frying, packaging and storage of potato crisps from four Kenyan potato cultivars. American Journal of Food Technology. 2011;**6**:772-780

[112] Cho KS, Jeong HJ, Cho JH, Park YE, Hong SY, Won HS, et al. Vitamin C content of potato clones from Korean breeding lines and compositional changes during growth and after storage. Horticulture, Environment, and Biotechnology. 2013;**54**:70-75

[113] Kleinkopf GE. Early season storage. American Potato Journal. 1995;**72**:449-462

Section 2

# Breeding Research of Root Vegetables

**Chapter 6**

# Advances and Milestones of Radish Breeding: An Update

*Anand Kumar and Prashant Kaushik*


## Abstract

Radish is a member of the Cruciferae family. The important traits for radish breeding include high yield, early maturity, late bolting, pungency, cold-hardiness, drought resistance, heat tolerance and soil adaptability. For successful radish production, one needs to the understand nature and behaviour of the flower and very important to identify the S haplotypes of parental lines to produce $F_1$ hybrids based on self-incompatibility to get rid of laborious hand emasculation in radish. Therefore, further breeding programmes depend on inter-specific and intra-specific hybridization, which is vital in genomic studies and crop improvement by introducing desirable agronomic characters. It is essential to acquire detailed genetic information on chromosomes and inheritance. Genomics is now at the core of radish breeding to study the underlying differences in genotypes. Moreover, researchers have produced transgenic radishes with various agronomic characteristics over the last decade.


**Keywords:** radish, breeding, inter-specific hybridization, molecular breeding, genomics, genetic engineering

## 1. Introduction

Radish is an annual herbaceous vegetable known as *Raphanus sativus* [1], and it is a diploid containing two sets of 18 chromosomes (2n = 18) [2]. Radish belongs to the Cruciferae family and is eaten fresh as grated radish, a garnish and a salad [3]. Radish is regularly served in eastern Asian cuisine; radish has also featured in food worldwide [4]. There is a focus on developing high-quality radish varieties ideal for tropical and subtropical temperatures [5]. Breeding work has been performed on numerous agronomical traits including tolerance to pathogens and consumption adaptability. Traits for radish breeding include high yield, early maturity, late bolting, pungency, cold-hardiness, drought resistance, heat tolerance and soil adaptability [3]. There is a positive correlation between the radish's consistency and its amount of sugar, pungency, elaboration of the cell, water content and pore extent [6, 7]. Although, the main endeavour has been to modify the radish cultivation to various growing seasons [8]. It is essential to acquire detailed genetic information on chromosomes and information on inheritance for multiple genes responsible for agronomical, biochemical traits and resistance to biotic and abiotic stresses for carrying out a successful radish breeding [9–11]. Marketing assessment and consumer preference are primarily associated with physical attractiveness such as length, shape, size and skin colour [12]. A primary colour changes into white and different pink, red, purple, yellow and green.

The anthocyanin pelargonidin is the colour-causing pigment in red colour radish varieties, and they have a mild flavour (not as pungent) and are around 40 cm in length [13]. Quality-related traits are remarkably heritable. They are often strongly influenced by cultivation methods. The swollen tap roots of radishes may be oval, tapered or cylindrical [14]. Moreover, mechanical harvesting often includes cylindrical root cultivars [15]. Rich in antler velvet, radish roots produce useful phytochemicals. They have cancer-preventive properties and a significant contributor to the taste and flavour of Brassica vegetables [16]. In addition, radishes provide complex carbohydrates, dietary fibre often organic nutrients and minerals to humans [17].

Omics approaches using Next-Generation Sequencing (NGS) methods provide a large amount of genomic data that enable the identification of novel genes and sequences. In addition, genome-wide study results reveal the genetic causes of diverse characteristics [18, 19]. Furthermore, less study has been published that discusses the historical milestones and technological advancements in radish breeding. As a result, we have gathered information on different aspects of radish breeding and its numerous accomplishments in this section. We believe that this work will prove to be a valuable resource for vegetable breeders in the future.

## 2. Breeding goals

Radish is high in its nutrition content, health benefits conferred by its chemical compounds and a significant contribution to the human being [20]. Root length is an important trait of radish for consumers, and preference always goes to radish on length, diameter and colour; visual indicators such as the colour of the label and where it is presented are crucial to buyers' selection phase. Too many experiments have concentrated on gene mapping to classify significant colour-gene associations in various vegetables for this cause [21]. Productive radish root colour is vital to radish crop productivity. The discovery and detection of considerable plant genes involved in radish colouring would aid in advancing colour genetics [22]. The range of potential skin inheritance trends has been investigated extensively in radish. It was reported that 609 chemical compounds are present within 23 categories of the most studied varieties of Radish, such as red (30%), white (13%) and black (6%) [23]. This study also found that nutrients, antioxidants and phytochemicals are mainly identified in roots, sprouts and leaves, which could be considered an important part of a healthy diet [23]. In addition, researchers have focused on red radish with red flesh because it contains large amounts of a natural red pigment widely used in foods, wine and cosmetics. The uniformity of various colours, sizes and yields are the factors becoming a high-priority goal in radish breeding [24]. Radish has a wide variety of colours that affect its appearance and its nutritional quality [25]. However, the detection, identification and quantification of flavonoids in multicolour radish are rarely explored. At the same time, it was also identified that red and purple radishes contained similar anthocyanin compounds responsible for colour pigmentation, including red cyanidin, callistephin and pelargonin [26]. Purple ZJL contains cyanidin *o*-syringic acid and cyanin, whereas callistephin and pelargonin contain more amount in dark red TXH. The metabolites in coloured radishes that differed from SZB genes are broadly involved in the plant secondary metabolites biosynthetic pathway, such as flavonoid, flavone, isoflavonoid and phenylpropanoid biosynthesis. This approach would be useful for cultivating important and valuable new radish varieties [26, 27]. These results explain anthocyanin synthesis in radish and provide potential genetic clues for improving anthocyanins in radish roots [28].

Fusarium wilt (FW) is a soil-borne vascular wilt disease caused by fungal pathogen *Fusarium oxysporum* f. sp. Raphanin, causes severe yield losses in radish production [29]. The most effective method to control the FW is using resistant varieties in crop improvement. Fusarium resistance is highly studied among 'Motohashi-' or 'Kuroba-mino' lines of the Minowase variety, and Tosai' is the strongest line among Nerami varieties [30]. A pathogen could damage harassment of yield, and in root colour, these varieties would not be preferred for the consumer. Bioactive compounds in *R. sativus* (radish) are being studied to treat several diseases. Therefore, radish has attracted scientific attention due to its nutritional and phytochemical composition, which reduces the risk of developing many cancers and cardiovascular diseases. Further, the important goals are provided in **Figure 1**.

Moreover, salinisation is considered as one of the significant soil pollutions in the environment affecting plant growth and soil fertility globally [31]. This scenario alarms an urgent need to enrich the soil or to identify stress-tolerant plants. It is reported that antioxidant enzymes (HOD-Hydrogen Peroxide; SOD-Superoxide; LOD-Lipid Peroxidation; CAT-Catalase) play a major role in reducing the effects of salts in plants by monitoring the oxidative stress in them [32]. To study the salt tolerance of Japanese wild radishes called 'Hamadaikon' (*R. sativus* f. raphanistroides Makino) and its characteristics such as seed germination, plant height, root length and fresh weight were examined under the salinity condition. It was found that higher germination and growth in NaCl were shown at 25°C than those at 20°C [33]. Hence, wild radishes could be considered for salt tolerance breeding. Moreover, halopriming is a seed priming technique in which the seeds were soaked in various salt solutions to enhance germination and seedling emergence uniformly under adverse conditions. The effect of halopriming on germination, initial growth and development of radish

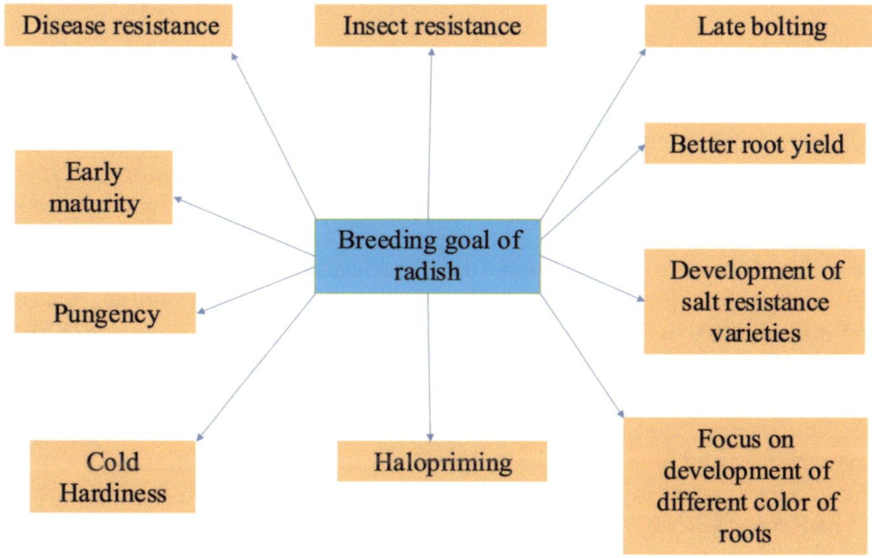


**Figure 1.**
*Breeding goals of radish (Raphanus sativus L.).*

under salt stress conditions was studied. It was found that the best outcome was achieved by priming with $CaCl_2$ for germination characteristics and vigour and with KCl for initial development [34].

## 3. Botany of radish

Radish (*R. sativus* L.) is an entomophilous flower classified as an allogamous plant [35]. Regular flowering appears from three florets on the tip of each branch of the panicle, and each flower is effective in producing a pod up to 1–3 inches long and consists of 1–6 seeds. The radish flowers open in the morning with fresh corolla and remain until the next day [36]. Kremer also reported that pollen receptivity of the flower limits up to few hours a day. Its flowers are 1.5–2 cm in width, whitish to pinkish and purplish colour with purple veins and have four erected sepals and clawed petals, six stamens and 3–4 cm long style [37, 38]. Siliqua or seedpod, a type of seed capsule of radish, is 1.5 cm wide and 3–7 cm long, consisting of 6–12 seeds/pod with a long conical seedless beak [13]. The inflorescence of radish is a typical elongated, erected, an oblong raceme of Cruciferae. The main objective of the investigation was the cross-pollination of radish by [39] found that the 'Icicle' and 'Scarlet Globe' cvs were self-incompatible pollinated with the help of honeybees [40]. The studies indicated that the seed yield is greatly influenced by the number of honeybees striking the radish flowers. Radchenko [41] also reported that honeybees were the main pollinators of radish flowers, approximately 77–99% in total, increasing the crop yield by 22% and enhancing the seed quality. Therefore, radish is considered almost entirely insect-pollinated. During fruit maturation, seeds' colour is somewhat yellow and turns reddish-brown with age. The mature radish leaves are alternate, arranged in a rosette pattern and have a lyrate shape set apart pinnately with an enhanced terminal lobe and minor lateral lobes. A longer root form, including oriental radishes, daikon or mooli and winter radishes, grows up to 60 cm (24 in) long with foliage about 60 cm (24 in) high with a spread of 45 cm (18 in) [42].

## 4. S haplotype

Radish is a self-incompatible crop exhibiting the high heterosis, the production of $F_1$ hybrids based on self-incompatibility is desired to eliminate laborious hand emasculation in radish [43]. The main aim of a plant breeder is to identify breeding lines of S haplotypes. The plant breeder can avoid cross-influencing of the parental lines [44]. The S haplotype of each parental line needs to show the compatible reaction between parental lines. Therefore, for producing $F_1$ hybrid breeding, it is very important to identify the S haplotypes of parental lines [45]. The abundance of S haplotype determines a specific S haplotype by using traditional methods, including test cross method, pollination, isoelectric focusing, immunoblot analysis and the pollen tube fluorescence analysis [46, 47]. The S alleles are highly variable in S haplotype. Moreover, Nikura and Matsuura identified 37 alleles in radish [48].

Several S haplotypes in *Raphinus sativus* were identified based on polymorphism in SLG, SRK and SCR/SP11 sequence and S haplotypes are numbered as S-1, S-2, S-3, etc. [48]. Although radish belongs to a genus different from Brassica, nucleotide sequences of SP11, SRK and SLG alleles of radish and Brassica are intermingled in phylogenetic trees of SP11, SRK and SLG, respectively, indicating that diversification of these alleles predates speciation of these genera [44]. SP11, SRK and SLG alleles of some S haplotypes in radish are highly similar to those of some S haplotypes in Brassica, and one S haplotype in radish has been revealed to have the same recognition specificity as that of one S haplotype in Brassica rapa [44]. Comparison of nucleotide sequences of SP11 and SRK alleles and recognition specificities between similar S haplotypes of radish and Brassica may provide valuable information for understanding the molecular structures of SP11 and SRK proteins. However, researchers' numbering of S haplotypes in radish varies, and nucleotide sequence information on S haplotypes is thus confusing [43].

Besides, analysis of SLG and SRK is utilised to identify S haplotype in Raphanus and Brassica by using methods of polymerase chain reaction–restriction fragment length polymorphism (PCR-RFLP) [49, 50]. However, the own limitation of PCR-RFLP, first, is challenging to design a universal primer that can amplify SLG and SRK alleles; second, the presence of multiple genes homologous to the SLG or SRK genes in Brassicaceae plants aggravates PCR amplification of specific SLG or SRK alleles [51–53]. Additional advanced radish cultivars (cultivars with improved yield and higher quality) were also produced by the Ogura CMS method to assist in radish hybrid [54]. This variety shows that bulk selection, mixed mass pedigree selection or bud pollination will take 8–12 years to produce a new variety, new varieties must be produced from other genetic means [55].

## 5. Inter-specific hybridization

Inter-specific and intra-specific hybridization has a vital role in the genomic study and crop improvement by introducing desirable agronomic characteristics and specific traits such as disease, insects and stress resistance from wild species to cultivated ones [56, 57]. The study indicated that the average podding rate of the cross between radish and turnip (67.03%) was much higher than that of the reciprocal cross between a turnip and radish (55.04%) [57], and it was also reported that the average seed-setting rate and hybrid acquisition rate of the radish and turnip based on cross pattern (e.g. 2.25 and 0% respectively), however, seed production of the $F_1$ hybrids and their $F_2$ progeny was up to 0.4 and 2%, respectively, as compared with wild radish [58]. Therefore, the study indicated a low hybridization affinity between radish and Chinese kale, but incompatibility still prevailed [57].

Similarly, the radish-wild mustard inter-specific hybrid was studied. It was found that production was higher with radish pollen competition, i.e. 42 and three interspecific hybrid seeds per 1000 seeds were observed [59]. Another study indicated that the modified flower culture method is the best method for hybridization between radish and (transgenic) oilseed rape (*Raphanobrassica* hybrids) without labour-intensive production in vitro ovule or embryo rescue techniques. This is a potential approach for breeding programmes by introducing useful radish genes, e.g. nematode resistance genes, into oilseed rape [60]. Moreover, clubroot is a common disease of cabbages, cauliflower, radishes, turnips and other plants of the family Brassicaceae caused by *Plasmodiophora brassicae* [61]. Radish is a close relative of the brassica family, and it was found that a synthesised allotetraploid *Brassicoraphanus* (RRCC, 2n = 36) between *R. sativus* cv. HQ-04 (2n = 18, RR) and *Brassica oleracea* var. *alboglabra* (L.H Bailey) (2n = 18, CC) proved resistant to multiple clubroot disease pathogen *P. brassicae* causing club root disease [62]. However, the spontaneous hybridization event between *Brassica napus* (oilseed rape) and *Raphanus raphanistrum* (wild radish) was screened. It was found that hybrids with wild radish as the seed parent contribute to herbicide resistance belonging to rape. Another study indicated that wild radish in an oilseed rape field produced as many as three interspecific hybrids per 100 plants and was the first ever such report of such a spontaneous event [58].

## 6. Molecular markers to QTLian breeding

Several economically important traits in radish are being identified. These traits are yield, insect resistance and disease resistance. Yield is a complex trait governed by polygenic characters. Identifying these traits using conventional breeding/traditional breeding is problematic because these traits depend on phenotypic expression and have environmental and genotypic interaction. The new tool molecular breeding overcomes these problems, identification of quantitative trait is being utilised with the help of DNA markers [7] and linkage mapping [63]. There are several DNA markers being used in breeding programmes, such as restriction fragment length polymorphism (RFLPs), random amplified polymorphism DNA (RAPD), simple sequence repeats (SSRs), single-nucleotide polymorphism (SNPs) [63–67]. Molecular markers such as RAPD have been used to establish the origin of hortensis var. sativus and var. niger, which formed from distinct progenitors and came from different sources [68]. Several Asian varieties show more incredible skin and flesh colour, size, length and weight of roots, var. hortensis' genetic variability is also not a surprise.

A genome-wide association study (GWAS) analysis identified 44 single-nucleotide polymorphisms (SNPs) and 20 putative candidate genes significantly associated with FW resistance. A total of four QTLs were identified from the $F_2$ population derived from an FW resistant line and a susceptible line, one of which was co-located with the SNPs on chromosome 7. These markers are emerging tools for molecular breeding programmes and marker-assisted selection to develop FW-resistant varieties of *R. sativus* [69]. Moreover, for the identification of the disease Fusarium wilt, Yu et al. [70] constructed a genetic linkage map on the $F_2$ population, they observed a total of eight loci conferring FW resistance that were distributed on 4LGs, namely 2, 3, 6 and 7 of the Raphanus genome. Synteny analysis using the linked markers QTL showed homology to A. thaliana chromosome 3, which contains disease-resistance gene clusters, suggesting the conservation of resistance genes between them. The list of significant QTLs is identified in the radish, and their location is provided in **Table 1**.

Moreover, identification of root shape and red pigmentation is performed, and it was observed that three quantitative trait loci for root shape, namely LG3, LG8 and LG9, two QTLs for root diameter, namely LG4 and LG8 and one for red pigmentation are identified with the help of using AFLP, SSR and SLG-CAPS [65]. Kamei et al. [71] constructed a genetic linkage map using AFLP and SSR markers and concluded that CR is governed by the single gene or closely linked gene loci, namely Crs1, Crs2 and Crr3. A genetic map was constructed using an $F_2$ population by using markers SRAP, RAPD, SSR, ISSR, RAMP and RGA markers, and they found that a novel QTL qRCD9 is responsible for controlling root CD [63]. Resistance against cyst nematode (*Heterodera schachtii*) was identified using RAPD, dpRAPD, AFLP and SSR markers [66]. To identify quantitative traits in radish for morphological characters, namely ovule number per silique, seed number per silique, plant shape, pubescence, whole plant weight (g), upper part weight (g), whole root weight (g), main root weight (g) using recombinant inbred lines, they identified 8 and 10 quantitative traits in 2008 and 2009 respectively [67]. In the identified QTL regions, nine SNP markers were

| Locus ID | Locus Name | Trait | Cross | Population | Marker Name | Max LOD | References |
|---|---|---|---|---|---|---|---|
| t3726. T000001 | qFW1 | Fusarium wilt resistance | 835 x B2 | F2 | nu_ mBRPGM1376 | 3.72 | 70 |
| t3726. T000002 | qFW1 | Fusarium wilt resistance | 835 x B2 | F2 | ACMP0606 | 4.34 | 70 |
| t3726. T000003 | qFW1 | Fusarium wilt resistance | 835 x B2 | F2 | nu_ mBRPGM1208 | 9.42 | 70 |
| t3726. T000004 | qFW1 | Fusarium wilt resistance | 835 x B2 | F2 | ACMP0357 | 10.08 | 70 |
| t3726. T000005 | qFW1 | Fusarium wilt resistance | 835 x B2 | F2 | nia032a | 3.31 | 70 |
| t3726. T000006 | qFW2 | Fusarium wilt resistance | 835 x B2 | F2 | nu_ mBRPGM1376 | 3.29 | 70 |
| t3726. T000007 | qFW3 | Fusarium wilt resistance | 835 x B2 | F2 | ACMP0590 | 5.62 | 70 |
| t3726. T000008 | qFW4 | Fusarium wilt resistance | 835 x B2 | F2 | nu_ mBRPGM0432 | 3.76 | 70 |
| t3726. T000009 | qFW5 | Fusarium wilt resistance | 835 x B2 | F2 | nu_ mBRPGM0432 | 4.84 | 70 |
| t3726. T000010 | qFW6 | Fusarium wilt resistance | 835 x B2 | F2 | nu_ mBRPGM0432 | 4.23 | 70 |
| t3726. T000011 | qFW7 | Fusarium wilt resistance | 835 x B2 | F2 | ACMP0606 | 7.94 | 70 |
| t3726. T000012 | qFW8 | Fusarium wilt resistance | 835 x B2 | F2 | ACMP0606 | 8.84 | 70 |
| t3726. T000013 | | Root shape (length/ width) | Huang-he hong wan x Utsugi-gensuke | F2 | BRMS-303-2 | 2.42 | 65 |
| t3726. T000014 | | Thickening | Huang-he hong wan x Utsugi-gensuke | F2 | ACA/CTT5 | 3.03 | 65 |
| t3726. T000015 | | Thickening | Huang-he hong wan x Utsugi-gensuke | F2 | AAG/CTA12 | 4.51 | 65 |
| t3726. T000016 | | Root shape (length/ width) | Huang-he hong wan x Utsugi-gensuke | F2 | ACA/CTG5 | 2.88 | 65 |
| t3726. T000017 | | Red pigmentation | Huang-he hong wan x Utsugi-gensuke | F2 | SLG | 9.58 | 65 |
| t3726. T000017 | | Red pigmentation | Huang-he hong wan x Utsugi-gensuke | F2 | SLG | 9.58 | 65 |

| Locus ID | Locus Name | Trait | Cross | Population | Marker Name | Max LOD | References |
|---|---|---|---|---|---|---|---|
| t3726. T000018 | | CR (clubroot resistance), Crs1 (clubroot resistance locus of Rap | Koga-benimaru x Utsugi-gensuke | F2 | BN142 | 12.64 | 71 |
| t3726. T000019 | qRCd1 | RCd (mg $kg^{-1}$), root Cd concentration | Nau-Dysx x Nau-Yh | F2 | Ni2E05_525 | 5.24 | 63 |
| t3726. T000019 | qRCd1 | RCd (mg $kg^{-1}$), root Cd concentration | Nau-Dysx x Nau-Yh | F2 | Ni2E05_525 | 5.24 | 63 |
| t3726. T000020 | qRCd4 | RCd (mg $kg^{-1}$), root Cd concentration | Nau-Dysx x Nau-Yh | F2 | BRMS129_510 | 4.36 | 63 |
| t3726. T000020 | qRCd4 | RCd (mg $kg^{-1}$), root Cd concentration | Nau-Dysx x Nau-Yh | F2 | BRMS129_510 | 4.36 | 63 |
| t3726. T000021 | qRCd6 | RCd (mg $kg^{-1}$), root Cd concentration | Nau-Dysx x Nau-Yh | F2 | EM6fc3_331 | 5.28 | 63 |
| t3726. T000021 | qRCd6 | RCd (mg $kg^{-1}$), root Cd concentration | Nau-Dysx x Nau-Yh | F2 | EM6fc3_331 | 5.28 | 63 |
| t3726. T000022 | qRCd9 | RCd (mg $kg^{-1}$), root Cd concentration | Nau-Dysx x Nau-Yh | F2 | EM5me6_286 | 23.64 | 63 |
| t3726. T000022 | qRCd9 | RCd (mg $kg^{-1}$), root Cd concentration | Nau-Dysx x Nau-Yh | F2 | EM5me6_286 | 23.64 | 63 |
| t3726. T000023 | qRDW5 | RDW (g), root dry weight | Nau-Dysx x Nau-Yh | F2 | BRMS058_550 | 7.28 | 63 |
| t3726. T000024 | qRDW6 | RDW (g), root dry weight | Nau-Dysx x Nau-Yh | F2 | NAUrp705_644 | 3.96 | 63 |
| t3726. T000025 | qRDW9 | RDW (g), root dry weight | Nau-Dysx x Nau-Yh | F2 | EM3me6_291 | 3.58 | 63 |
| t3726. T000026 | qRL1 | RL (cm), root length | Nau-Dysx x Nau-Yh | F2 | Na10F06_545 | 3.38 | 63 |
| t3726. T000027 | qRL3.1 | RL (cm), root length | Nau-Dysx x Nau-Yh | F2 | RamRM24-568_618 | 3.26 | 63 |

| Locus ID | Locus Name | Trait | Cross | Population | Marker Name | Max LOD | References |
|---|---|---|---|---|---|---|---|
| t3726. T000028 | qRL3.2 | RL (cm), root length | Nau-Dysx x Nau-Yh | F2 | PM2fc8_314 | 3.64 | 63 |
| t3726. T000029 | qRL5 | RL (cm), root length | Nau-Dysx x Nau-Yh | F2 | NAUrp782_643 | 4.89 | 63 |
| t3726. T000030 | qRL7 | RL (cm), root length | Nau-Dysx x Nau-Yh | F2 | EM4odd48_365 | 4.06 | 63 |
| t3726. T000031 | qSCd1 | SCd (mg $kg^{-1}$), shoot Cd concentration | Nau-Dysx x Nau-Yh | F2 | NAUrp362_706 | 4.37 | 63 |
| t3726. T000031 | qSCd1 | SCd (mg $kg^{-1}$), shoot Cd concentration | Nau-Dysx x Nau-Yh | F2 | NAUrp362_706 | 4.37 | 63 |
| t3726. T000032 | qSCd3 | SCd (mg $kg^{-1}$), shoot Cd concentration | Nau-Dysx x Nau-Yh | F2 | EM16ga18_383 | 7.64 | 63 |
| t3726. T000032 | qSCd3 | SCd (mg $kg^{-1}$), shoot Cd concentration | Nau-Dysx x Nau-Yh | F2 | EM16ga18_383 | 7.64 | 63 |
| t3726. T000033 | qSDW2 | SDW (g), shoot dry weight | Nau-Dysx x Nau-Yh | F2 | NAUJKC19_565 | 4.74 | 63 |
| t3726. T000034 | qSDW6 | SDW (g), shoot dry weight | Nau-Dysx x Nau-Yh | F2 | Na10F08_665 | 3.78 | 63 |
| t3726. T000035 | qSDW9 | SDW (g), shoot dry weight | Nau-Dysx x Nau-Yh | F2 | Ol14E06_720 | 4.62 | 63 |
| t3726. T000036 | qSH2 | SH (cm), shoot height | Nau-Dysx x Nau-Yh | F2 | EM16ga18_384 | 4.25 | 63 |
| t3726. T000037 | qSH5 | SH (cm), shoot height | Nau-Dysx x Nau-Yh | F2 | RGA12F12R_300 | 3.64 | 63 |
| t3726. T000038 | qTDW1.1 | TDW (g), total dry weight | Nau-Dysx x Nau-Yh | F2 | RamM2-706_616 | 3.54 | 63 |
| t3726. T000039 | qTDW1.2 | TDW (g), total dry weight | Nau-Dysx x Nau-Yh | F2 | PM2em11_367 | 5.6 | 63 |
| t3726. T000040 | qTDW5 | TDW (g), total dry weight | Nau-Dysx x Nau-Yh | F2 | PM17em10_327 | 4.43 | 63 |
| t3726. T000041 | qTDW6 | TDW (g), total dry weight | Nau-Dysx x Nau-Yh | F2 | NAUrp586_752 | 3.14 | 63 |
| t3726. T000042 | qTDW7 | TDW (g), total dry weight | Nau-Dysx x Nau-Yh | F2 | PM18odd44_362 | 4.84 | 63 |

| Locus ID | Locus Name | Trait | Cross | Population | Marker Name | Max LOD | References |
|---|---|---|---|---|---|---|---|
| t3726. T000043 | qTDW9 | TDW (g), total dry weight | Nau-Dysx x Nau-Yh | F2 | PM36em8_423 | 6.2 | 63 |
| t3726. T000044 | Hs1_Rph | BCN-resistance, resistance against the beet cyst nematode (H. sc | Pegletta x Siletta Nova | F2 | E41M59–297 | 22.6 | 66 |
| t3726. T000045 | | PS, Plant shape | rat-tail radish x Haru-S | RIL | RsHH019 | 3.8 | 67 |
| t3726. T000046 | | MW, Main root weight (g) | rat-tail radish x Haru-S | RIL | REL-13 | 6.9 | 67 |
| t3726. T000047 | | Pubescence | rat-tail radish x Haru-S | RIL | RM_1 | 10.1 | 67 |
| t3726. T000048 | | MW, Main root weight (g) | rat-tail radish x Haru-S | RIL | RES-1 | 11.7 | 67 |
| t3726. T000049 | | WW, Whole plant weight (g) | rat-tail radish x Haru-S | RIL | AtSTS-1015 | 3.8 | 67 |
| t3726. T000050 | | Pubescence | rat-tail radish x Haru-S | RIL | RsSR104 | 13.6 | 67 |
| t3726. T000051 | GSL-QTL-1 | 4MTB-GSL contents, 4-methylthio-3-butenyl glucosinolate contents | TBS x AZ26H | F2 | RS2CL6432s | 5.87 | 72 |
| t3726. T000052 | GSL-QTL-1 | 4MTB-GSL contents, 4-methylthio-3-butenyl glucosinolate contents | TBS x AZ26H | F2 | S2CL4585s | 3.62 | 72 |
| t3726. T000053 | GSL-QTL-1 | 4MTB-GSL contents, 4-methylthio-3-butenyl glucosinolate contents | TBS x AZ26H | F2 | RS2CL3356s | 3.85 | 72 |
| t3726. T000054 | GSL-QTL-1 | 4MTB-GSL contents, 4-methylthio-3-butenyl glucosinolate contents | TBS x AZ26H | F2 | RS2CL4290s | 7.36 | 72 |

| Locus ID | Locus Name | Trait | Cross | Population | Marker Name | Max LOD | References |
|---|---|---|---|---|---|---|---|
| t3726. T000055 | GSL-QTL-2 | 4MTB-GSL contents, 4-methylthio-3-butenyl glucosinolate contents | TBS x AZ26H | F2 | RS2CL6432s | 19.1 | 72 |
| t3726. T000056 | GSL-QTL-3 | 4MTB-GSL contents, 4-methylthio-3-butenyl glucosinolate contents | TBS x AZ26H | F2 | RS2CL6594s | 5.62 | 72 |
| t3726. T000057 | GSL-QTL-4 | 4MTB-GSL contents, 4-methylthio-3-butenyl glucosinolate contents | TBS x AZ26H | F2 | RS2CL6594s | 1.54 | 72 |
| t3726. T000058 | GSL-QTL-5 | 4MTB-GSL contents, 4-methylthio-3-butenyl glucosinolate contents | TBS x AZ26H | F2 | S2CL4585s | 5.19 | 72 |

**Table 1.**
*List of QTLs identified for important traits in radish.*

newly produced. Nucleotide sequences and expression of these genes suggested their possible function in 4MTB-GSL biosynthesis in radish roots. [72]. Whereas recently, it was discovered that the R2R3-MYB transcription factor responsible for anthocyanin pigment 2 (PAP2) production is located on chromosome 2. The amino acid sequence encoded by the RsPAP2 gene was entirely distinguishable from other previously published RsMYB genes responsible for the red skin colour of radish [73].

## 7. Genomics

Genomics is now at the core of crop improvement, and the radish crop has been exploited to study the underlying differences in genotypes. The rapid development of genomic data boosted the discoveries regarding the genetic basis of plant traits, such as increased yield, flowering or disease resistance [74]. Various studies of the radish have investigated their genomes' arrangement and the reorganisation of chromosomes during polyploidy events [75], of which draft genomic sequences have been assembled. Another study reported an Asian radish cultivar, WK10039 which was sequenced entirely by combining 454, Illumina and PacBio sequencing systems and bacterial artificial chromosome clones obtained through end sequencing was fully sequenced by using the end sequencing method and sequencing equipment from the ABI firm [76]; over the last decade, a variety of genomic studies on the cultivated radish have been performed [77]. Moreover, a chromosome-scale genome assembly (rs1.0) of WK10039, an Asian radish cultivar, was constructed compared with assemblies documented previously [78]. It revealed more details than those previously recorded (having greater coverage of the genome, a greater number of contigs and chromosome anchoring) [79]. However, Radish Base is a genomic and genetic database containing radish mitochondrial genome sequences [80]. This database presently includes the mitochondrial genomes of two newly sequenced radish species, one from the normal cytoplasm and the other from the male-sterile cytoplasm of Ogura [81]. The previous study published the bioinformatics analysis in radish and identified 20 *COL* transcription factors in the radish genome among 54,357 coding genes [82]. Every *COL* gene in the 'Aokubi daikon' cultivar matched the *COL* gene in the 'kazusa' cultivar. A total of 20 radish *COL* genes were also searched in the cultivar 'WK10039' [82]. Besides, in the radish genome, 35 unique RsOFPs and five RsOFP-likes (with no/partial OVATE domain) were identified by BLASTP, and analysis of exon-intron organisation revealed that most genes were intron-less containing maximum coding sequences in the genome [82].

Based on 17-mer analysis, the estimated size of the genome came out to be 530 Mb. A 387.73 Mb was assembled into 44,820 high-quality scaffolds using SOAP denovo [83, 84] and SSPACE [85]. The assembly in this study showed excellent results with fosmid clones (98.86% covered). The assembly showed a much higher quality than the draft genome of *Raphanus raphanistrum* (254 Mb contigs) [86] and two assemblies (116.0 and 179.8 Mb) of *R. sativus* 'Aokubi' [87] which was released previously. After de novo assembly of the 'Okute-Sakurajima' genome, an estimated haploid genome size of 498.5 Mb was found. The de novo assembly showed a substantially heterozygous genome [88]. Subsequent long-read sequencing produced 36.0 Gb data (60.7 coverage of the estimated genome size) in 2.3 million reads with an N50 length of 29.1 kb. After two rounds of data polishing, the long-read assembly consisting of 504.5 Mb primary contigs (including 1437 sequences with an N50 length of 1.2 Mb) and 263.5 Mb alternative contigs consists of the other haplotypes with different alleles, also known as haploid sequences (including 2373 sequences with an N50 length of 154.6 kb) [88].

A study performed on the radish genome after polyploidy has shown fundamental information about the radish genome production and evolution, which provides valuable insights into radish genetics and breeding. The detailed data and genomic methods obtained through these investigations support a greater understanding of the radish triplicated genome composition. Additionally, these methods help radish breeding by promoting marker-assisted collection, comparative genomic studies and the transmission of knowledge from the reference data to other radish accessions [89]. Consequently, a portal that is home to considerable quantities of genomic information and various links to specific genome analysis methods is precious in radish research and breeding.

## 8. Genetic engineering

Genetic engineering has pivotal role in agriculture by improving the characteristics in the crops and satisfying the need of poor nourished countries. The developments in gene technology and metabolic engineering systems accelerated the production of valuable germplasms [90]. Progress is being achieved in plant methods by improving the traits; researchers have successfully produced transgenic radishes with various agronomic characteristics [91–93]. Gene transmission is done with the help of pathogen, known as agrobacterium, which is extensively used for plant hairy root lines, which appear to yield better than other forms of root systems [94]. Herbaceous hairy roots have advantageous due to their longevity, pace of growth and capacity to assist plants in growing from the root up [95]. The hairy roots are produced in nutrient solution with the help of increasing agrobacterium that contains unusual properties, including biochemically and bio-transforming different metabolites. It is best to use Agrobacterium to produce secondary metabolites since they help to enhance growth regulators [96]. Working on the hairy roots, new sources of natural compounds [97]. In addition, chromosomal disruption or amplification may affect the fertility of cultivated plants. Antibiotics, herbicides, metabolic analogues and non-toxic agents all facilitate transformed cells for survival. Kanamycin and hygromycin B hamper radish regeneration [98].

Recent advancements in plant biotechnology indicate that radish could be genetically modified via a process called 'floral-dipping'. This technique involves co-suppression of the photoperiodic gene GIGANTEA in radish and contributes to the plant's ability to delay bolting and blooming. It can be used to boost a crop's medicinal value [98]. The prospects for improving transformation efficiency and selecting new traits for generating late-flowering radish are published [68]. In 2001, it was demonstrated that plants derived from plants dipped into an Agrobacterium suspension containing both the beta-glucuronidase (gusA) gene and the herbicide resistance gene (bar) between the flanking T-DNA border sequences could be used to generate transgenic radish (*R. sativus* L. longipinnatus Bailey) [91]. In the end, Southern blotting results revealed that both the gusA and bar genes integrated into the genome of transformed plants and segregated as dominant Mendelian traits [91]. A study revealed that The *RHA2b* gene from radish encodes a transcription factor involved in abscisic acid (ABA) signal transduction and is responsible for seed dormancy and pre-harvest sprouting [99]. The study performed the experimentation in which The *RsRHA2b* gene was cloned and transferred into Zhengmai 9023 via Agrobacterium-mediated stem apex transformation [99]. The agrobacterium-mediated transformation became a more appropriate method for genetic transformation [100]. Using adventitious shoot growth on hypocotyl explants for Agrobacterium-mediated radish genetic transformation was investigated using transgenic radish (*Raphanus sativa* L., cv. Jin Ju Dae Pyong) grown on Murashige and Skoog medium [101]. Besides, northern blot results revealed that the GUS gene transcript was detected in a few regenerated plants, confirming genetic transformation. In addition, the techniques available for introducing pharmaceutical proteins into radish for on-site delivery of edible proteins into it are discussed by Curtis in his study [98]. The concerns of releasing transgenic radish to the field in pollen-mediated gene transfer have also been explored. Risks that might exist and the introduction of transgenic radish to the field are sometimes brought up in discussions about transgenic crops [91, 102, 103].

## 9. Conclusion and future directions

For successful production of radish yield, inter- and intra-specific hybridizations are vital to genetic research and crop improvement because they enable the introduction of desirable agronomic traits into the population. The production of yields, early maturity and late bolting, pungency, cold-hardiness, drought resistance, heat tolerance and soil adaptability are just a few of the essential radish breeding traits. The radish genome contains self-incompatibility alleles, allowing for the generation of $F_1$ hybrids without the labour-intensive and hand emasculation required in radish. When generating $F_1$ combinations, it is critical to determine the S haplotypes of the parental lines to avoid hand emasculation. Collecting complete genetic data on chromosomes and information on inheritance is critical. To better understand and forecast resistance, yield characteristics and fruit quality, researchers must understand the regulatory factors synchronising at various developmental stages for each attribute discussed. It remains necessary to develop a robust and long-lasting strategy for plant disease resistance, which is currently under consideration. This is because diseases are capable of evading resistance by generating novel bacterial strains.

Speed breeding is one such strategy; as genome sequencing costs continue to decline, RAD-sequencing and DNA microarrays will become more common, enabling faster genome mapping and tagging of new quantitative trait loci. These quantitative trait loci (QTLs) may incorporate resistance into high-yielding radish genotypes and combine them with significant resistance genes to increase the number of resistant radish genotypes. Additionally, GWAS (genome-wide association studies) can map characteristics to specific candidate genes on a genome-wide scale to improve crop production and quality in radish. The discovery of significant genetic and metabolic diversity paves the way to develop controlled harvest variations in agriculture and genetic enhancement via breeding.


## Funding

This research has received no external funding.


## Conflict of interest

Authors declare no conflict of interest.


## Author details

Anand Kumar[1] and Prashant Kaushik[2,3*]

1 Department of Genetics and Plant Breeding, Chandra Shekhar Azad University of Agriculture and Technology, Kanpur, India

2 Kikugawa Research Station, Yokohama Ueki, Kikugawa City, Japan

3 Institute of Conservation and Improvement of Valencian Agrodiversity, Polytechnic University of Valencia, Valencia, Spain

*Address all correspondence to: prakau@doctor.upv.es

## References

[1] Arro J, Labate JA. Genetic variation in a radish (Raphanus sativus L.) geodiversity collection. Genetic Resources and Crop Evolution. 2021;**69**: 163-171

[2] Richharia RH. Cytological investigation of *Raphanus sativus*, Brassica oleracea, and their $F_1$ and $F_2$ hybrids. Journal of Genetics. 1937;**1**: 19-44

[3] Kaneko Y, Matsuzawa Y. Radish: *Raphanus sativus* L. In: Genetic Improvement of Vegetable Crops. 1993. pp. 487-510

[4] Patra JK, Das G, Paramithiotis S, Shin HS. Kimchi and other widely consumed traditional fermented foods of Korea: A review. Frontiers in Microbiology. 2016;**7**:1493

[5] Ebert AW. Ex situ conservation of plant genetic resources of major vegetables. In: Conservation of Tropical Plant Species. 2013. pp. 373-417

[6] Singh S, Selvakumar R, Mangal M, Kalia P. Breeding and genomic investigations for quality and nutraceutical traits in vegetable crops-a review. Indian Journal of Horticulture. 2020;**1**:1-40

[7] Kaneko Y, Kimizuka-Takagi C, Bang SW, Matsuzawa Y. Radish. In: Vegetables. 2007. pp. 141-160

[8] Simard MJ, Légère A. Synchrony of flowering between canola and wild radish (*Raphanus raphanistrum*). Weed Science. 2004;**6**:905-912

[9] Kumar R, Solankey SS, Singh M. Breeding for drought tolerance in vegetables. Vegetable Science. 2012;**39**:1-15

[10] Dixon GR. Vegetable brassicas and related crucifers (No. 14). CABI. 2007.

[11] Porteus MH, Carroll D. Gene targeting using zinc finger nucleases. Nature Biotechnology. 2005;**8**:967-973

[12] Petropoulos SA, Sampaio SL, Di Gioia F, Tzortzakis N, Rouphael Y, Kyriacou MC, et al. Grown to be blue—Antioxidant properties and health effects of colored vegetables. Part I: Root vegetables. Antioxidants. 2019;**12**:617

[13] Singh A, Sharma S. Radish. In: Antioxidants in Vegetables and Nuts-Properties and Health Benefits. Springer; 2020. pp. 209-235

[14] Rubatzky VE, Yamaguchi M. Cole crops, other brassica, and crucifer vegetables. In: World Vegetables. 1997. pp. 371-417

[15] Bhardwaj RK, Kumari R, Vikram A. Efficient methods for the improvement of temperate root vegetables. In: Accelerated Plant Breeding. Vol. 2. 2020. pp. 155-196

[16] Cartea ME, Velasco P, Obregón S, Padilla G, de Haro A. Seasonal variation in glucosinolate content in Brassica oleracea crops grown in Northwestern Spain. Phytochemistry. 2008;**2**:403-410

[17] El-Ramady HR, Domokos-Szabolcsy É, Abdalla NA, Taha HS, Fári M. Postharvest management of fruits and vegetables storage. In: Sustainable Agriculture Reviews. 2015. pp. 65-152

[18] Kobayashi H, Shirasawa K, Fukino N, Hirakawa H, Akanuma T, Kitashiba H. Identification of genome-wide single-nucleotide polymorphisms among geographically diverse radish accessions. DNA Research. 2020;**27**:1-8

[19] Available from: https://www.ncbi.nlm.nih.gov/pmc/articles/PMC4885859

[20] Al-Shehbaz IA. Brassicaceae (mustard family). *e LS*. 2001.

[21] Yu D, Gu X, Zhang S, Dong S, Miao H, Gebretsadik K, et al. Molecular basis of heterosis and related breeding strategies reveal its importance in vegetable breeding. Horticulture Research. 2021;**8**:1-17

[22] Schreiner M, Huyskens-Keil S, Peters P, Schonhof I, Krumbein A, Widell S. Seasonal climate effects on root colour and compounds of red radish. Journal of the Science of Food and Agriculture. 2002;**11**:1325-1333

[23] Gamba M, Asllanaj E, Raguindin PF, Glisic M, Franco OH, Minder B, et al. Nutritional and phytochemical characterization of radish (*Raphanus sativus*): A systematic review. Trends in Food Science & Technology. 2021;**113**:205-218

[24] Elsayed AYA, Hamdino MI, Ahmed A, Shabana AI. Expected genetic gain in radish (*Raphanus sativus* var. red radicula) submitted to different procedures of selection. Egyptian Journal of Plant Breeding. 2016;**20**:313-328

[25] Zhang J, Zhao J, Tan Q, Qiu X, Mei S. Comparative transcriptome analysis reveals key genes associated with pigmentation in radish (*Raphanus sativus* L.) skin and flesh. Scientific Reports. 2021;**1**:1-11

[26] Zhang J, Qiu X, Tan Q, Xiao Q, Mei S. A comparative metabolomics study of flavonoids in radish with different skin and flesh colors (*Raphanus sativus* L.). Journal of Agricultural and Food Chemistry. 2020;**49**:14463-14470

[27] Dongarwar LN, Kashiwar SR, Ghawade SM, Dongarwar UR. Varietal Performance of Radish (*Raphanus sativus* L.) Varieties in Black Soils of Vidharbha-Maharashtra, India. International Journal of Current Microbiology and Applied Sciences. 2018;**1**:491-501

[28] Giusti MM, Wrolstad RE. Characterization of red radish anthocyanins. Journal of Food Science. 1996;**61**:322-326

[29] Bosland PW. Fusarium oxysporum, a pathogen of many plant species. Genetics of Plant Pathogenic Fungi. 1988;**6**:281-288

[30] Hida K, Ashizawa M. Breeding of radishes for Fusarium resistance. JARQ. 1985;**3**:190-195

[31] Shrivastava P, Kumar R. Soil salinity: A serious environmental issue and plant growth promoting bacteria as one of the tools for its alleviation. Saudi Journal of Biological Sciences. 2015;**2**:123-131

[32] Vishnu PS, Elavarasan R, Selvan T. Studies on radish salinity tolerance and their growth response analysis. International Journal of Advanced Research in Biological Sciences. 2020;**7**:30-39

[33] Sugimoto K. Evaluation of salt tolerance in Japanese wild radishes (*Raphanus sativus* f. raphanistroides Makino). Vol. 39. Bull. of Minamik. Univ.; 2009. pp. 79-88

[34] Kanjevac MM, Bojović BM, Todorović MS, Stanković MS. Effect of seed halopriming on improving salt tolerance in *Raphanus sativus* L. Kragujevac Journal of Science. 2021;**43**:87-98

[35] Bateman AJ. Self-incompatibility systems in angiosperms. *3. Cruciferae* (No. RESEARCH). 1955

[36] Kremer JC. Influence of honeybee habits on radish seed yield.

Michigan University Agricultural Experiment Station Quarterly Bulletin. 1945;**27**:413-420

[37] Nishio T. Economic and academic importance of radish. In: The Radish Genome. 2017. pp. 1-10

[38] Nayik GA, Gull A. Antioxidants in Vegetables and Nuts-Properties and Health Benefits. Springer; 2020

[39] Crane MB, Mather K. The natural cross-pollination of crop plants with particular reference to the radish. Annals of Applied Biology. 1943;**4**:301-308

[40] Young HJ, Stanton ML. Influences of floral variation on pollen removal and seed production in wild radish. Ecology. 1990;**2**:536-547

[41] Radchenko TG. Role of honeybees as pollinators in increasing the seed crop from cabbage and radish. Bdzhilnistvo. 1966;**2**:72-74

[42] Brickell C. The Royal Horticultural Society Encyclopedia of Gardening (Print). London: Dorling Kindersley; 1992. pp. 356-357

[43] Nishio T, Sakamoto K. Polymorphism of self-incompatibility genes. In: The Radish Genome. 2017. pp. 177-188

[44] Wang R, Mei Y, Xu L, Zhu X, Wang Y, Guo J, et al. Genome-wide characterization of differentially expressed genes provides insights into regulatory network of heat stress response in radish (*Raphanus sativus* L.). Functional & Integrative Genomics. 2018;**18**:225-239

[45] Ockendon DJ. The S-allele collection of Brassica oleracea. In III International Symposium on Brassicas and XII Crucifer Genetics Workshop. 2000;**539**:25-30

[46] Ruffio-Châble V, Hervé Y, Dumas C, Gaude T. Distribution of S-haplotypes and its relationship with self-incompatibility in Brassica oleracea. Part 1. In inbred lines of cauliflower (B. oleracea var 'botrytis'). Theoretical and Applied Genetics. 1997;**94**:338-346

[47] Georgia AVS, Ockendon DJ, Gabrielson RL, Maguire. Self-incompatibility alleles in broccoli. Horticultural Science. 1982;**17**:748-749

[48] Niikura S, Maguire JD, Matsuura S. Genetic variation of the self-incompatibility alleles (S-alleles) in the cultivated radish (*Raphanus sativus* L.) by the PCR-RFLP method. Acta Horticulturae. 2001;**546**:359-366

[49] Nishio T, Kusaba M, Sakamoto K, Ockendon DJ. Polymorphism of the kinase domain of the S-locus receptor kinase gene (SRK) in Brassica oleracea L. Theoretical and Applied Genetics. 1997;**95**:335-342

[50] Okamoto S, Sato Y, Sakamoto K, Nishio T. Distribution of similar self-incompatibility (S) haplotypes in different genera Raphanus and Brassica. Sexual Plant Reproduction. 2004;**17**:33-39

[51] Lalonde BA, Nasrallah ME, Dwyer KG, Chen CH, Barlow B, Nasrallah JB. A highly conserved Brassica gene with homology to the S-locus-specific glycoprotein structural gene. The Plant Cell. 1989;**1**:249-258

[52] Boyes DC, Nasrallah JB. Physical linkage of the SLG and SRK genes at the self-incompatibility locus of Brassica oleracea. Molecular & General Genetics. 1993;**236**:369-373

[53] Kumar V, Trick M. Sequence complexity of the S receptor kinase gene family in Brassica. Molecular and General Genetics. 1993;**241**:440-446

[54] Bhardwaj RK, Kumari R, Vikram A. Efficient methods for the improvement of temperate root vegetables. In: Accelerated Plant Breeding. 2020. pp. 155-196

[55] Cox K, Beaton C. Fruit and Vegetables for Scotland: What to Grow and How to Grow It. Birlinn Ltd.; 2020

[56] Kaneko Y, Bang SW. Interspecific and intergeneric hybridization and chromosomal engineering of Brassicaceae crops. Breeding Science. 2014;**64**:14-22

[57] Jin P, Zhu Z, Guo X, Chen F, Wu Y, Chen J, et al. Production and characterization of intergeneric hybrids by crossing radish with turnip and with Chinese kale. Euphytica. 2020;**216**:1-15

[58] Darmency H, Lefol E, Fleury A. Spontaneous hybridizations between oilseed rape and wild radish. Molecular Ecology. 1998;**7**:1467-1473

[59] Eber F, Boucherie R, Broucqsault LM, Bouchet Y, Chèvre AM. Spontaneous hybridisation between vegetable crops and weeds. I. Garden radish (*Raphanus sativus* L.) and wild mustard (Sinapis arvensis L.). Agronomie. 1998;**18**:489-497

[60] Metz PL, Nap JP, Stiekema WJ. Hybridization of radish (*Raphanus sativus* L.) and oilseed rape (Brassica napus L.) through a flower-culture method. Euphytica. 1995;**83**:159-168

[61] Bhattacharya I, Dutta S, Mondal S, Mondal B. Clubroot disease on Brassica crops in India. Canadian Journal of Plant Pathology. 2014;**36**:154-160

[62] Zhan Z, Nwafor CC, Hou Z, Gong J, Zhu B, Jiang Y, et al. Cytological and morphological analysis of hybrids between Brassicoraphanus, and Brassica napus for introgression of clubroot resistant trait into Brassica napus L. PLoS One. 2017;**12**:e0177470

[63] Xu L, Wang L, Gong Y, Dai W, Wang Y, Zhu X, et al. Genetic linkage map construction and QTL mapping of cadmium accumulation in radish (*Raphanus sativus* L.). Theoretical and Applied Genetics. 2012;**125**:659-670

[64] Muminović J, Merz A, Melchinger AE, Lübberstedt T. Genetic structure and diversity among radish varieties as inferred from AFLP and ISSR analyses. Journal of the American Society for Horticultural Science. 2005;**130**:79-87

[65] Tsuro M, Suwabe K, Kubo N, Matsumoto S, Hirai M. Mapping of QTLs controlling root shape and red pigmentation in radish, *Raphanus sativus* L. Breeding Science. 2008;**58**:55-61

[66] Budahn H, Peterka H, Mousa M, Ding Y, Zhang S, Li J. Molecular mapping in oil radish (*Raphanus sativus* L.) and QTL analysis of resistance against beet cyst nematode (*Heterodera schachtii*). Theoretical and Applied Genetics. 2009;**118**:775-782

[67] Hashida T, Nakatsuji R, Budahn H, Schrader O, Peterka H, Fujimura T, et al. Construction of a chromosome-assigned, sequence-tagged linkage map for the radish, *Raphanus sativus* L. and QTL analysis of morphological traits. Breeding Science. 2013;**63**:218-226

[68] Curtis IS. Genetic engineering of radish: current achievements and future goals. Plant Cell Reports. 2011;**30**:733-744

[69] Lee O, Koo H, Yu JW, Park HY. Genotyping-by-Sequencing-Based Genome-Wide Association Studies of

Fusarium Wilt Resistance in Radishes (*Raphanus sativus* L.). Genes. 2021;**12**:858

[70] Yu X, Choi SR, Ramchiary N, Miao X, Lee SH, Sun HJ, et al. Comparative mapping of *Raphanus sativus* genome using Brassica markers and quantitative trait loci analysis for the Fusarium wilt resistance trait. Theoretical and Applied Genetics. 2013;**126**:2553-2562

[71] Kamei A, Tsuro M, Kubo N, Hayashi T, Wang N, Fujimura T, et al. QTL mapping of clubroot resistance in radish (*Raphanus sativus* L.). Theoretical and Applied Genetics. 2010;**120**:1021-1027

[72] Zou Z, Ishida M, Li F, Kakizaki T, Suzuki S, Kitashiba H, et al. QTL analysis using SNP markers developed by next-generation sequencing for identification of candidate genes controlling 4-methylthio-3-butenyl glucosinolate contents in roots of radish, *Raphanus sativus* L. PLoS One. 2013;**8**:e53541

[73] Fan L, Wang Y, Xu L, Tang M, Zhang X, Ying J, et al. A genome-wide association study uncovers a critical role of the RsPAP2 gene in red-skinned *Raphanus sativus* L. Horticulture Research. 2020;**7**:1-13

[74] Tuberosa R, Salvi S. Genomics-based approaches to improve drought tolerance of crops. Trends in Plant Science. 2006;**11**:405-412

[75] Li F, Hasegawa Y, Saito M, Shirasawa S, Fukushima A, Ito T, et al. Extensive chromosome homoeology among Brassiceae species were revealed by comparative genetic mapping with high-density EST-based SNP markers in radish (*Raphanus sativus* L.). DNA Research. 2011;**18**:401-411

[76] Jeong YM, Chung WH, Choi AY, Mun JH, Kim N, Yu HJ. The complete mitochondrial genome of cultivated radish WK10039 (*Raphanus sativus* L.). Mitochondrial DNA Part A. 2016;**27**:941-942

[77] Jeong YM, Chung WH, Mun JH, Kim N, Yu HJ. De novo assembly and characterization of the complete chloroplast genome of radish (*Raphanus sativus* L.). Gene. 2014;**551**:39-48

[78] Mitsui Y, Michihiko S, Kenji K, Nobukazu N, Mari S, Misaki I, et al. The radish genome and comprehensive gene expression profile of tuberous root formation and development. Scientific Reports. 2015;**5**:1-14

[79] Lee YJ, Mun JH, Jeong YM, Joo SH, Yu HJ. Assembly of a radish core collection for evaluation and preservation of genetic diversity. Horticulture, Environment, and Biotechnology. 2018;**59**:711-721

[80] Shen D, Sun H, Huang M, Zheng Y, Li X, Fei Z. RadishBase: A database for genomics and genetics of radish. Plant and Cell Physiology. 2013;**54**:33

[81] Tanaka Y, Tsuda M, Yasumoto K, Yamagishi H, Terachi T. A complete mitochondrial genome sequence of Ogura-type male-sterile cytoplasm and its comparative analysis with that of normal cytoplasm in radish (*Raphanus sativus* L.). BMC Genomics. 2012;**13**:1-12

[82] Hu T, Wei Q, Wang W, Hu H, Mao W, Zhu Q, et al. Genome-wide identification and characterization of CONSTANS-like gene family in radish (*Raphanus sativus*). PLoS One. 2018;**13**:e0204137

[83] Xiaohui Z, Zhen Y, Shiyong M, Yang Q, Xinhua Y, Xiaohua C, et al.

A de novo Genome of a Chinese Radish Cultivar. Horticultural Plant Journal. 2015;**1**:155-164

[84] Li Z, Zhou M, Zhang Z, Ren L, Du L, Zhang B, et al. Expression of a radish defensin in transgenic wheat confers increased resistance to Fusarium graminearum and Rhizoctonia cerealis. Functional & Integrative Genomics. 2011;**11**:63-70

[85] Boetzer M, Henkel CV, Jansen HJ, Butler D, Pirovano W. Scaffolding pre-assembled contigs using SSPACE. Bioinformatics. 2011;**27**:578-579

[86] Moghe GD, Hufnagel DE, Tang H, Xiao Y, Dworkin I, Town CD, et al. Consequences of whole-genome triplication as revealed by comparative genomic analyses of the wild radish *Raphanus raphanistrum* and three other Brassicaceae species. The Plant Cell. 2014;**26**:1925-1937

[87] Kitashiba H, Li F, Hirakawa H, Kawanabe T, Zou Z, Hasegawa Y, et al. Draft sequences of the radish (*Raphanus sativus* L.) genome. DNA Research. 2014;**21**:481-490

[88] Shirasawa K, Hirakawa H, Fukino N, Kitashiba H, Isobe S. Genome sequence and analysis of a Japanese radish (*Raphanus sativus*) cultivar named 'Sakurajima Daikon'possessing giant root. DNA Research. 2020;**27**:010

[89] Varshney RK, Bohra A, Yu J, Graner A, Zhang Q, Sorrells ME. Designing future crops: Genomics-assisted breeding comes of age. Trends in Plant Science. 2021;**26**:631-649

[90] Wang J, Qiu Y, Wang X, Yue Z, Yang X, Chen X, et al. Insights into the species-specific metabolic engineering of glucosinolates in radish (*Raphanus sativus* L.) based on comparative genomic analysis. Scientific Reports. 2017;**1**:1-9

[91] Curtis IS, Nam HG. Transgenic radish (*Raphanus sativus* L. longipinnatus Bailey) by floral-dip method–plant development and surfactant are important in optimizing transformation efficiency. Transgenic Research. 2001;**10**:363-371

[92] Tzfira T, Citovsky V. Agrobacterium: From Biology to Biotechnology. Springer; 2008

[93] Lacroix B, Citovsky V. Pathways of DNA transfer to plants from Agrobacterium tumefaciens and related bacterial species. Annual Review of Phytopathology. 2019;**25**:231-251

[94] Ali MA, Azeem F, Abbas A, Joyia FA, Li H, Dababat AA. Transgenic strategies for enhancement of nematode resistance in plants. Frontiers in Plant Science. 2017;**8**:750

[95] Gelvin S. Agrobacterium in the genomics age. Plant Physiology. 2009;**150**:1665-1676

[96] Giri A, Narasu ML. Transgenic hairy roots: recent trends and applications. Biotechnology Advances. 2000;**18**:1-22

[97] Berkov S, Pavlov A, Kovatcheva P, Stanimirova P, Philipov S. Alkaloid spectrum in diploid and tetraploid hairy root cultures of Datura stramonium. Zeitschrift fur Naturforschung C. 2003;**58**:42-46

[98] Curtis IS. The noble radish: Past, present and future. Trends in Plant Science. 2003;**8**:305-307

[99] Li S, Zhang N, Li J, Xiang J. A new male sterile line of "Duanye-13" radish (*Raphanus sativus* L.) produced by ethyl methanesulfonate mutagenesis.

Genetic Resources and Crop Evolution. 2019;**66**:981-987

[100] Hayta S, Smedley MA, Demir SU, Blundell R, Hinchliffe A, Atkinson N, et al. An efficient and reproducible Agrobacterium-mediated transformation method for hexaploid wheat (*Triticum aestivum* L.). Plant Methods. 2019;**15**:1-15

[101] Cho MA, Min SR, Ko SM, Liu JR, Choi PS. Agrobacterium-mediated genetic transformation of radish (*Raphanus sativus* L.). Plant Biotechnology. 2008;**25**:205-208

[102] Rissler J, Mellon M, Mellon MG. The ecological risks of engineered crops. MIT Press; 1996

[103] Snow AA, Palma PM. Commercialization of transgenic plants: potential ecological risks. BioScience. 1997;**47**:86-96

Chapter 7

# Transgenic Approaches for Nutritional Enhancement of Potato

*Sagar S. Datir and Sharon Regan*

## Abstract

Potatoes provide an excellent source of carbohydrates, minerals, vitamins, carotenoids, anthocyanins, and several other metabolites which play an important role in human nutrition. These bioactive compounds are effective in preventing diseases like cancer, diabetes, and heart-related issues. In addition to their industrial uses, potatoes are a major focus of genetic engineering programs for the modification of nutritional properties. Several important candidate genes operating in phenylpropanoid mechanism, ascorbic acid biosynthesis pathway, carbohydrate metabolism, steroidal glycoalkaloid biosynthesis pathway, and other-related metabolic steps have been cloned and characterized at the biochemical and molecular levels. Overexpression and down regulation of genes operating in these pathways has revealed important insights into improved nutritional quality. Expression of a transgene has successfully resulted in increasing carotenoids, anthocyanins, and vitamin content in transgenic tubers. Reduction in glycoalkaloid content, enzymatic browning, flesh color, and chipping quality has been achieved via modification of the genes involved in the respective biochemical pathway in potatoes. Transgenic approaches not only resulted in improved quality but also helped in understanding the biochemical and molecular mechanisms associated with the regulation of genes in these pathways. Although the commercialization of transgenic potatoes is still hindered by consumers approval and ethical restrictions, the Clustered Regularly Interspaced Short Palindromic Repeats (CRISPR)/Cas9 system holds promise as a non-transgenic alternative for developing nutritionally enhanced potatoes.

**Keywords:** anthocyanin, carotenoids, glycoalkaloids, potato, transgenic, vitamins

## 1. Introduction

Potatoes form an integral component of human food diet and rank as the third most important staple food crop in the world after wheat and rice [1]. The tubers are important dietary source of carbohydrates, proteins, carotenoids, antioxidants, minerals, phenolics, anti-nutrients, and vitamins [2–5]. These bioactive compounds are known to prevent and combat chronic diseases such as hypertension, cancer, diabetes, and heart disease [6–9]. Potatoes also contains recommended amounts of minerals such as potassium, magnesium, iron, and bioactive compounds such as chlorogenic acid, the flavonoids apigenin, rutin, and kaempferol 3-O-rutinoside, polyamines and alkaloids such as calystegines, solanine, tomatine, and chaconine [10, 11]. It is widely used as a raw marketable product and in industry for making processed food stuffs such as chips and french fries etc. [12]. Although, they provide most of the calories and protein needed, potatoes are not nutritionally complete foods [13, 14]. To overcome the challenges of poverty and hunger worldwide, potato is considered as one of the promising crops for nutritional enhancement [15]. Traditional processing and preparation methods such as peeling, roasting, microwaving, boiling, frying, and baking alter nutritional quality of potatoes including loss of key micronutrients, adsorption of fat, and conversion of naturally resistant starch into highly digestible starch [16, 17]. Biotechnology could enhance micronutrient content and has the potential to reduce malnutrition especially among poor people from developing countries [13]. This approach will only be successful if clear advantages are demonstrated to both growers and consumers and the necessary safety precautions are addressed [18].

Potatoes are an excellent model crop for the genetic modification of metabolic pathways. Large-scale metabolomic studies have identified significant variation in nutrient content, minerals, and bioactive compounds in potatoes leading to the selection of specific cultivars for consumption of raw or processed potatoes [19–21]. Metabolite and mineral variation in raw and cooked potatoes can be used to predict the nutrient and bioactive content in cooked potato tuber for improved health traits [22]. Although conventional breeding efforts are underway for producing potato

**Figure 1.**
*Important quality traits in potato. Figure highlights the important tuber quality traits. A very special thanks to Prof. Dr. David G. Holm, Department of Horticulture and Landscape Architecture, Colorado State University, USA and Dr. Sanjay Gupta, Department of Soil, Water and Climate, University of Minnesota, USA for providing the potato photographs used in the figure. Also, I would like to extend my thanks to https://www.idahopacific.com/potato-granules and Waltz (2015; 10.1038/nbt0115-12) from which the photographs have been obtained and modified.*

cultivars with altered nutritional properties [23, 24], the efforts are complicated by the tetrasomic inheritance and high level of heterozygosity of potatoes [25]. Transgenic technology and genome editing tools [26] offer significant opportunities for tailored improvement of nutritional quality traits in potato, such as starch and sugar content, chipping quality, flesh color, and taste, as well as ascorbic acid, anthocyanin, carotenoid, and glycoalkaloid content [27–32].

**Figure 1** outlines the important quality traits that have been modified by transgenic technologies or other genome editing tools in potato. These technologies include production of transgenic potatoes by overexpression or antisense repression of genes, RNA interference (RNAi) and gene editing tools such as Transcription Activator-Like Effector Nucleases (TALENs), and Clustered Regularly Interspaced Short Palindromic Repeats (CRISPR/Cas9) [26, 27, 31, 33, 34]. TALENs and CRISPR/Cas9 systems make site-specific gene modification by creating double-stranded DNA break. While TALEN recognizes the target site based on DNA protein interaction, the CRISPR system is based on site specific RNA protein interactions. CRISPR/Cas9 is a targeted mutagenesis technique to generate knockout mutations via non-homologous end-joining as well as gene targeting to edit an endogenous gene by homologous recombination [26, 27, 31, 33, 34]. CRISPR/cas9 avoids foreign DNA insertions in the plant genome, an important criterion in the development of crop varieties not subjected to the cumbersome GMO regulation process [26–28, 34]. Moreover, the availability of the potato genome sequence [35] has facilitated the development of comparative genomic analyses and functional studies of candidate genes to improve several important traits in potato [36]. In this review we discuss the past developments, and future perspectives of nutritional enhancement in potato using transgenic technologies.

## 2. Reducing the glycoalkaloid content in transgenic tubers

α-Chaconine and α-solanine are the two main glycoalkaloids that are toxic secondary metabolites present in the tubers of cultivated potato (*Solanum tuberosum* L.) [37, 38]. Symptoms of glycoalkaloid poisoning include abdominal pain, vomiting, and diarrhea in humans. Due to their toxic properties, 200 mg/kg fresh weight is the safety limit for total glycoalkaloid content in the tubers of released commercial potato cultivars [39, 40] because glycoalkaloids cannot be destroyed during food-processing treatments, such as boiling, baking or frying and even at high temperatures [41]. The general pathway of glycoalkaloid metabolism and important candidate genes used in the modification of glycoalkaloid content are highlighted in **Figure 2** and **Table 1**).

The biosynthesis of γ-solanine is catalyzed by the enzyme UDP-galactose:solanidine galactosyltransferase (SGT) from galactose and solanidine [44]. Transgenic potato lines were produced using an antisense version of a cDNA encoding *SGT* under the control of Cauliflower Mosaic Virus 35S (CaMV35S) promoter or a tuber-specific granule bound starch synthase (GBSS) promoter. Transgenic lines produced from potato cv. 'Lenape' expressing antisense *SGT* exhibited significantly lower steroidal glycoalkaloids in the tubers. In another study [45], antisense suppression of the genes that encode the enzyme for the biosynthesis of γ-solanine from UDP-galactose and solanidine (SGT1), γ-chaconine from UDP-glucose and solanidine (SGT2), and α-solanine and α-chaconine from UDP-rhamnose, β-solanine and β-chaconine (*SGT3*) were down-regulated under the control of GBSS6 promoter. Down-regulation of *SGT1* reduced the concentration of α-solanine without affecting

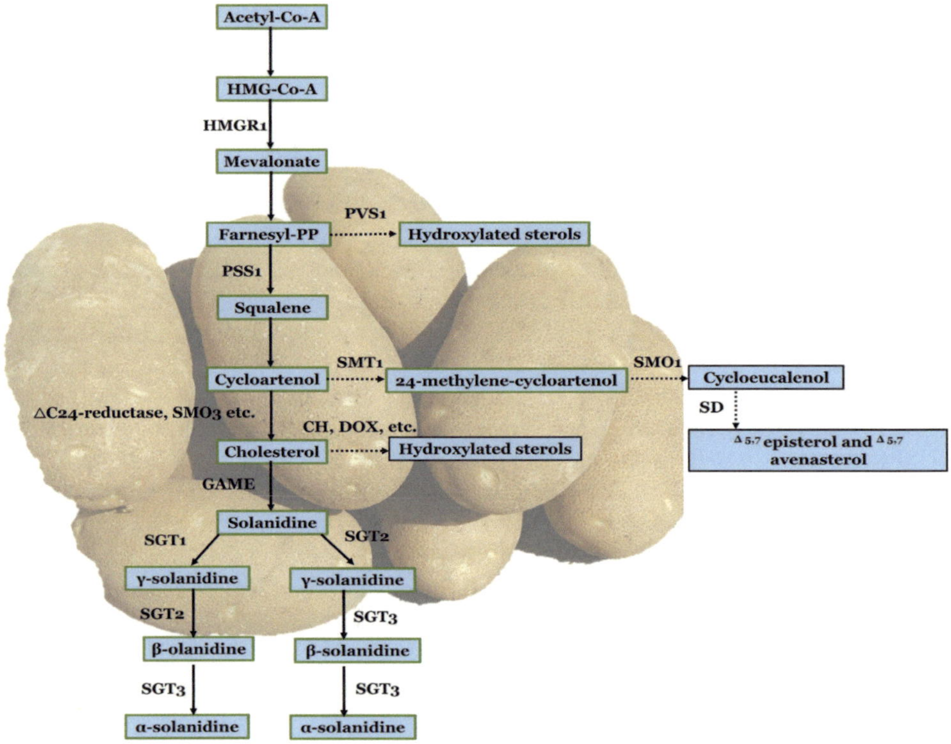


**Figure 2.**
*Glycoalkaloid biosynthetic pathway. The glycoalkaloid biosynthetic pathway starts from Acetyl-Co-A. The representative enzymes are HMGR1, 3-Hydroxy-3-methylglutaryl coenzyme A reductase; PVS1, vetispiradiene sesquiterpene cyclase; PSS1, squalene synthase; SMT1, sterol C24-methyltransferase type1; CH, cholestrole hydroxylase; SGT1, solanidine galactosyltransferase; SGT2, solanidine glucosyltransferas; SGT3, glycosterol rhamnosyltransferase; SMO, C-4 sterol methyl oxidase; SD, sterol C-5(6) desaturase; SSR, sterol side chain reductase; and GAME, glycoalkaloid metabolism genes. The figure is adapted and modified from Arnqvist et al. [27], Khan et al. [42], Sonawane et al. [43].*

the levels of α-chaconine. In contrast, down-regulation of *SGT2* resulted in reduction of α-chaconine and increased levels of α-solanine. Down-regulation of *SGT3* reduced concentrations of both α-chaconine and α-solanine [45]. Antisense manipulation of the *SGT* caused reduced glycoalkaloids content thereby decreasing toxicity in potato tubers.

RNAi, TALENs and CRISPR/Cas9-based systems have been used to reduce glycoalkaloid content [46, 52]. Glycoalkaloid biosynthesis is carried out by *PGA1* and *PGA2* encoding cytochrome P450 monooxygenases (CYP72A208 and CYP72A188) respectively. Transgenic lines using RNAi expressing either *PGA1* or *PGA2* showed very little steroidal glycoalkaloids accumulation and no effect on vegetative growth and tuber production [46]. Cholesterol and sterol side chain reductase 2 (SSR2) is a key enzyme in the biosynthesis of cholesterol and related steroidal glycoalkaloids [47, 56]. TALEN induced mutations in the *SSR2* gene have lower levels of cholesterol and steroids without affecting the plant growth [48–50]. CRISPR/Cas9-edited *StSSR2* resulted in a significant reduction of steroidal glycoalkaloids content [57]. CRISPR-Cas9-based knockout of *CYP88B1*, resulted in reduced levels of α-solanine and α-chaconine [51]. Likewise, knockout of dioxygenase *St16DOX*, responsible for

| Gene name | Abbreviation | Quality trait | Reference |
|---|---|---|---|
| *UDP-galactose:solanidine galactosyltransferase* | *SGT* | Glycoalkaloid | McCue et al. [44], Shepherd et al. [45] |
| *Cytochrome P450 monooxygenases* | *CYP72A208 and CYP72A188* | Glycoalkaloid | Umemoto et al. [46] |
| *Sterol side chain reductase 2* | *SSR2* | Glycoalkaloid | Heftmann et al. [47], Sawai et al. [48], Itkin et al. [49], Nahar et al. [50] |
| *Cytochrome P450 monooxygenase* | *CYP88B1* | Glycoalkaloid | Akiyama et al. [51] |
| *Dioxygenase* | *DOX* | Glycoalkaloid | Nakayasu et al. [52] |
| *Sterol 24-C-methyltransferase, sterol desaturase and C-4 sterol methyl oxidase* | *SMT1, SD and SMO* | Glycoalkaloid | Arnqvist et al. [27], Kaminski et al. [53] |
| *GLYCOALKALOID METABOLISM 9* | *GAME9* | Glycoalkaloid | Cárdenas et al. [54], Ginzberg et al. [55] |

**Table 1.**
*Genes used in the modification of glycoalkaloid contents in transgenic potatoes.*

steroid 16α-hydroxylation abolished *St16DOX* expression and prevented glycoalkaloids production [52]. These studies indicate that suppression or knockout of genes involved in glycoalkaloid biosynthesis is a viable strategy to manipulate the steroidal glycoalkaloid levels in potato.

In contrast to gene suppression strategies, other studies have shown that the overexpression of genes involved in steroidal glycoalkaloid biosynthesis pathway are also an effective strategy to reduce cholesterol and glycoalkaloid levels [27]. Steroidal glycoalkaloids are thought to be synthesized from cholesterol that is converted to solanidine, and then by two separate pathways to α-solanine and α-chaconine. Altered glycoalkaloid content has been associated with overexpression of genes such as sterol *24-C-methyltransferase* (*SMT1*), *sterol desaturase* (*SD*) and *C-4 sterol methyl oxidase* (*SMO*) [53]. *SMTs* are involved in the biosynthesis of sterols and other products [58]. Overexpressing soybean *SMT* (*GmSMT1*) increased total sterols accompanied by a decrease in cholesterol and glycoalkaloids in leaves and tubers [27]. *Glycoalkaloid metabolism9* (GAME9) is an APETALA2/Ethylene Response Factor associated with a major quantitative trait for steroidal glycoalkaloid content in tubers [54, 59]. Overexpression of *GAME9* altered the levels of steroidal glycoalkaloids leaves and tuber skin [54, 55].

## 3. Genetic modification for vitamin C content

A whole, baked potato is an excellent source of vitamin C, vitamin B6, niacin and folate [60–62]. High consumption of potato tubers has been correlated with increased antioxidant level in blood and tissues and increased protection against oxidative stress [60]. However, the level of vitamin C is reduced if the potato is frozen, stored under refrigerated conditions, boiled or fried [63–65]. Thus far, there has been limited success in increasing vitamin C content in transgenic potato tubers [66, 67].

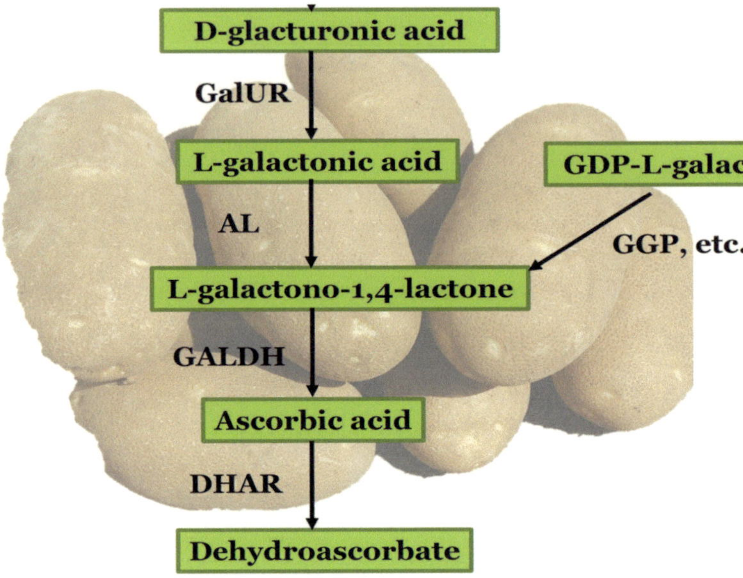


**Figure 3.**
*L-ascorbic acid biosynthesis pathway. The representative enzymes are GGP, GDP-L-galactose phosphorylase; GalUR, D-galacturonic acid reductase; DHAR, dehydroascorbate reductase; GALDH, L-galactono-1,4-lactone dehydrogenase; and AL, aldono lactonase. The figure is adapted and modified from Hemavathi et al. [68], Venkatesh and Park [69].*

The pathway of ascorbic acid (vitamin C) production (**Figure 3**) includes the reduction of D-galacturonic acid to L-galactonic acid by D- galacturonic acid reductase (GalUR), followed by conversion to L-galactono-1,4-lactone by aldonolactonase. The L-galactono-1,4-lactone is then oxidized to ascorbic acid by L-galactono-1,4-lactone dehydrogenase (GALDH) [70]. **Table 2** outlines the key genes used in the modification of ascorbic acid content. Overexpression of the strawberry *GalUR* resulted in increased ascorbic acid levels in potato [3]. Dehydroascorbate reductase (DHAR) plays an important role in maintaining the normal level of ascorbic acid by recycling oxidized ascorbic acid. Transgenic potato over-expressing the cytosolic *DHAR* significantly increased DHAR activity and ascorbic acid content in potato leaves and tubers, whereas chloroplastic *DHAR* overexpression only increased DHAR activity and ascorbic acid content in leaves [29]. Overexpression of the *Arabidopsis GDP-L-galactose phosphorylase* (*GGP*) resulted in significantly enhanced ascorbic acid content. Overall studies suggest that genetic alteration of specific vitamin-related genes could be excellent targets for improving the nutritional content of potato.

| Gene name | Abbreviation | Quality trait | Reference |
|---|---|---|---|
| *D- galacturonic acid reductase* | *GalUR* | Ascorbic acid/vitamin C | Hemavathi et al. [3] |
| *L-galactono-1,4-lactone dehydrogenase* | *GALDH* | Ascorbic acid/vitamin C | Linster et al. [70] |
| *Dehydroascorbate reductase* | *DHAR* | Ascorbic acid/vitamin C | Qin et al. [29] |
| *GDP-L-galactose phosphorylase* | *GGP* | Ascorbic acid/vitamin C | Bulley et al. [67] |

**Table 2.**
*Genes used in the modification of ascorbic acid content in transgenic potatoes.*

## 4. Enhancing carotenoid content in transgenic potato

Carotenoids are yellow to red pigments that play an essential role in human nutrition, with the most important carotenoid being β-carotene, a major source of provitamin A. Deficiency of vitamin A is a major global micronutrient problem that causes blindness and weakens the immune system [71, 72]. Carotenoids enhance may improve the immune system, reduce cardiovascular disease and cancer and help prevent atherosclerosis [73–75]. For these reasons, there is considerable interest in developing potatoes with increased levels of carotenoids [76–79].

Xanthophylls lutein, zeaxanthin, violaxanthin and neoxanthin are the major carotenoids present in the tubers of cultivated potato while that of β-carotene is found in low levels [80, 81]. The Candidate genes used in the modification of carotenoid content in transgenic potatoes are highlighted in **Figure 4** and **Table 3**. Lycopene is produced from phytoene by phytoene desaturase (CRTI), and cyclized by lycopene β-cyclase (LCY-β) to form β-carotene. While α-carotene is produced by both LCY-β and LCY-ε). The hydroxylation of α-carotene yields lutein. Two subsequent hydroxylations of β-carotene, catalyzed by carotenoid β-hydroxylase (CHYB) produce zeaxanthin. Zeaxanthin can be epoxidized by zeaxanthin epoxidase (ZEP) to form violaxanthin, which can be used by violaxanthin de-epoxidase (VDE) to regenerate zeaxanthin. The final step in carotenoid biosynthesis is the conversion of violaxanthin

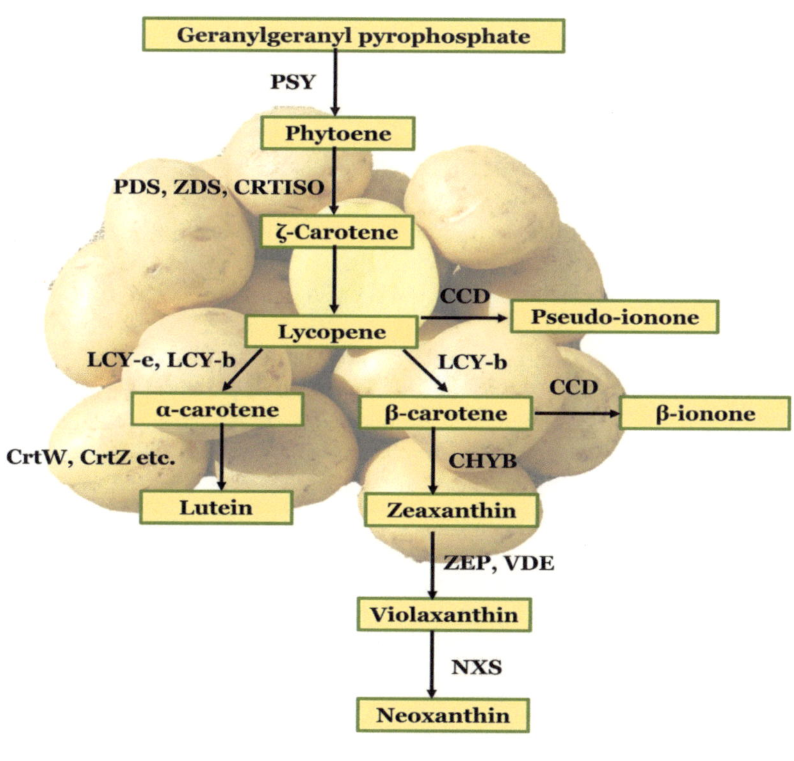


**Figure 4.**
*Carotenoid biosynthesis pathway. The representative enzymes are PSY, phytoene synthase; PDS, phytoene desaturase; ZDS, ζ-carotene desaturase; CRTISO, carotenoids isomerase; LCY-e, lycopene e-cyclase; LCY-b, lycopene β-cyclase; CHB/BCH, β-carotene hydroxylase; ZEP, zeaxanthin epoxidase; VDE, violaxanthin de-epoxidase; NXS, neoxanthin synthase; CCD, carotenoid cleavage dioxygenase; Crt-ketolase. The figure is adapted and modified from Zhou et al. [31].*

| Gene name | Abbreviation | Quality trait | Reference |
|---|---|---|---|
| *Phytoene synthase* | *PSY* | Carotenoid | Cazzonelli and Pogson [82] |
| *Phytoene synthase CrtB* | *CrtB* | Carotenoid | Diretto et al. [83], Ducreux et al. [84] |
| *Lycopene epsilon cyclase* | *LCY-e* | Carotenoid | Diretto et al. [85] |
| *β-carotene hydroxylases* | *CHY* | Carotenoid | Diretto et al. [86] |
| *β-carotene hydroxylase gene* | *CHB* | Carotenoid | Van Eck et al. [75] |
| *Carotenoid cleavage dioxygenases* | *CCDs* | Carotenoid | Campbell et al. [87] |
| *Ketolase crtO, crtW, bkt* | *crtO β-carotene, crtW, bkt* | Carotenoid | Gerjets and Sandmann [88], Lu et al. [89] |
| *Cauliflower Or* | *Or* | Carotenoid | Lu et al. [89], Lopez et al. [90], Goo et al. [91] |

**Table 3.**
*Genes used in the modification of carotenoid content in transgenic potatoes.*

to neoxanthin by neoxanthin synthase [92]. Introduction of various carotenogenesis-related genes has resulted in increased production of specific carotenoids and total carotenoid content using overexpression [84, 93], co-transformation with more than two genes [83], antisense [94], and RNAi technology [75, 87].

Phytoene synthase (PSY) is the rate-limiting step in the carotenoid biosynthetic pathway [82] and manipulation of *PSY* expression resulted in enhanced carotenoid synthesis in tubers [84]. The hydroxylation of β-carotene is a second important regulatory step in carotenogenesis [95]. Tuber-specific overexpression of the bacterial phytoene synthase (*CrtB*) gene caused a 7-fold increase in total carotenoids [84]. Expression of three genes from the bacterium *Erwinia herbicol* encoding phytoene synthase (CrtB), phytoene desaturase (CrtI) and lycopene betacyclase (CrtY), under the tuber-specific patatin promoter resulted with deep yellow flesh and increased levels of β-carotene, α-carotene, lutein and violaxanthin [83]. Silencing of LCY-e resulted in increased carotenoid levels, with up to 14-fold more β-carotene in tubers [85]. Silencing of the genes encoding β-carotene hydroxylases CHY1 and CHY2 using the tuber-specific patatin promoter increased β-carotene levels along with increased levels of phytofluene, violaxanthin, neoxanthin, lutein and total carotenoids [86]. Both the overexpression and silencing of the major genes in carotenoid biosynthesis pathway produced increased carotenoids in transgenic potato tubers.

Zeaxanthin have become increasingly important due to their benefits in the prevention of degenerative diseases [96]. Zeaxanthin is an immediate biochemical derivative of β-carotene thus tubers that accumulate high levels of zeaxanthin produce β-carotene that subsequently serves as the substrate for zeaxanthin synthesis [75]. RNAi was used to silence the β-carotene hydroxylase gene (*BCH/CHB*), which converts β-carotene to zeaxanthin under the control of GBSS or CaMV35S promoters [75]. Transgenic lines with silenced *CHB* expression showed altered carotenoid profiles. Transformants derived from the GBSS promoter contained more β-carotene than CaMV35S transformants, demonstrating that silencing *CHB* has the potential to increase the content of carotenoids in potato for mitigating vitamin A deficiency [75]. Oxidative cleavage of carotenoids is catalyzed by carotenoid cleavage dioxygenases

(CCDs). Down-regulation of *CCD4* though RNAi resulted in a 2-5-fold higher carotenoid content due to elevated violaxanthin content. Down-regulation of zeaxanthin epoxidase under the control of GBSS promoter resulted in zeaxanthin-rich potato lines, increased total carotenoids, and reduced the amount of lutein [94]. Thus, RNAi is a useful strategy to improve the carotenoid content in potato tubers thereby alleviating the vitamin A deficiency.

Astaxanthin, is an important ketocarotenoid associated with the reduction in oral cancer and mammary tumor growth [97, 98] and increasing astaxanthin and other ketocarotenoids levels has been studied in potato [88]. Astaxanthin is derived from β-carotene by 3-hydroxylation and 4-ketolation at both ionone end groups [99]. The hydroxylation reaction is widespread in many organisms, but ketolation is restricted to a few bacteria, fungi, and some unicellular green algae [100]. Previous studies used the transgenic expression of ketolase genes to produce ketocarotenoids in potato [88] as well as other Solanaceae members such as tomato [101] and tobacco [102]. A transgenic potato cultivar that accumulates increased zeaxanthin due to inactivated zeaxanthin epoxidase was co-transformed with the *ketolase* (*crtO β-carotene*) gene from the cyanobacterium *Synechocystis* under the control of a constitutive promoter. The resulting transgenic potato plants accumulated more ketocarotenoids in leaves, as well as more 3′-hydroxyechinenone, 4-ketozeaxanthin and astaxanthin in the tuber [88]. Likewise, overexpression of the *crtW* gene from the marine bacterium *Brevundimonas* under the control of GBSS promoter resulted in enhanced astaxanthin content in transgenic potato tubers [101, 103]. These and other studies reveal that transgenic potato lines can be produced with increased carotenoid content using bacterial and algal genes.

In another approach, increased carotenoid content was observed in tubers when cauliflower *Or* gene was overexpressed in potato [89, 90]. Lu et al. [89] showed that the cauliflower *Or* gene encoding a DnaJ cysteine-rich domain-containing protein that mediates high levels of β-carotene accumulation can be used to increase total carotenoid and β-carotene levels in potato tubers. Overexpression of the *Or* gene induced the formation of chromoplasts and resulted in high levels of carotenoids in transgenic tubers [90, 91, 103, 104]. This increase was found to be associated with the Or-regulated stability of PSY protein in these tubers, thus facilitating continuous carotenoid synthesis in the transgenic tubers [104]. Thus, overall studies indicated that genetic manipulation of *Or* genes can be an effective strategy to achieve increased carotenoid content and high quality potato tubers.

Overall, these studies indicate that the suppression and overexpression of various genes involved in carotenoid biosynthesis under different promoters resulted in altered carotenoid content in transgenic potato tubers. Although the transgenic strategy may be an effective way for increasing carotenoids production in potato, CRISPR/Cas9 gene editing might be challenging in tailoring efficient and non-transgenic potato cultivars with improved nutritional quality.

## 5. Manipulation for high levels of anthocyanins

Anthocyanins are water-soluble flavonoids that not only contribute to color of the fresh potatoes, but high anthocyanin content also enhances the antioxidant benefits on human health [105–108]. The major anthocyanidins in purple potato are cyanidin, petunidin, pelargonidin, peonidin, and malvidin, while red potatoes contain cyanidin, pelargonidin, and peonidin [109–111]. Transcriptome analysis has identified

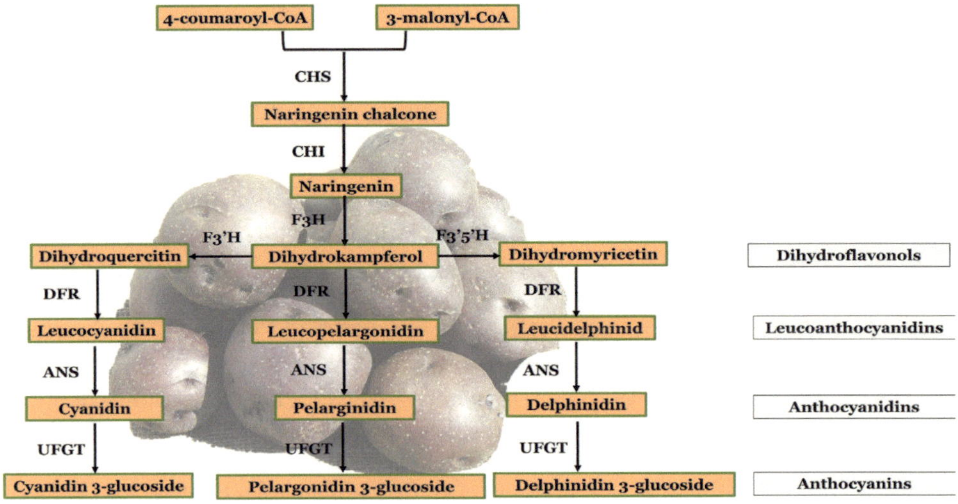


**Figure 5.**
*Anthocyanin biosynthetic pathway. The representative enzymes are CHS, chalcone, synthase; CHI, chalcone isomerase; F3H, flavanone 3-hydroxylase; F3′H, flavonoid 3′-hydroxylase; F3′5′H, flavonoid 3′,5′-hydroxylase; DFR, dihydroflavonol 4-reductase; ANS, anthocyanidin synthase and UFGT, flavonoid 3-O-glucosyltransferase. The figure is adapted and modified from Mattoo et al. [115].*

104 potentially important genes that may play an important role in anthocyanin biosynthesis in potato [112]. Several studies [113, 114] have investigated the impact of altered expression of some of the key genes in the anthocyanin biosynthesis pathway (**Figure 5** and **Table 4**).

The biosynthesis of anthocyanins (**Figure 5**) initiates from 4-coumaroyl-CoA and malonyl CoA catalyzed by the enzyme chalcone synthase (CHS) to synthesize naringenin chalcone which is then converted to naringenin by chalcone isomerase (CHI). Naringenin is converted to dihydrokaempferol, catalyzed by flavanonone 3′-hydroxylase (F3H). Flavonoid 3′-hydroxylase (F3′H) thereafter hydroxylates dihydrokaempferol (DHK) into dihydroquercitin (DHQ) or to dihydro-myricetin (DHM) which is catalyzed by flavonoid 3′,5′ –hydroxylase (F3′5′H). All three dihydroflavonols DHK, DHM and DHQ are independently converted to colorless leucoanthocyanidins by the enzyme dihydroflavonol 4-reductase (DFR). The next enzymatic reaction involves the enzyme anthocyanidin synthase (ANS), which converts all three

| Gene name | Abbreviation | Quality trait | Reference |
|---|---|---|---|
| *Dihydroflavonol 4-reductase* | *DFR* | Anthocyanin | Stobiecki et al. [113], Zhang et al. [116] |
| *Flavonoid 3',5'-hydroxylase* | *F3'5'H* | Anthocyanin | Jung et al. [114] |
| *Anthocyanidin 3-O-glucosyltransferase* | *3GT/UFGT* | Anthocyanin | Yoshihara et al. [117] |
| *UDP-glucose: flavonoid-3-O-glucosyltransferase* | *3GT* | Anthocyanin | Wei et al. [30] |
| *MYB transcription factor* | *MYB* | Anthocyanin | Liu et al. [118], Kranz et al. [119] |

**Table 4.**
*Genes used in the modification of anthocyanin content in transgenic potatoes.*

leucoathocyanidins to colored anthocyanidins [120]. Stobiecki et al. [113] demonstrated that overexpression of DFR gene under the control of the CaMV35S promoter showed increased tuber anthocyanin content along with a 4-fold increase in petunidin and pelargonidin derivatives in red skinned potato cv. 'Desiree'. However, overexpression of DFR in the potato cv 'Prince Hairy', which has a white tuber, resulted in change in the flower color from light blue to purple, but no change in tuber color [116]. Likewise, the overexpression of F3′5′H in the cv. 'Desiree' resulted in plants with purple-colored tubers and stems [114].

In addition to genes involved in the biosynthesis of anthocyanins, several transcription factors have been associated with anthocyanin biosynthesis [118, 119, 121, 122]. Stushnoff et al. [121] studied the gene expression pattern associated with the accumulation of purple tuber anthocyanins using microarray. A total of 27 genes were identified those were differentially expressed in purple and white tuber tissues. One of these genes coded for a novel single-domain MYB transcription factor (StMYBA1) that has been shown to influence anthocyanin-pigment production in potato [121, 122]. *StMYBA1* from potato was transformed into tobacco under the control of the CaMV 35S promoter and the resultant transformants showed anthocyanin accumulation in all tissues of transgenic tobacco lines [118]. In other studies, it has been demonstrated that there are two distinct classes of MYB transcription factors which negatively regulate anthocyanin accumulation: R3 MYB and R2R3 MYB repressors [119]. R2R3 MYB transcription factors; StAN1 and StbHLH1 are responsible for the coordinated regulation of the skin and flesh pigmentation, as well as anthocyanin biosynthetic pathway genes in white regions in potato [122–124]. Therefore, manipulation of specific transcription factor genes associated with anthocyanin synthesis through transgenic or CRISPR tools may be useful in enhancing nutritionally important traits of pigmented tuber flesh.

## 6. Improving the cooking and processing quality

Although potatoes are increasingly consumed in the form of processed food products such as chips, french fries, dehydrated products, etc., they do not deliver the same nutritional value as fresh potatoes [125–127]. Potatoes are rich in carbohydrates (starch, sucrose, glucose, and fructose) and non-starch polysaccharides from cell wall components [128]. Potato starch is composed of 20–30% amylose and 70–80% amylopectin [129, 130]. Both starch and sugars play an important role during growth and development (biosynthesis of starch) and during postharvest storage of potatoes (breakdown of starch) [131]. The amount of starch and sugars present in potato tubers is an important criterion for selection of potato cultivars for commercial purpose. Likewise, the amount of reducing sugars (RS) present in potato tubers affects the processing quality of fried products. The final color in fried potato products results from the heating reactions that occur between RS and amino acids such as asparagine [132]. Similarly, cooking and baking qualities that are important for consumers are related to color, texture, and flavor. Therefore, potatoes are selected with reduced sugar content, good processing quality, and with the absence of cooking defects such as enzymatic browning and stem-end blackening [133]. Engineering for modified starch and sugar content as well as alteration in cooking quality has resulted in enhanced nutritional quality of transgenic potato tubers [134–138].

### 6.1 Metabolic engineering for modifying starch content

**Figure 6** and **Table 5** highlights the information on important candidate genes used in the modification of starch content in transgenic potato tubers. ADP-glucose pyrophosphorylase (AGPase), starch synthase (SS) and starch branching enzyme (SBE) are involved in the process of starch synthesis while amylases (AMY) and starch phosphorylase (SP) are responsible for its breakdown [152]. Amylose is synthesized by the granule bound starch synthase (GBSS), whereas soluble starch synthases (isoforms) and SBEI/SBEII, with various debranching enzymes (DBE), kinases and

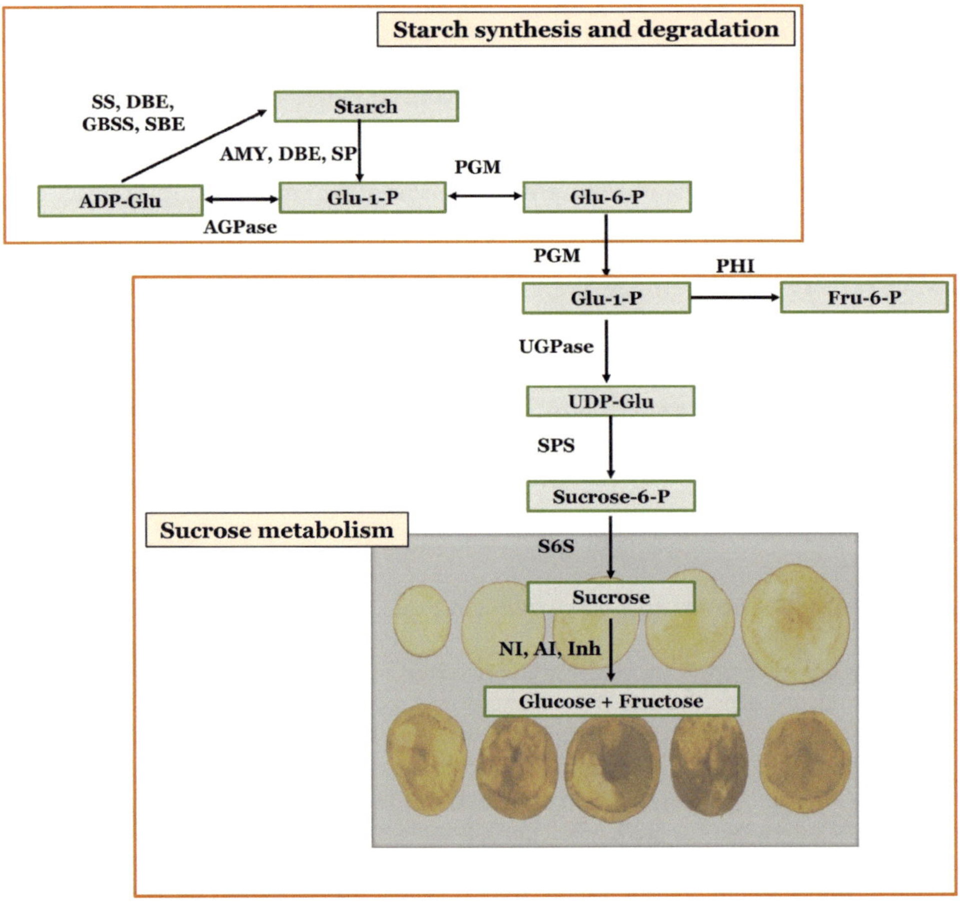


**Figure 6.**
*Simplified schematic overview of starch and sugar metabolism. The figure explains the simplified pathway of starch and sugar metabolism in potato tubers. Only the key enzymes involved in the pathway are shown in the figure. Starch synthesis involves AGPase, ADP-glucose pyrophosphorylase; SS, starch synthase; GBSS, granule bound starch synthase; and SBE, starch branching enzyme. This process involves more than one SS and SBE (not shown in figure). Further starch is degraded into Glucose-1-Phosphate (Glu-1-P) by Amy, amylase; SP, starch phosphorylase which forms Glucose-6-Phosphate (Glu-6-P) via PGM, phosphoglucomutase. Fructose-6-Phosphate (Fru-6-P) is formed from PHI, Glc-6-P via phosphohexoisomerase. Sucrose is formed in the cytoplasm from UDP-Glucose. SPS and S6S are sucrose phosphate synthase and sucrose 6 phosphate phosphatase. UGPase, UDP-glucose pyrophosphorylase; SPS, sucrose phosphate synthase; and S6P, sucrose 6-phosphate phosphatase are involved in the process. Sucrose is transported into the vacuole and converted by AI, acid invertase leading to the formation of RS glucose and fructose. AI is inhibited by vacuolar invertase inhibitor (INH) in the vacuole. NI, neutral invertase converts sucrose into glucose and fructose in the cytoplasm. The figure is adapted and modified from Sowokinos [139].*

| Gene name | Abbreviation | Quality trait | Reference |
|---|---|---|---|
| *Granule bound starch synthase* | *GBSS* | Starch | Visser [140], Kuipers et al. [141], Andersson et al. [142] |
| *ADP glucose pyrophosphorylase* | *AGPase* | Starch | Müller-Röber et al. [143] |
| *Glucose-6-phosphate/ phosphate translocator and the adenylate translocator (nucleotide translocator)* | *GPT and NTT* | Starch | Claudia et al. [144] |
| *R1 protein* | *R1 protein* | Starch | Lorberth [145] |
| *Starch branching enzyme* | *SBE* | Starch | Tuncel et al. [146] |
| *Vacuolar invertase* | *VI* | Sucrose | Bhaskar et al. [147], Liu [148], Clasen [149] |
| *RING finger* | *SbRFP1* | Sucrose | Scheidig [150] |
| *Vacuolar invertase inhibitor* | *INH2* | Sucrose | Liu et al. [138], McKenzie et al. [151] |

**Table 5.**
*Genes used in the modification of carbohydrate metabolism and enzymatic browning in transgenic potatoes.*

other enzymes are also involved in amylopectin synthesis [153, 154]. The modification of starch properties is important to food processing industries and the demand for amylose free potatoes wherein higher amylopectin content is desired [122, 155]. Food products containing high amylose content and long chains of amylopectin contribute to the formation of resistant starch that is responsible for a lower glycemic index after intake with enhanced health benefits by promoting the growth of healthy gut fora and lowering both the caloric intake and cholesterol levels in the blood [156–158].

Transgenic studies focused on the modification of starch content and its properties using the genes involved starch biosynthesis in potatoes [140, 141, 143–145, 159–163]. AGPase is the key enzyme involved in the synthesis of starch in amyloplasts which consists of two regulatory subunits and two slightly smaller catalytic subunits. The function of the large subunit of AGPase is to modulate the regulatory properties of the small subunit (sAGP) and the function of sAGP is primarily catalysis [152, 159, 160]. An antisense inhibition of sAGP under the control of CaMV 35S promoter resulted in starch reduction and a lower amylose content in transgenic tubers [143].

Nutritional quality enrichment by starch modification can be efficiently performed using TALENs technology [164] as well as CRISPR/Cas9 system [142]. Site-specific mutations induced in *GBSS* gene (CaMv35S promoter) using Emerald-Gateway TALEN system resulted in 63 nucleotide deletion suggesting that the system can be utilized for nutritional enhancement in potato [164]. All four *GBSS* alleles were knockout by CRISPR/Cas9 technology in cv. 'Kuras' protoplasts resulted in complete loss of GBSS activity and amylose-free high amylopectin starch in regenerated potato microtubers [142]. In another study, overexpression of SBEII using hybrid cDNA/gDNA intragene construct containing a single intron increased short-chain branching of amylopectin and altered the physicochemical properties of starch in potato tuber [165]. CRISPR-Cas9 was used to crete mutations in the two *SBEs* (*SBE1* or *SBE2* alone or in combination) in potato. Results revealed that lines mutated in SBE1 did not have an altered starch structure, while tuber cells from SBE2 mutated lines displayed an increased number of starch granules. One line had a strong reduction in both SBEs, resulting in starch with an altered granule phenotype, longer amylopectin chains and reduction in a degree of branching [146]. The quality characteristics of potatoes for table and processing purposes are largely dependent on the starch, dry matter, sugar concentration and tubers free from any deformities such as enzymatic discoloration. Overall studies revealed that, transgenic, intragenic as well as CRISPR tools proved their usefulness in modification of various starch synthesizing enzymes using gene silencing or gene knockout approaches. Alteration in starch synthesis genes can lead to modification in starch properties without affecting the yield and dry matter content of the transgenic tubers.

### 6.2 Postharvest tuber quality for enhanced cold-induced sweetening

After harvest, storage of potato tubers under cold conditions (below 8–10°C) is required for year-round processing as well as to mitigate the possibility of sprouting and diseases. Stored tubers loose some of their starch content and accumulate high RS in a process known as cold-induced sweetening (CIS). Upon frying at high temperatures, these RS interact with asparagine to produce dark-colored fried products [166, 167]. Therefore, minimizing the accumulation of RS in tubers is of high importance to potato processing industry. The amount of RS in a tuber is regulated by a balance between the activity of invertases (cell wall invertase, vacuolar invertase and neutral invertase), which convert sucrose into RS and invertase inhibitors which limit the activity of invertase via protein-protein interactions [28, 136, 168]. **Figure 6** and **Table 5** highlights the information on important candidate genes used in the modification of sugar content in transgenic potatoes. Among various invertases, the acid invertase (AI) is the key enzymes involved in the conversion of sucrose into RS and transgenic approaches confirmed that overexpression of the vacuolar invertase inhibitor (*INH2*) gene reduces the expression of vacuolar invertase gene, AI activity and RS in transgenic potato tubers under the control of CaMV 35S or class I patatin promoter [136–138, 166]. In another study, overexpression of the vacuolar invertase inhibitor isoforms from potato resulted in decreased vacuolar invertase activity, low RS and low acrylamide content with improved chip quality in cold-stored transgenic potato [138, 151]. Suppression of the AI activity by silencing of the vacuolar invertase gene using RNAi or TALENS resulted in a very strong decrease in RS accumulation, light colored chips and low acrylamide in cold-stored transgenic tubers of different potato cultivars [147, 169]. Knockout of the vacuolar invertase was performed in 'Ranger Russet' potatoes using the TALENs technology. Five regenerated plants contained knockouts of all four invertase alleles with no detectable RS, light brown chip color and lower levels of acrylamide [148].

Other than, the vacuolar invertase and its inhibitor gene, Zhang et al. [170] hypothesized that *RING finger gene* (*SbRFP1*) could be a potential target for manipulation of the CIS in potato tubers. RING finger proteins constitute a large protein family in higher plants involved in cold response [170]. When a novel *SbRFP1* was overexpressed (CaMV35S promoter) in potato microtubers (cv. 'E-potato 3'), it resulted in inhibition of beta-amylase and invertase activity. As a result, starch and sucrose degradation slowed down and the accumulation of RS in cold stored tubers was prevented [170]. Study suggested that other potential genes could be involved in CIS in potato tubers and transgenic manipulation can be performed to overcome this persistent problem.

So far, much attention has been given to the manipulation of the acid invertase activity via suppression and overexpression of the vacuolar invertase and invertase inhibitor genes to control the RS content in transgenic potato tubers respectively. In contrast to the acid invertase, the potential involvement of neutral invertase which is involved in the conversion of sucrose into RS has not been demonstrated and well-studied in potato tubers. Datir and Regan [150] identified 8 neutral invertase genes from potato. Based on their expression pattern and enzymatic pattern in cold-stored potato tubers, they concluded that neutral invertases also may play a role in sucrose degradation. Therefore, genetic manipulation of neutral invertase may result in decreased RS along with improved processing quality of potato tubers.

### 6.3 Reducing the enzymatic browning

Enzymatic browning or discoloration of potato tubers occurs when phenolic compounds are oxidized by the enzyme polyphenol oxidase (PPO). This results in negative effects on color, taste, flavor, and nutritional value and forms undesired dark pigments that result in considerable economic losses to the potato food and processing industry [171, 172]. To prevent the quality loss and increase consumers acceptance, sulfiting agents (sulfur dioxide, sodium etc.) can be used to prevent the enzymatic browning, however, there are concerns about the health risks of sulfites [173, 174]. For this reason, there is a need to develop the alternative technologies for developing potatoes that are resistant to enzymatic browning [175]. Silencing of the *PPO* gene resulted in significantly reduced enzymatic browning in the tubers of transgenic lines [175, 176]. Two potato cultivars 'Van Gogh' and 'Diamant' transformed with antisense *PPO* placed under patatin and GBSS promoters abolished the expression of PPO in transgenic tubers [176]. In another study, the PPO activity was inhibited by expression of a sense as well as antisense *PPO* RNAs from a tomato *PPO* cDNA under the control of the CaMV 35S promoter in cv. 'Russet Burbank'. Transgenic lines expressing *PPO* by sense and antisense approaches resulted in reduction in black spot susceptibility, decreased PPO activity and reduced enzymatic browning [175].

Recent studies have demonstrated that genome editing using CRISPR/Cas9 system can be successfully used to reduce the enzymatic browning in potato [177, 178]. The Cas9 nuclease guided by two RNA molecule/s (sgRNA/s) introduced a double stranded break in the PPO gene. The system that introduced the mutations in the *PPO* gene was delivered into the protoplasts of cv 'Desiree'. 24% of CRISPR/Cas9-edited lines carried mutations in all four alleles of *PPO* without any off-target mutations in other *PPO* genes. Mutations induced in the four alleles of *StPPO2* gene showed 69% reduction in tuber PPO activity and a 73% reduction in enzymatic browning, compared to the control [177]. These studies revealed that CRISPR/Cas9 system represents an important step towards the development of potato varieties that maintain the organoleptic, antioxidant and nutritional properties during harvest and post-harvest procedures, without the utilization of potentially harmful browning controlling agents.

## 7. Conclusions

Improvement in nutritional properties such as cooking, baking, and processing qualities such as dry matter content, vitamins, starch, sugar content, flavors, colors, and glycoalkaloids is one of the most important aspects of potato production. Genetic manipulation of these quality traits has significantly increased our understanding of the genes and their network involved in controlling these traits. However, transgenic potatoes still lack consumer and producer acceptance and are not widely used compared to other crops. Transgenics created doubts on the transformation processes and hence food safety evaluation tests are necessary to detect the any unintended effects in transgenic lines. Release of the Potato Genome Sequencing Consortium, genetic mapping, and genome-wide association-based studies, and the recent progress in CRISPR/Cas9-based tools have paved the way for development of potato cultivars with improved nutritional properties. Using intragenic or cisgenic and CRISPR-based approaches as well as proper assessment under field trials may alleviate the concerns by consumers, producers, and processors. Also, the unintended effects can be overcome by CRISPR/Cas9-based tools to produce transgene-free potatoes.

## Acknowledgements

Authors are grateful to Prof. Dr. David G. Holm, Department of Horticulture and Landscape Architecture, Colorado State University, USA and Dr. Sanjay Gupta, University of Minnesota, USA for providing the potato photographs used in **Figures 1–5**. This research was supported by a Natural Science and Engineering Research Council grant to SR.

## Conflict of interest

"The authors declare no conflict of interest."

## Acronyms and abbreviations

| | |
|---|---|
| amiRNAs | artificial microRNAs |
| CaMV35S | cauliflower mosaic virus 35S |
| CRISPR | Clustered Regularly Interspaced Short Palindromic Repeats |
| QTL | quantitative trait loci |
| RNAi | RNA interference |
| TALENS | Transcription Activator-Like Effector Nucleases |

## Author details

Sagar S. Datir and Sharon Regan*
Department of Biology, Queen's University, Kingston, Ontario, Canada

*Address all correspondence to: sharon.regan@queensu.ca

## References

[1] Birch PR, Bryan G, Fenton B, Gilroy EM, Hein I, Jones JT, et al. Crops that feed the world 8: Potato: Are the trends of increased global production sustainable? Food Security. 2012;**4**:477-508. DOI: 10.1007/s12571-012-0220-1

[2] Burlingame B, Mouille B, Charrondiere R. Nutrients, bioactive non-nutrients and anti-nutrients in potatoes. Journal of Food Composition and Analysis. 2009;**22**:494-502

[3] Hemavathi UCP, Young KE, Akula N, Kim HS, Heung JJ, Oh MO, et al. Over-expression of strawberry D-galacturonic acid reductase in potato leads to accumulation of vitamin C with enhanced abiotic stress tolerance. Plant Science. 2009;**177**:659-667. DOI: 10.1016/j.plantsci.2009.08.004

[4] Zaheer K, Akhtar MH. Potato production, usage, and nutrition—A review. Critical Reviews in Food Science and Nutrition. 2016;**56**:711-721. DOI: 10.1080/10408398.2012.724479

[5] Hellmann H, Goyer A, Navarre DA. Antioxidants in potatoes: A functional view on one of the major food crops worldwide: A review. Molecules. 2021;**26**:2446. DOI: 10.3390/molecules26092446

[6] Shepherd T, Dobson G, Verrall S, Conner S, Griffiths DW, McNicol J, et al. Potato metabolomics by GC-MS: What are the limiting factors? Metabolomics. 2007;**3**:475-488. DOI: 10.1007/s11306-007-0058-2

[7] Thompson MD, Thompson HJ, McGinley JN, Neil ES, Rush DK, Holm DG, et al. Functional food characteristics of potato cultivars (*Solanum tuberosum* L.): Phytochemical composition and inhibition of 1-methyl-1- nitrosourea induced breast cancer in rats. Journal of Food Composition and Analysis. 2009;**22**:571-576. DOI: 10.1016/j.jfca.2008.09.002

[8] Madiwale GP, Reddivari L, Stone M, Holm DG, Vanamala J. Combined effects of storage and processing on the bioactive compounds and pro-apoptotic properties of color-fleshed potatoes in human colon cancer cells. Journal of Agricultural Food Chemistry. 2012;**60**:11088-11096. DOI: 10.1021/jf303528p

[9] Tian J, Chen J, Ye X, Chen S. Health benefits of the potato affected by domestic cooking: A review. Food Chemistry. 2016;**202**:165-175. DOI: 10.1016/j.foodchem.2016.01.120

[10] Gibson S, Kurilich A. The nutritional value of potatoes and potato products in the UK diet. Nutrition Bulletin. 2013;**38**:389-399. DOI: 10.1111/nbu.12057

[11] Lemos MA, Aliyu MM, Hungerford G. Influence of cooking on the levels of bioactive compounds in purple majesty potato observed via chemical and spectroscopic means. Food Chemistry. 2015;**173**:462-467. DOI: 10.1016/j.foodchem.2014.10.064

[12] Scott G, Suarez V. The rise of Asia as the Centre of global potato production and some implications for industry. Potato Journal. 2012;**39**:1-22. DOI: 10.7748/ns2012.01.26.18.1.p7241

[13] Hirschi KD. Nutrient biofortification of food crops. Annual Review of Nutrition. 2009;**29**:401-421. DOI: 10.1146/annurev-nutr-080508-141143

[14] Patil VU, Singh R, Vanishree G, Dutt S, Kawar PG, Bhardwaj V, et al. Genetic engineering for enhanced nutritional quality in potato—A review. Potato Journal. 2016;**43**:1-21

[15] Bakhsh A. Development of efficient, reproducible, and stable agrobacterium-mediated genetic transformation of five potato cultivars. Food Technology and Biotechnology. 2020;**58**:57-63. DOI: 10.17113/ftb.58.01.20.6187

[16] García-Alonso A, Goñi I. Effect of processing on potato starch: In vitro availability and glycaemic index. Die Nahrung. 2000;**44**(1):19-22. DOI: 10.1002/(SICI)1521-3803(20000101)44:1<19::AID-FOOD19>3.0.CO;2-E

[17] Tyburczy C, Delmonte P, Fardin-Kia AR, Mossoba MM, Kramer JKG, Rader JI. Profile of trans fatty acids (FAs) including trans polyunsaturated fats in representative fast-food samples. Journal of Agricultural Food Chemistry. 2012;**60**:4567-4577. DOI: 10.1021/jf300585s

[18] Dias JS, Ortiz R. Transgenic vegetable breeding for nutritional quality and health benefits. Food and Nutrition Sciences. 2012;**3**:1209-1219. DOI: 10.4236/fns.2012.39159

[19] Datir SS, Yousf S, Sharma S, Kochle M, Ravikumar A, Chugh J. Cold storage reveals distinct metabolic perturbations in processing and non-processing cultivars of potato (*Solanum tuberosum* L.). Scientific Reports. 2020;**10**(1):6268. DOI: 10.1038/s41598-020-63329-5

[20] Dobson G, Shepherd T, Verrall SR, Griffiths WD, Ramsay G, McNicol JW, et al. A metabolomics study of cultivated potato (*Solanum tuberosum*) groups Andigena, Phureja, Stenotomum, and tuberosum using gas chromatography-mass spectrometry. Journal of Agricultural Food Chemistry. 2010;**58**:1214-1223. DOI: 10.1021/jf903104b

[21] Defernez M, Gunning YM, Parr AJ, Shepherd LV, Davies HV, Colquhoun IJ. NMR and HPLC-UV profiling of potatoes with genetic modifications to metabolic pathways. Journal of Agricultural Food Chemistry. 2004;**52**:6075-6085. DOI: 10.1021/jf049522e

[22] Chaparro JM, Holm DG, Broeckling CD, Prenni JE, Heuberger AL. Metabolomics and ionomics of potato tuber reveals an influence of cultivar and market class on human nutrients and bioactive compounds. Frontiers in Nutrition. 2018;**5**:36. DOI: 10.3389/fnut.2018.00036

[23] Xiong X, Tai GCC, Seabrook JEA. Effectiveness of selection for quality traits during the early stage in the potato breeding population. Plant Breeding. 2002;**121**:441-444

[24] Hamernik AJ, Hanneman RE Jr, Jansky SH. Introgression of wild species germplasm with extreme resistance to cold sweetening into the cultivated potato. Crop Science. 2009;**49**:529-542. DOI: 10.2135/cropsci2008.04.0209

[25] Bradshaw JE. Genetics of Agri horticultural traits. In: Gopal J, Khurana SMP, editors. Handbook of Potato Production, Improvement, and Postharvest Management. Food Products Press: New York; 2006. pp. 41-75

[26] Bhardwaj A, Nain V. TALENs—An indispensable tool in the era of CRISPR: A mini review. Journal of Genetic Engineering and Biotechnology. 2021;**19**(125). DOI: 10.1186/s43141-021-00225-z

[27] Arnqvist L, Dutta PC, Jonsson L, Sitbon F. Reduction of cholesterol and glycoalkaloid levels in transgenic potato plants by overexpression of a type 1 sterol methyltransferase cDNA. Plant Physiology. 2003;**131**:1792-1799. DOI: 10.1104/pp.102.018788

[28] Brummell DA, Chen RK, Harris JH, Zhang H, Hamiaux C, Kralicek AA, et al. Induction of vacuolar invertase inhibitor mRNA in potato tubers contributes to cold-induced sweetening resistance and includes spliced hybrid mRNA variants. Journal of Experimental Botany. 2011;**62**:3519-3534. DOI: 10.1093/jxb/err043

[29] Qin AG, Shi QH, Yu XC. Ascorbic acid contents in transgenic potato plants overexpressing two dehydroascorbate reductase genes. Molecular Biology Reports. 2011;**38**:1557-1566. DOI: 10.1007/s11033-010-0264-2

[30] Wei Q, Wang QY, Feng ZH, Wang W, Zhang YF, Yang Q. Increased accumulation of anthocyanins in transgenic potato tubers by overexpressing the *3GT* gene. Plant Biotechnology Reports. 2012;**6**:69-75. DOI: 10.1007/s11816-011-0201-4

[31] Zhou X, McQuinn R, Fei Z, Wolters AA, Van Eck J, Brown C, et al. Regulatory control of high levels of carotenoid accumulation in potato tubers. Plant Cell and Environment. 2011;**34**:1020-1030. DOI: 10.1111/j.1365-3040.2011.02301.x

[32] Datir SS. Invertase inhibitors in potato: Towards a biochemical and molecular understanding of cold-induced sweetening. Critical Reviews in Food Science and Nutrition. 2020:1-15. DOI: 10.1080/10408398.2020.1808876

[33] Wiberley-Bradford AE, Busse JS, Jiang J, Bethke PC. Sugar metabolism, chip color, invertase activity, and gene expression during long-term cold storage of potato (*Solanum tuberosum*) tubers from wild-type and vacuolar invertase silencing lines of Katahdin. BMC Research Notes. 2014;**7**:801. DOI: 10.1186/1756-0500-7- 801

[34] Wang S, Zhang S, Wang W, Xiong X, Meng F, Cui X. Efficient targeted mutagenesis in potato by the CRISPR/Cas9 system. Plant Cell Reports. 2015;**34**:1473-1476. DOI: 10.1007/s00299-015- 1816-7

[35] Xu X, Pan S, Cheng S, Zhang B, Mu D, Ni P, et al. Genome sequence and analysis of the tuber crop potato. Nature. 2011;**475**:189-195. DOI: 10.1038/nature10158

[36] Nahirñak V, Almasia NI, González MN, Massa GA, Décima Oneto CA, Feingold SE, et al. State of the art of genetic engineering in potato: From the first report to its future potential. Frontiers in Plant Science. 2022;**12**:768233. DOI: 10.3389/fpls.2021.768233

[37] Van Gelder WMJ, Vinke JH, Scheffer JJC. Steroidal glycoalkaloids in tubers and leaves of *Solanum* species used in potato breeding. Euphytica. 1988;**39**:147-158. DOI: 10.1007/BF00043378

[38] Friedman M, McDonald GM. Potato glycoalkaloids: Chemistry, analysis, safety, and plant physiology. Critical Revies in Plant Sciences. 1997;**16**:55-132. DOI: 10.1080/07352689709701946

[39] Smith DB, Roddick JG, Jones JL. Potato glycoalkaloids: Some unanswered questions. Trends in Food Science and Technology. 1996;**7**:126-131. DOI: 10.1016/0924-2244(96)10013-3

[40] Valkonen JTP, Keskitalo M, Vasara T, Pietila L, Raman KV.

Potato glycoalkaloids: A burden or a blessing? Critical Reviews in Plant Science. 1996;**15**:1-20. DOI: 10.1080/07352689609701934

[41] Salunkhe DK, Kadam SS. In: Salunkhe DK, Kadam SS, editors. Handbook of Vegetable Science and Technology. Production, Composition, Storage and Processing. CRC Press; 1998

[42] Khan MS, Minur I, Khan I. The potential of unintended effects in potato glycoalkaloids. African Journal of Biotechnology. 2013;**12**:754-766

[43] Sonawane PD, Pollier J, Panda S, Szymanski J, Massalha H, Yona M, et al. Plant cholesterol biosynthetic pathway overlaps with phytosterol metabolism. Nature Plants. 2016;**22**(3):16205. DOI: 10.1038/nplants.2016.205. Erratum in: Nat Plants. 2017;**3**:17101

[44] McCue KF, Shepherd LVT, Allen PV, Maccree MM, Rockhold DR, Corsini DL, et al. Metabolic compensation of steroidal glycoalkaloid biosynthesis in transgenic potato tubers: Using reverse genetics to confirm the in vivo enzyme function of a steroidal alkaloid galactosyltransferase. Plant Science. 2005;**168**:367-373. DOI: 10.1016/j.plantsci.2004.08.006

[45] Shepherd LV, Hackett CA, Alexander CJ, McNicol JW, Sungurtas JA, Stewart D, et al. Modifying glycoalkaloid content in transgenic potato—Metabolome impacts. Food Chemistry. 2015;**187**:437-443. DOI: 10.1016/j.foodchem.2015.04.111

[46] Umemoto N, Nakayasu M, OhyamaK,Yotsu-YamashitaM,MizutaniM, Seki H, et al. Two cytochrome P450 monooxygenases catalyze early hydroxylation steps in the potato steroid glycoalkaloid biosynthetic pathway. Plant Physiology. 2016;**171**:2458-2467. DOI: 10.1104/pp.16.00137

[47] Heftmann E, Lieber ER, Bennett RD. Biosynthesis of tomatidine from cholesterol in *Lycopersicon pimpinellifolium*. Phytochemistry. 1967;**6**:225-229

[48] Sawai S, Ohyama K, Yasumoto S, Seki H, Sakuma T, Yamamoto T, et al. Sterol side chain reductase 2 is a key enzyme in the biosynthesis of cholesterol, the common precursor of toxic steroidal glycoalkaloids in potato. The Plant Cell. 2014;**26**:3763-3774. DOI: 10.1105/tpc.114.130096

[49] Itkin M, Heinig U, Tzfadia O, Bhide AJ, Shinde B, Cardenas PD, et al. Biosynthesis of antinutritional alkaloids in solanaceous crops is mediated by clustered genes. Science. 2013;**341**:175-179. DOI: 10.1126/science

[50] Nahar N, Westerberg E, Arif U, Huchelmann A, Guasca AO, Beste L, et al. Transcript profiling of two potato cultivars during glycoalkaloid-inducing treatments shows differential expression of genes in sterol and glycoalkaloid metabolism. Scientific Reports. 2017;7:43268. DOI: 10.1038/srep43268

[51] Akiyama R, Nakayasu M, Umemoto N, Muranaka T, Mizutani M. Molecular breeding of SGA-free potatoes accumulating pharmaceutically useful saponins. Regulation of Plant Growth & Development. 2017;**52**(2):92-98. DOI: 10.18978/jscrp.52.2_92

[52] Nakayasu M, Akiyama R, Lee HJ, Osakabe K, Osakabe Y, Watanabe B, et al. Generation of α-solanine-free hairy roots of potato by CRISPR/Cas9 mediated genome editing of the St16DOX gene. Plant Physiology and Biochemistry. 2018;**131**:70-77. DOI: 10.1016/j.plaphy.2018.04.026

[53] Kaminski KP, Kørup K, Andersen MN, Sønderkær M,

Andersen MS, Kirk HG, et al. Next generation sequencing bulk segregant analysis of potato support that differential flux into the cholesterol and stigmasterol metabolite pools is important for steroidal glycoalkaloid content. Potato Research. 2016;**59**:81-97. DOI: 10.1007/s11540-015-9314-4

[54] Cárdenas PD, Sonawane PD, Pollier J, Vanden Bossche R, Dewangan V, Weithorn E, et al. GAME9 regulates the biosynthesis of steroidal alkaloids and upstream isoprenoids in the plant mevalonate pathway. Nature Communications. 2016;7:10654. DOI: 10.1038/ncomms10654

[55] Ginzberg I, Thippeswamy M, Fogelman E, Demirel U, Mweetwa AM, Tokuhisa J, et al. Induction of potato steroidal glycoalkaloid biosynthetic pathway by overexpression of cDNA encoding primary metabolism HMG-CoA reductase and squalene synthase. Planta. 2012;**235**(6):1341-1353. DOI: 10.1007/s00425-011-1578-6

[56] Moehs CP, Allen PV, Friedman M, Belknap WR. Cloning and expression of solanidine UDP-glucose glucosyltransferase from potato. Plant Journal. 1997;**11**:227-236. DOI: 10.1046/j.1365-313x.1997.11020227.x

[57] Zheng Z, Ye G, Zhou Y, Pu X, Su W, Wang J. Editing sterol side chain reductase 2 gene (StSSR2) via CRISPR/Cas9 reduces the total steroidal glycoalkaloids in potato. All Life. 2021;**14**:401-413. DOI: 10.1080/26895293.2021.1925358

[58] Bouvier-Navé P, Husselstein T, Benveniste P. Two families of sterol methyltransferases are involved in the first and the second methylation steps of plant sterol biosynthesis. European Journal of Biochemistry. 1998;**256**:88-96

[59] Sørensen KK, Kirk HG, Olsson K, Labouriau R, Christiansen J. A major QTL and an SSR marker associated with glycoalkaloid content in potato tubers from *Solanum tuberosum* x *S. sparsipilum* located on chromosome I. Theoretical and Applied Genetics. 2008;**117**:1-9

[60] Camire ME, Kubow S, Dnnelly DJ. Potatoes and human health. Critical Reviews in Food Science and Nutrition. 2009;**49**:823-840. DOI: 10.1080/10408390903041996

[61] Han J, Kosukue N, Young K, Lee K, Friedman M. Distribution of ascorbic acid in potato tubers and in home-processed and commercial potato foods. Journal of Agricultural Food Chemistry. 2004;**52**:6516-6521. DOI: 10.1021/jf0493270

[62] USDA Economic Research Service. USDA Economic Research Service—Potatoes. 2014. Available from: http://www.ers.usda.gov/topics/crops/vegetables-pulses/potatoes.aspx [Accessed: May 10, 2022]

[63] Mazza G, Hung J, Dench MJ. Processing nutritional quality changes in potato tubers during growth and long-term storage. Canadian Institute of Food Science and Technology Journal. 1983;**16**:39-44. DOI: 10.1016/S0315-5463(83)72017-1

[64] Linnemann AR, van Es A, Hartmans KJ. Changes in the content of L-ascorbic acid, glucose, fructose, sucrose and total glycoalkaloids in potatoes (cv. Bintje) stored at 7, 16 and 28 C. Potato Research. 1985;**28**:271-278. DOI: 10.1007/BF02357581

[65] Ikanone CEO, Oyekan PO. Effect of boiling and frying on the total carbohydrate, vitamin c and mineral contents of Irish (*Solanun tuberosum*) and sweet (*Ipomea batatas*) potato tubers. Nigerian Food Journal. 2014;**32**:33-39

[66] Upadhyaya CP, Venkatesh J, Gururani MA, Asnin L, Sharma K, Ajappala H, et al. Transgenic potato overproducing L-ascorbic acid resisted an increase in methylglyoxal under salinity stress via maintaining higher reduced glutathione level and glyoxalase enzyme activity. Biotechnology Letters. 2011;**33**:2297-2307. DOI: 10.1007/s10529-011-0684-7

[67] Bulley S, Wright M, Rommens C, Yan H, Rassam M, Lin-Wang K, et al. Enhancing ascorbate in fruits and tubers through over-expression of the l-galactose pathway gene GDP-l-galactose phosphorylase. Plant Biotechnology Journal. 2012;**10**:390-397. DOI: 10.1111/j.1467-7652.2011.00668.x

[68] Hemavathi UCP, Akula N, Young KE, Chun SC, Kim DH, Park SE. Enhanced ascorbic acid accumulation in transgenic potato confers tolerance to various abiotic stresses. Biotechnology Letters. 2010;**32**:321-330. DOI: 10.1007/s10529-009-0140-0

[69] Venkatesh J, Park SW. Role of L-ascorbate in alleviating abiotic stresses in crop plants. Botanical Studies. 2014;**55**(1):38. DOI: 10.1186/1999-3110-55-38

[70] Linster CL, Van Schaftingen E, Vitamin C. Biosynthesis, recycling and degradation in mammals. FEBS Letters. 2007;**274**(1):1-22. DOI: 10.1111/j.1742-4658.2006.05607.x

[71] West KP Jr. Extent of vitamin A deficiency among preschool children and women of reproductive age. Journal of Nutrition. 2002;**132**:285. DOI: 10.1093/jn/132.9.2857S

[72] West CE, Eilander A, van Lieshout M. Consequences of revised estimates of carotenoid bioefficacy for dietary control of vitamin A deficiency in developing countries. Journal of Nutrition. 2002;**132**:2920-2926. DOI: 10.1093/jn/132.9.2920S

[73] Fraser PD, Bramley PM. The biosynthesis and nutritional uses of carotenoids. Progress in Lipid Research. 2004;**43**:228-265. DOI: 10.1016/j.plipres.2003.10.002

[74] Bonierbale M, Grüneberg W, Amoros W, Burgos G, Salas E, Porras E, et al. Total and individual carotenoid profiles in Solanum phureja cultivated potatoes: II. Development and application of near-infrared reflectance spectroscopy (NIRS) calibrations for germplasm characterization. Journal of Food Composition and Analysis. 2009;**22**:509-516

[75] Van Eck J, Conlin B, Garvin DF, Mason H, Navarre DA, Brown CR. Enhancing beta-carotene content in potato by RNAi-mediated silencing of the beta-carotene hydroxylase gene. American Journal of Potato Research. 2007;**84**:331-342

[76] Hejtmánková K, Kotíková Z, Hamouz K, Pivec V, Vacek J, Lachman J. Influence of flesh colour, year and growing area on carotenoid and anthocyanin content in potato tubers. Journal of Food Composition and Analysis. 2013;**32**:20-27. DOI: 10.1016/j.jfca.2013.07.001

[77] Lachman J, Hamouz K, Orsák M, Kotíková Z. Carotenoids in potatoes—A short overview. Plant and Soil Environment. 2016;**62**:474-481

[78] Stahl W, Sies H. Bioactivity and protective effects of natural carotenoids. Biochimica et Biophysica Acta. 2005;**1740**:101-107. DOI: 10.1016/j.bbadis.2004.12.006

[79] Yan J, Kandianis C, Harjes C, Harjes CE, Bai L, Kim EH, et al. Rare genetic variation at *Zea mays* crtRB1 increases b-carotene in maize grain. Nature Genetics. 2010;**42**:322-327. DOI: 10.1038/ng.551

[80] Brown CR, Edwards CG, Yang CP, Dean BB. Orange flesh trait in potato: Inheritance and carotenoid content. Journal of American Society for Horticultural Science. 1993;**118**:145-150 10.21273/JASHS.118.1.145

[81] Iwanzik W, Tevini M, Stute R, Hilbert R. Carotinoidgehalt und -zusammensetzung verschiedener deutscher Kartoffelsorten und deren Bedeutung fur die Fleischfarbe der Knolle. Potato Research. 1983;**26**: 149-162

[82] Cazzonelli CI, Pogson BJ. Source to sink: Regulation of carotenoid biosynthesis in plants. Trends Plant Science. 2010;**15**:266-274. DOI: 10.1016/j.tplants.2010.02.003

[83] Diretto G, Al-Babili S, Tavazza R, Papacchioli V, Beyer P, Giuliano G. Metabolic engineering of potato carotenoid content through tuber-specific overexpression of a bacterial mini-pathway. PLoS One. 2007a;**2**(4):e350. DOI: 10.1371/journal.pone.0000350

[84] Ducreux LJ, Morris WL, Hedley PE, Shepherd T, Davies HV, Millam S, et al. Metabolic engineering of high carotenoid potato tubers containing enhanced levels of beta-carotene and lutein. Journal of Experimental Botany. 2005;**56**(409):81-89. DOI: 10.1093/jxb/eri016

[85] Diretto G, Tavazza R, Welsch R, Pizzichini D, Mourgues F, Papacchioli V, et al. Metabolic engineering of potato tuber carotenoids through tuber-specific silencing of lycopene epsilon cyclase. BMC Plant Biology. 2006;**6**:13. DOI: 10.1186/1471-2229-6-13

[86] Diretto G, Welsch R, Tavazza R, Mourgues F, Pizzichini D, Beyer P, et al. Silencing of beta-carotene hydroxylase increases total carotenoid and beta-carotene levels in potato tubers. BMC Plant Biology. 2007;**7**:11. DOI: 10.1186/1471-2229-7-11

[87] Campbell R, Ducreux LJ, Morris WL, Morris JA, Suttle JC, Ramsay G, et al. The metabolic and developmental roles of carotenoid cleavage dioxygenase4 from potato. Plant Physiology. 2010;**154**:656-664. DOI: 10.1104/pp.110.158733

[88] Gerjets T, Sandmann G. Ketocarotenoid formation in transgenic potato. Journal of Experimental Botany. 2006;**57**:3639-3645. DOI: 10.1093/jxb/erl103

[89] Lu S, Van Eck J, Zhou X, Lopez AB, O'Halloran DM, Cosman KM, et al. The cauliflower or gene encodes a DnaJ cysteine-rich domain-containing protein that mediates high levels of beta-carotene accumulation. The Plant Cell. 2006;**18**(12):3594-3605. DOI: 10.1105/tpc.106.046417

[90] Lopez AB, Van Eck J, Conlin BJ, Paolillo DJ, O'Neill J, Li L. Effect of the cauliflower or transgene on carotenoid accumulation and chromoplast formation in transgenic potato tubers. Journal of Experimental Botany. 2008;**59**:213-223. DOI: 10.1093/jxb/erm299

[91] Goo YM, Han EH, Jeong JC, Kwak SS, Yu J, Kim YH, et al. Overexpression of the sweet potato IbOr gene results in the increased accumulation of carotenoid and confers tolerance to environmental

stresses in transgenic potato. Comptes Rendus Biologies. 2015;**338**(1):12-20. DOI: 10.1016/j.crvi.2014.10.006

[92] Al-Babili S, Hugueney P, Schledz M, Welsch R, Frohnmeyer H, Laule O, et al. Identification of a novel gene coding for neoxanthin synthase from Solanum tuberosum. FEBS Letters. 2000;**485**(2-3):168-172. DOI: 10.1016/s0014-5793(00)02193-1

[93] Morris WL, Ducreux LJ, Fraser PD, Millam S, Taylor MA. Engineering ketocarotenoid biosynthesis in potato tubers. Metabolic Engineering. 2006;**8**:253-263. DOI: 10.1016/j.ymben.2006.01.001

[94] Römer S, Lubeck J, Kauder F, Steiger S, Adomat C, Sandmann G. Genetic engineering of a zeaxanthin-rich potato by antisense inactivation and co-suppression of carotenoid epoxidation. Metabolic Engineering. 2002;**4**:263-262. DOI: 10.1006/mben.2002.0234

[95] Fiore A, Dall'osto L, Fraser PD, Bassi R, Giuliano G. Elucidation of the beta-carotene hydroxylation pathway in *Arabidopsis thaliana*. FEBS Letters. 2006;**580**:4718-4722. DOI: 10.1016/j.febslet.2006.07.055

[96] van den Berg H, Faulks R, Fernando Granado H, Hirschberg J, Olmedilla B, Sandmann G, et al. The potential for the improvement of carotenoid levels in foods and the likely systemic effects. Journal of the Science of Food and Agriculture. 2000;**80**:880-912. DOI: 10.1002/(SICI)1097-0010(20000515)80:7%3C880::AID-JSFA646%3E3.0.CO;2-1

[97] Tanaka T, Makita H, Ohnishi M, Mori H, Satoh K, Hara A. Chemoprevention of rat oral carcinogenesis by naturally occurring xanthophylls, astaxanthin and canthaxanthin. Cancer Research. 1995;**55**:4059-4064

[98] Chew BP, Park JS, Wong MW, Wong TS. A comparison of the anticancer activities of dietary beta-carotene, canthaxanthin and astaxanthin in mice in vivo. Anticancer Research. 1999;**19**:1849-1853

[99] Sandmann G. Carotenoid biosynthesis and biotechnological application. Archives of Biochemistry and Biophysics. 2001;**385**:4-12. DOI: 10.1006/abbi.2000.2170

[100] Johnson EA, An GH. Astaxanthin from microbial sources. Critical Reviews in Biotechnology. 1991;**11**:297-326. DOI: 10.3109/07388559109040622

[101] Huang JC, Zhong YJ, Liu J, Sandmann G, Chen F. Metabolic engineering of tomato for high-yield production of astaxanthin. Metabolic Engineering. 2013;**17**:59-67. DOI: 10.1016/j.ymben.2013.02.005

[102] Hasunuma T, Miyazawa S, Yoshimura S, Shinzaki Y, Tomizawa K, Shindo K, et al. Biosynthesis of astaxanthin in tobacco leaves by transplastomic engineering. The Plant Journal. 2008;**55**:857-868. DOI: 10.1111/j.1365-313X.2008.03559.x

[103] Nishida Y, Adachi K, Kasai H, Shizuri Y, Shindo K, Sawabe A, et al. Elucidation of a carotenoid biosynthesis gene cluster encoding a novel enzyme, 2,2′-beta-hydroxylase, from *Brevundimonas* sp. strain SD212 and combinatorial biosynthesis of new or rare xanthophylls. Applied and Environmental Microbiology. 2005;**71**(8):4286-4296. DOI: 10.1128/AEM.71.8.4286-4296.200

[104] Li L, Yang Y, Xu Q, Owsiany K, Welsch R, Chitchumroonchokchai C, et al. The or gene enhances carotenoid accumulation and stability during post-harvest storage of potato tubers. Molecular Plant. 2012;**5**:339-352. DOI: 10.1093/mp/ssr099

[105] Lukaszewicz M, Matysiak-Kata I, Skala J, Fecka I, Cisowski W, Szopa J. Antioxidant capacity manipulation in transgenic potato tuber by changes in phenolic compounds content. Journal of Agricultural Food Chemistry. 2004;**52**:1526-1533. DOI: 10.1021/jf034482k

[106] Stintzing FC, Carle R. Functional properties of anthocyanins and betalains in plants, food, and human nutrition. Trends in Food Science & Technology. 2004;**15**:19-38. DOI: 10.1016/j.tifs.2003.07.004

[107] Wegener CB, Jansen G, Jürgens HU, Schütze W. Special quality traits of coloured potato breeding clones: Anthocyanins, soluble phenols and antioxidant capacity. Journal of the Science of Food and Agriculture. 2008;**89**:206-215. DOI: 10.1002/jsfa.3426

[108] Zhao C, Guo H, Dong Z, Zhao Q. Pharmacological and nutritional activities of potato anthocyanins. African Journal of Pharmacy and Pharmacology. 2009;**3**:463-468

[109] Chen-Yi H, Murray JR, Ohmann SM, Tong CBS. Anthocyanin accumulation during potato tuber development. Journal of the American Society for Horticultural Science. 1997;**122**:20-23. DOI: 10.21273/JASHS.122.1.20

[110] Brown CR, Wrolstadt R, Durst R, Yang CP, Clevidence B. Breeding studies in poting high concentrations of anthocyanins. American Journal of Potato Research. 2003;**80**:241-249. DOI: 10.1007/BF02855360

[111] Giusti MM, Polit MF, Ayvaz H, Tay D, Manrique I. Characterization and quantitation of anthocyanins and other phenolics in native Andean potatoes. Journal of Agricultural Food Chemistry. 2014;**62**:4408-4416. DOI: 10.1021/jf500655n

[112] Tengkun N, Dongdong W, Xiaohui M, Yue C, Qin C. Analysis of key genes involved in potato anthocyanin biosynthesis based on genomics and transcriptomics data. Frontiers in Plant Science. 2019;**10**:603. DOI: 10.3389/fpls.2019.00603

[113] Stobiecki M, Matysiak KI, Franski R, Skala J, Szopa J. Monitoring changes in anthocyanin and steroid alkaloid glycoside content in lines of transgenic potato plants using liquid chromatography/mass spectrometry. Phytochemistry. 2003;**62**:959-969. DOI: 10.1016/s0031-9422(02)00720-3

[114] Jung C, Griffiths H, De Jong D, Cheng S, Bodis M, De Jong W. The potato P locus codes for flavonoid 30,50-hydroxylase. Theoretical and Applied Genetics. 2005;**110**:269-275. DOI: 10.1007/s00122-004-1829-z

[115] Mattoo AK, Dwivedi SL, Dutt S, Singh B, Garg M, Ortiz R. Anthocyanin-rich vegetables for human consumption-focus on potato, sweetpotato and tomato. International Journal of Molecular Sciences. 2022;**23**(5):2634. DOI: 10.3390/ijms23052634

[116] Zhang Y, Cheng S, De Jong D, Griffiths H, Halitschke R, De Jong W. The potato R locus codes for dihydroflavonol 4-reductase. Theoretical and Applied Genetics. 2009;**119**:931-993

[117] Yoshihara N, Imayama T, Fukuchi-Mizutani M, Okuhara H, Tanaka Y, Ino I, et al. cDNA cloning and characterization of UDP-glucose: Anthocyanidin 3-O-glucosyltransferase in Iris hollandica. Plant Science. 2005;**169**:496-501. DOI: 10.1016/j.plantsci.2005.04.007

[118] Liu Y, Wang L, Zhang J, Yu B, Wang J, Wang D. The MYB transcription factor *StMYBA1* from potato requires light to activate anthocyanin biosynthesis in transgenic tobacco. Journal of Plant Biotechnology. 2017;**60**:93-101. DOI: 10.1007/s12374-016-0199-9

[119] Kranz HD, Denekamp M, Greco R, Jin H, Leyva A, Meissner RC, et al. Towards functional characterisation of the members of the R2R3-MYB gene family from *Arabidopsis thaliana*. The Plant Journal. 1998;**16**:263-276. DOI: 10.1046/j.1365-313x.1998.00278.x

[120] Liu Y, Tikunov Y, Schouten RE, Marcelis LFM, Visser RGF, Bovy A. Anthocyanin biosynthesis and degradation mechanisms in solanaceous vegetables: A review. Frontiers in Chemistry. 2018;**6**:52. DOI: 10.3389/fchem.2018.00052

[121] Stushnoff C, Ducreux LJ, Hancock RD, Hedley PE, Holm DG, McDougall GJ, et al. Flavonoid profiling and transcriptome analysis reveals new gene-metabolite correlations in tubers of *Solanum tuberosum* L. Journal of Experimental Botany. 2010;**61**(4):1225-1238. DOI: 10.1093/jxb/erp394

[122] Yuhui L, Lin-Wang K, Espley RV, Wang L, Li Y, Liu Z, et al. StMYB44 negatively regulates anthocyanin biosynthesis at high temperatures in tuber flesh of potato. Journal of Experimental Botany. 2019;**70**:3809-3824. DOI: 10.1093/jxb/erz194

[123] Jung CS, Griffiths HM, De Jong DM, Cheng S, Bodis M, Kim TS, et al. The potato developer (D) locus encodes an R2R3 MYB transcription factor that regulates expression of multiple anthocyanin structural genes in tuber skin. Theoretical and Applied Genetics. 2009;**120**:45-57

[124] Liu Y, Lin-Wang K, Espley RV, Wang L, Yang H, Yu B, et al. Functional diversification of the potato R2R3 MYB anthocyanin activators AN1, MYBA1, and MYB113 and their interaction with basic helix-loop-helix cofactors. Journal of Experimental Botany. 2016;**67**:2159-2176. DOI: 10.1093/jxb/erw014

[125] Bradshaw JE, Ramsay G. Potato origin and production. In: Singh J, Kaur L, editors. Advances in Potato Chemistry and Technology. Academic Press: San Diego; 2009. pp. 1-26

[126] Erk T, Williamson G, Renouf M, Marmet C, Steiling H, Dionisi F, et al. Dose-dependent absorption of chlorogenic acids in the small intestine assessed by coffee consumption in ileostomists. Molecular Nutrition and Food Research. 2012;**10**:1488-1500. DOI: 10.1002/mnfr.201200222

[127] Furrer AN, Chegeni M, Ferruzzi MG. Impact of potato processing on nutrients, phytochemicals, and human health. Critical Reviews in Food Science and Nutrition. 2018;**58**:146-168. DOI: 10.1080/10408398.2016.1139542

[128] Lister CE, and Munro J. Nutrition and Health Qualities of Potatoes—A Future Focus. A report prepared for New Zealand Federation of Vegetable and Potato Growers. 2000. pp. 1-47. Available from: https://studylib.net/doc/8701427/nutrition-and-health-qualities-of-potatoes---a [Accessed: May 10, 2022]

[129] Hoover R. Composition, molecular structure, and physicochemical properties of tuber and root starches: A review. Carbohydrate Polymers. 2001;**45**:253-267. DOI: 10.1016/S0144-8617(00)00260-5

[130] Schwall GP, Safford R, Westcott RJ, Jeffcoat R, Tayal A, Shi Y-C, et al. Production of very-high-amylose potato starch by inhibition of SBE A and B. Nature Biotechnology. 2000;**18**:551-554. DOI: 10.1038/75427

[131] Bánfalvi Z, Molnár A, Lakatos L, Hesse H, Höefgen R. Differences in sucroseto-starch metabolism of *Solanum tuberosum* and *Solanum brevidens*. Plant Science. 1999;**147**:81-88

[132] Shallenberger RS, Smith O, Treadway RH. Food color changes, role of the sugars in the browning reaction in potato chips. Journal of Agricultural Food Chemistry. 1959;**7**:274-277. DOI: 10.1021/jf60098a010

[133] Burton WG. The distribution and composition of the dry matter in the potato tuber. The Potato. 1989;**3**:286-335

[134] Chen X, Salamini F, Gebhardt C. A potato molecular function map for carbohydrate metabolism and transport. Theoretical and Applied Genetics. 2001;**102**:284-295. DOI: 10.1007/s001220051645

[135] Menéndez CM, Ritter E, Schäfer-Pregl R, Walkemeier B, Kalde A, Salamini F, et al. Cold sweetening in diploid potato: Mapping quantitative trait loci and candidate genes. Genetics. 2002;**162**(3):1423-1434. DOI: 10.1093/genetics/162.3.1423

[136] Greiner S, Rausch T, Sonnewald U, Herbers K. Ectopic expression of a tobacco invertase inhibitor homolog prevents cold-induced sweetening of potato tubers. Nature Biotechnology. 1999;**17**:708-711. DOI: 10.1038/10924

[137] Cheng S-H, Liu J, Xie C-H, Song B-T, Li C-J. Role of tobacco vacuolar invertase regulated by patatin promoter in resistance of potato tubers to cold-sweetening. Chinese Journal of Agricultural Biotechnology. 2007;**4**:81-86. DOI: 10.1017/S1479236207001313

[138] Liu X, Cheng S, Liu J, Ou Y, Song B, Zhang C, et al. The potato protease inhibitor gene, St-Inh, plays roles in the cold-induced sweetening of potato tubers by modulating invertase activity. Postharvest Biology and Technology. 2013;**86**:265-271. DOI: 10.1016/j.postharvbio.2013.07.001

[139] Sowokinos JR. Biochemical and molecular control of cold-induced sweetening in potatoes. American Journal of Potato Research. 2001;**78**:221-236

[140] Visser RGF. Regeneration and transformation of potato by *Agrobacterium tumefaciens*. In: Lindsey K, editor. Plant Tissue Culture Manual. Dordrecht, B5: Springer; 1991. pp. 1-9. DOI: 10.1007/978-94-009-0103-2_16

[141] Kuipers AGJ, Soppe WJJ, Jacobsen E, Visser RGF. Field evaluation of transgenic potato plants expressing an antisense granule-bound starch synthase gene: Increase of the antisense effect during tuber growth. Plant Molecular Biology. 1994;**26**:1759-1773. DOI: 10.1007/BF00019490

[142] Andersson M, Turesson H, Nicolia A, Fält AS, Samuelsson M, Hofvander P. Efficient targeted multiallelic mutagenesis in tetraploid potato (*Solanum tuberosum*) by transient CRISPR-Cas9 expression in protoplasts. Plant Cell Reports. 2017;**36**:117-128. DOI: 10.1007/s00299-016-2062-3

[143] Müller-Röber B, Sonnewald U, Willmitzer L. Inhibition of the ADP-glucose pyrophosphorylase in transgenic potatoes leads to sugar-storing tubers and influences tuber formation and expression of tuber storage protein genes. The EMBO Journal. 1992;**11**(4):1229-1238

[144] Claudia J, Uwe S, Mohmmad R, Ulf-Ingo FL. Simultaneous boosting of source and sink capacities doubles tuber starch yield of potato plants. Plant Biotechnology Journal. 2012;**10**:1088-1098. DOI: 10.1111/j.1467-7652.2012.00736.x

[145] Barrera-Gavira JM, Pont SDA, Morris JA, Hedley PE, Stewart D, Taylor MA, et al. Senescent sweetening in potato (*Solanum tuberosum*) tubers is associated with a reduction in plastidial glucose-6-phosphate/phosphate translocator transcripts. Postharvest Biology and Technology. 2021;**181**:111637. DOI: 10.1016/j.postharvbio.2021.111637

[146] Tuncel A, Corbin KR, Ahn-Jarvis J, Harris S, Hawkins E, Smedley MA, et al. Cas9-mediated mutagenesis of potato starch-branching enzymes generates a range of tuber starch phenotypes. Plant Biotechnology Journal. 2019;**17**:2259-2271. DOI: 10.1111/pbi.13137

[147] Bhaskar PB, Wu L, Busse JS, Whitty BR, Hamernik AJ, Jansky SH, et al. Suppression of the vacuolar invertase gene prevents cold-induced sweetening in potato. Plant Physiology. 2010;**154**:939-948. DOI: 10.1104/pp.110.162545

[148] Clasen BM, Stoddard TJ, Luo S, Demorest ZL, Li J, Cedrone F, et al. Improving cold storage and processing traits in potato through targeted gene knockout. Plant Biotechnology Journal. 2016;**14**:169-176. DOI: 10.1111/pbi.12370

[149] Invalid references

[150] Datir SS, Regan S. Role of alkaline/neutral invertases in postharvest storage of potato. Post-Harvest Biology and Technology. 2022;**184**(9):111779. DOI: 10.1016/j.postharvbio.2021.111779

[151] McKenzie MJ, Chen RKY, Harris JC, Ashworth MJ, Brummell DA. Post-translational regulation of acid invertase activity by vacuolar invertase inhibitor affects resistance to coldinduced sweetening of potato tubers. Plant Cell & Environment. 2013;**36**:176-185. DOI: 10.1111/j.1365-3040.2012.02565.x

[152] Preiss J. Regulation of the biosynthesis and degradation of starch. Annual Review of Plant Physiology. 1982;**33**:431-454. DOI: 10.1146/annurev.pp.33.060182.002243

[153] Zeeman SC, Kossmann J, Smith AM. Starch: Its metabolism, evolution, and biotechnological modification in plants. Annual Review of Plant Biology. 2010;**61**:209-234. DOI: 10.1146/annurev-arplant-042809-112301

[154] Kötting O, Kossmann J, Zeeman SC, Lloyd JR. Regulation of starch metabolism: The age of enlightenment? Current Opinion in Plant Biology. 2010;**13**:321-329

[155] Jobling S. Improving starch for food and industrial applications. Current Opinion in Plant Biology. 2004;7:210-218. DOI: 10.1016/j.pbi.2003.12.001

[156] Lehmann U, Robin F. Slowly digestible starch: Its structure and health implications—A review. Trends Food Science and Technology. 2007;**18**:346-355. DOI: 10.1016/j.tifs.2007.02.009

[157] Zhao X, Andersson M, Andersson R. Resistant starch and other dietary fiber components in tubers from

a high-amylose potato. Food Chemistry. 2018;**251**:58-63. DOI: 10.1016/j.foodchem.2018.01.028

[158] Zhao X, Jayarathna S, Turesson H, Fält A-S, Nestor G, González MN, et al. Amylose starch with no detectable branching developed through DNA-free CRISPR-Cas9 mediated mutagenesis of two starch branching enzymes in potato. Scientific Reports. 2021;**11**:4311. DOI: 10.1038/s41598-021-83462-z

[159] Ballicora MA, Laughlin MJ, Fu Y, Okita TW, Barry GF, Preiss J. Adenosine 5′-diphosphate-glucose pyrophosphorylase from potato tuber. Significance of the N terminus of the small subunit for catalytic properties and heat stability. Plant Physiology. 1995;**109**:245-251. DOI: 10.1104/pp.109.1.245

[160] Okita TW, Nakata PA, Anderson JM, Sowokinos J, Morell M, Preiss J. The subunit structure of potato tuber ADPglucose pyrophosphorylase. Plant Physiology. 1990;**93**:785-790. DOI: 10.1104/pp.93.2.785

[161] Sweetlove LJ, Burrel MM, Ap Rees T. Characterization of transgenic potato (*Solanum tuberosum*) tubers with increased ADPglucose pyrophosphorylase. Biochemical Journal. 1996;**320**:487-492. DOI: 10.1042/bj3200487

[162] Preiss J. Biology and molecular biology of starch synthesis and its regulation. Oxford Surveys of Plant Molecular & Cell Biology. 1991;7: 59-114

[163] Zhang L, Rainer E, Häusler CG, Mohammad-Reza HI, Haferkamp H, Ekkehard N, et al. Overriding the co-limiting import of carbon and energy into tuber amyloplasts increases the starch content and yield of transgenic potato plants. Plant Biotechnology Journal. 2008;**6**:453-465. DOI: 10.1111/j.1467-7652.2008.00332.x

[164] Kusano H, Onodera H, Kihira M, Aoki H, Matsuzaki H, Shimada H. A simple gateway-assisted construction system of TALEN genes for plant genome editing. Scientific Reports. 2016;**6**:30234. DOI: 10.1038/srep30234

[165] Brummell DA, Watson LM, Zhou J, McKenzie MJ, Hallett IC, Simmons L, et al. Overexpression of *STARCH BRANCHING ENZYME II* increases short-chain branching of amylopectin and alters the physicochemical properties of starch from potato tuber. BMC Biotechnology. 2015;**15**:28. DOI: 10.1186/s12896-015-0143-y

[166] Dale MFB, Bradshaw JE. Progress in improving processing attributes in potato. Trends Plant Science. 2003;**8**:10-312. DOI: 10.1016/S1360-1385(03)00130-4

[167] Shepherd LVT, Bradshaw JE, Dale MFB, McNicol JW, Pont SDA, Mottram DS, et al. Variation in acrylamide producing potential in potato: Segregation of the trait in a breeding population. Food Chemistry. 2010;**123**:568-573. DOI: 10.1016/j.foodchem.2010.04.070

[168] Rausch T, Greiner S. Plant protein inhibitors of invertases. Biochimica et Biophysica Acta. 2004;**1696**:253-261. DOI: 10.1016/j.bbapap.2003.09.017

[169] Ye J, Shakya R, Shrestha P, Rommens CM. Tuber-specific silencing of the acid invertase gene substantially lowers the acrylamide forming potential of potato. Journal of Agricultural and Food Chemistry. 2010;**58**:12162-12167. DOI: 10.1021/jf1032262

[170] Zhang H, Liu X, Liu J, Ou Y, Lin Y, Li M, et al. A novel RING finger gene, SbRFP1, increases resistance to cold-induced sweetening of potato tubers. FEBS Letters. 2013;**587**:749-755. DOI: 10.1016/j.febslet.2013.01.066

[171] Llorente B, Vanina R, Guillermo D, Alonso HNT, Mirtha MF, Fernando FB. Improvement of aroma in transgenic potato because of impairing tuber browning. PLoS One. 2010;**5**(11):e14030. DOI: 10.1371/journal.pone.0014030

[172] Friedman M. Chemistry, biochemistry, and dietary role of potato polyphenols. A review. Journal of Agricultural and Food Chemistry. 1997;**45**:1523-1540. DOI: 10.1021/jf960900s

[173] Vaughn KC, Lax AR, Duke SO. Polyphenol oxidase—The chloroplast oxidase with no established function. Physiologia Plantarum. 1988;**72**:659-665. DOI: 10.1111/j.1399-3054.1988.tb09180.x

[174] McEvily AJ, Iyengar R, Otwell WS. Inhibition of enzymatic browning in foods and beverages. Critical Reviews in Food Science and Nutrition. 1992;**32**:253-273. DOI: 10.1080/10408399209527599

[175] Coetzer C, Corsini D, Love S, Pavek J, Tumer N. Control of enzymatic browning in potato (*Solanum tuberosum* L.) by sense and antisense RNA from tomato polyphenol oxidase. Journal of Agriculture and Food Chemistry. 2001;**49**:652-657. DOI: 10.1021/jf001217f

[176] Bachem CWB, Speckmann GJ, van der Linde PCG, Verhaggen FTM, Hunt MD, Zabeau M. Antisense expression of polyphenol-oxidase genes inhibits enzymatic browning of potato tubers. Biotechnology. 1994;**12**:1101-1127. DOI: 10.1038/nbt1194-1101

[177] González MN, Massa GA, Andersson M, Turesson H, Olsson N, Fält AS, et al. Reduced enzymatic browning in potato tubers by specific editing of a polyphenol oxidase gene *via* ribonucleoprotein complexes delivery of the CRISPR/Cas9 system. Frontiers in Plant Science. 2020;**10**:1649. DOI: 10.3389/fpls.2019.01649

[178] Maioli A, Gianoglio S, Moglia A, Acquadro A, Valentino D, Milani AM, et al. Simultaneous CRISPR/Cas9 editing of three PPO genes reduces fruit flesh browning in *Solanum melongena* L. Frontiers in Plant Science. 2020;**11**:607161. DOI: 10.3389/fpls.2020.607161

Section 3

# Production of Vegetables

Chapter 8

# Red Beetroot (*Beta Vulgaris* L.)

*Dóra Székely and Mónika Máté*


## Abstract

Beetroot has long been a known and consumed vegetable, it was cultivated by the ancient Egyptians, Greeks, and Romans. Beetroot is a type of vegetable belonging to the beet (*Beta*) genus, which also includes chard, sugar beet, and fodder beet. Beetroot is easy to grow, as it is not one of the vegetables with special needs. The characteristic color of beets is due to red pigments known as betacyanins. Extremely rich in valuable ingredients, it is an extremely good immune booster due to its vitamin A, B, and C content. It is rich in antioxidants and also contains pantothenic acid, lycopene, biotin, silicon, potassium, magnesium, sodium, calcium, zinc, copper, manganese, and iron. Thanks to its content, it even has many health-protective effects, thanks to which it is highly recommended to consume. Beetroot is a deliciously sweet, albeit slightly earthy, superfood. It can be consumed in many ways; raw, as vegetable juice, boiled or fried, fermented, dried, but also as a food supplement in powdered form, and it can also be used as a natural colorant to color different food products.


**Keywords:** cultivation, nutrient content, traditional and innovative processing

## 1. Introduction

Red beetroots are becoming more and more important vegetable nowadays due to their many positive nutritional and physiological properties. In addition to its significant potassium and magnesium content, it is associated with low sodium concentrations, which have a beneficial effect on the ionic balance of the human body. The betacyanins in it reduce oxidative stress and the harmful effects of free radicals, have antibacterial and antiviral properties, inhibit the proliferation of cancer cells, and are involved in the prevention of cardiovascular disease. The plant itself can occur almost anywhere, as it is easy to grow and requires no additional care other than hoeing or thinning [1].

Beetroot is a real superfood, which is proven by its wide range of health-protective effects in addition to the diversity of its active ingredients.

The antioxidants in beetroots contribute to the prevention of the formation of tumors and effectively treat existing tumor cells. According to numerous studies, it can be used mainly for colon, prostate, and breast cancer, but it also supports healing in the case of pancreatic cancer [2–4].

It has a detoxifying effect, which is due to the large amount of fiber and saponin in the gut, supports the body's natural detoxification processes, and cleans the intestines and blood. It improves digestion and the regular, thorough emptying of accumulated waste, which is a prerequisite for the proper absorption of nutrients. In this way, beets help prevent the development of deficiency diseases. It optimizes the functioning of the liver, it is great for various liver diseases, such as hepatitis or cirrhosis. Beetroot is excellent at reducing inflammation of the intestinal tract, relieving unpleasant symptoms, and speeding up the healing process [3].

Significant amounts of pickles and juice are produced from red beetroot raw material. In addition, dried and concentrated red beetroot juice is used in many foods to increase the intensity of the red color as a natural colorant. Baby food companies prefer to associate beetroots with other vegetables, fruits, or meat. In this case, the manufacturer must pay great attention to the quality of the raw material, as the nitrite and nitrate content of beetroots can be quite high, which is less manageable for the young, developing organism. Furthermore, dietary supplement tablets and syrups containing red beetroots are also known in the market. In addition to the ones listed so far, the consumption of beets as a dried product seems to be a promising solution, as drying is one of the oldest preservation methods that can increase the shelf life of foods without the addition of chemicals [1].

## 2. Taxonomic classification, origin

Beets (*Beta vulgaris* L. ssp. *Esculenta* convar. *Crassa* provar. *conditiva*) belong to the Conditiva group, including members of the *Caryophyllidae* subclass, the *Caryophillanae* main order, the *Chenopodiales* order, and member of *Chenopodiaceae* family [5]. The beetroot is related to sugar beet, fodder beet, and chard [6].

Beetroots have been known and cultivated since ancient times in both white and red versions. *Beta maritima* is the ancestor of all beet cultivars grown today, including beetroots, which can be originated around the Mediterranean [7]. This species of sea beet has been found since ancient times on the coasts of Europe and North Africa, the Middle East, and parts of Asia. Sea beet leaves have probably been collected since humans first began experimenting with edible green plant parts. It has been used by the peoples since 1000 BC, in the Roman Empire, its leaf was used as food, while its root was used as medicine [8]. However, the usefulness of the tubers was only discovered later. Sea beet was first domesticated in the eastern Mediterranean and the Middle East [9]. It later became popular mainly in India, where not only were its nutritionally important properties exploited, but it was also often used for healing purposes. It was known and consumed by the Greeks and Romans as Sicilian beets.

## 3. Botanical characterization

The beetroot is a biennial plant, in the first year it develops a carrot body and a rosette, then after the winter dormancy, the seed stalk and flower appear next year, followed by the seed. The bare, simple, glossy leaves can vary in color from dark green to dark red. The length of the petiole and the color content depend significantly on the variety grown. The thinner, dark purple petiole is observed in the smaller-leafed cultivars, while the petioles of the longer-leaved plants are thicker and orange color with purple stripes. The angular, branching soft stem develops in the second year on which a clumpy inflorescence is located [10]. The flowers with five petals are small [11], bivalve, but the stamens ripen earlier than the pistil, so they are foreign pollinators, so fertilization is done by the wind and insects, respectively. The seeds retain their

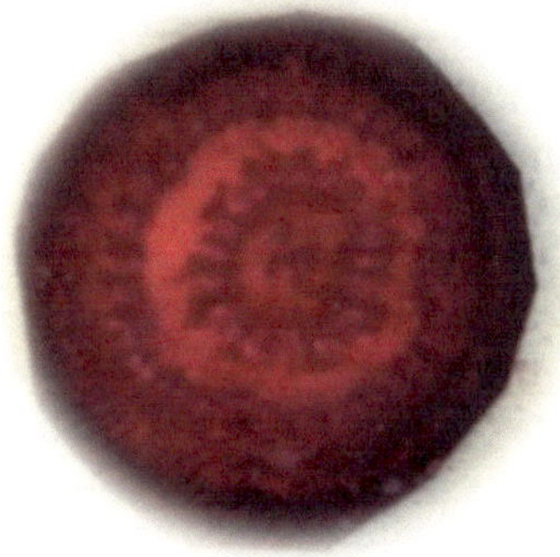

**Figure 1.**
*Ring system caused by secondary thickening.*

ability to germinate for 3–4 years, under favorable conditions for up to 8–10 years. The thousand-seed weight of beets is 13–22 g, depending on the variety [10].

It has a taproot penetrating the soil, on the side of which are the densely spaced, thin side roots 1–2 cm long. The most commonly consumed part is the beetroot body, which can be divided into cylindrical, round and flat groups according to its shape. Spherical cultivars are only attached to the soil by thin taproots, so they are preferred during harvest because they can be harvested with less soil contamination and without damage. By the end of the growing season, one-third of the carcass body is above ground, so picking can be done easily by machine and by hand. The carcass body is characterized by secondary thickening. In the cross-section of the roots of older plants, the tree and spleen elements form concentric circles with a cambium zone between them (**Figure 1**). The visibility of the rings is caused by the fact that the cells of the spleen have the highest content of red dye characteristic of beets [6].

## 4. Ecological demand

Beetroots cannot be grown on extreme soil types such as sandy, saline, and stony soils. The best quality and quantity of beetroots can be harvested from loamy, sandy loam, and humus-rich sandy soils. Frequent watering causes loosening of the loose, sandy soil, which strengthens the root of the beet body, degrades the quality of the harvested beetroots and makes them more difficult to clean. The optimal soil pH for the cultivation of the plant is between 6.5–7.5 pH [12]. After sowing the beetroot seeds, the harvest takes place 75–90 days in summer and 100–120 days in winter [13].

Beetroots can be classified as plants with medium water requirements. During germination and in the initial developmental stage of the plant, it requires a higher amount of continuous soil moisture, which is important for even germination (homogeneous stock) and initial development. Except for this period, beetroots are less sensitive to water shortages compared to other root vegetables.

In terms of light requirements, beets are a medium-demand vegetable. It develops well even in weaker, diffused light, but in this case, the beet body has less color and sugar content.

Beetroots have a medium heat demand, which fluctuates significantly during development. According to Markov–Haen's law, the optimum temperature for beetroots is 19° C. Germination starts at 5–6° C, but germination is fastest at 25–26° C [12]. Lower temperatures promote the formation of deep red pigments [13]. It is most sensitive to the cold at a young age but develops well during the growing season at much higher and lower than optimal temperatures. If exposed to low temperatures for a long time, it produces seed stalks in the first year and disturbs the development of the beet body. Due to its sensitivity to frost, it should be harvested before frost.

Like all other plants, beetroots have a certain need for nutrients. Knowing this, it is possible to determine which fertilizer and in what amount can be used to obtain the best quality crop. Beetroot is a medium-nutrient vegetable in terms of phosphorus and nitrogen, but it needs more potassium during its development. Per one ton of crop, the specific nutrient requirements of beetroots from the mentioned macronutrients are as follows: 2.4 kg of nitrogen, 1.4 kg of phosphorus, and 6 kg of potassium are needed to grow vegetables of the desired quality [12]. Particular attention should be paid to the addition of nitrogen, because in the event of an overdose, the quality of the beet body will deteriorate, resulting in poorer storage of nitrate and a reduction in color and dry matter content. In addition, the plant's resistance to disease is reduced by excess nitrogen. Of the trace elements, it is particularly sensitive to manganese deficiency [10].

Both organic and fertilizer can be used to meet the nutrient needs of plants. In the case of beetroots, the use of organic fertilizer is not recommended, as fresh stable manure results in an overdeveloped, deformed crop with an unpleasant taste [14].

It is important that the chemical, biological and physical contamination of the plant is as low as possible in order to validate its positive effects. Of these, the risk of chemical contamination is particularly noticeable in the case of beetroots, as the presence of root vegetables can accumulate a large amount of residues of fertilizers and fungicides (pesticides). That is particularly dangerous because the human body can convert the absorbed material into a substance (that is more dangerous than the original active ingredient instead of emptying it) or store it. Harmful substances accumulated or transformed in this way can cause allergic reactions, immune disorders, gene mutations, and possibly carcinogenic effects even years later [15].

As the most common chemical contaminants in the case of root vegetables, including beetroots, are the accumulation of nitrite and nitrate, efforts should be made to keep their content to a minimum. Inorganic nitrogen from fertilizers, most of which is not incorporated into tissues, can accumulate in plants in significant amounts in the form of nitrite and nitrate. The degree of accumulation can be influenced by number of factors, such as the lettuce has higher nitrate content during cultivation if less sunlight or molybdenum and iron are obtained, which are essential micronutrients [16].

## 5. Composition values

Its beneficial dietary and medicinal effects are due to the high content of red color content (betacyanins), high vitamins (C, B), minerals, and fiber. Examination of its composition has accelerated since the discovery of antitumor, which has been the subject of numerous studies worldwide [12]. The processing of beets and the consumption of products made from them has increased rapidly since it was recognized as an extremely rich source of antioxidants [17].

### 5.1 Nutrient content

Numerous studies report that the nutrient content of fresh beetroots is influenced by the variety, growing, and harvesting conditions alike [17]. **Table 1** shows that

| Energy components | Unit | RODLER [18] | NEELWARNE [19] | SOUCI et al., [20] | USDA [21] |
|---|---|---|---|---|---|
| Protein | g | 1.3 | 1.61 | 1.53 | 1.61 |
| Fat | g | 0.1 | 0.17 | 0.10 | 0.17 |
| Carbohydrate | g | 5.9 | 9.56 | 6.76 | 9.56 |
| Energy content | kJ<br>kcal | 130<br>31 | 45 | 175<br>41 | 180<br>43 |
| Ash content | g | 0.9 | — | — | 1.08 |
| Water content | g | 90.9 | — | 86.2 | 87.58 |
| Fiber | g | — | 2.8 | — | — |

**Table 1.**
*Energy components of beet per 100 g.*

beetroots are a good source of carbohydrates and protein. Because it contains little fat and no cholesterol at all, it results in a low-calorie intake. This is one of the reasons why it fits well into a weight loss diet. Its relatively high carbohydrate and sugar content does not affect this either, as the body immediately converts its easy-to-use sugar content into energy [9]. In terms of sugar content, it is a beneficial property for athletes to contain the highest amounts of sucrose, as it is beneficial for them to consume low concentrations of fructose and high sucrose, thereby increasing their physical capacity [22, 23]. A significant amount of sucrose content is also confirmed by a study conducted by WRUSS et al. [24], which examined the sugar content of seven popular beetroot varieties. The average sugar content of the investigated beetroot variety was 77.5 g/l, which contained 94.8% sucrose, 3.3% glucose, and less than 1.9% fructose.

Beetroots contain significant amounts of essential and non-essential amino acids (**Table 2**).

## 5.2 Macroelements

Beetroots contain large amounts of metallic macronutrients. These are potassium, sodium, magnesium, and calcium (**Table 3**). Magnesium is an activator of

| Amino acid | Quantity mg/100 g | Amino acid | Quantity mg/100 g |
|---|---|---|---|
| tryptophan | 0.019 | cystine | 0.019 |
| isoleucine | 0.048 | arginine | 0.042 |
| leucine | 0.068 | histidine | 0.021 |
| lysine | 0.058 | alanine | 0.060 |
| threonine | 0.047 | glutamic acid | 0.428 |
| methionine | 0.018 | glycine | 0.031 |
| phenylalanine | 0.046 | proline | 0.042 |
| tyrosine | 0.038 | aspartic acid | 0.116 |
| valine | 0.056 | serine | 0.059 |

**Table 2.**
*Amount of each amino acid present in beetroots [25].*

| Minerals | Unit | RODLER [18] | NEELWARNE [19] | YASHWANT, [13] | SOUCI et al., [20] | USDA [21] |
|---|---|---|---|---|---|---|
| Phosphorus | mg | 87 | — | 38 | 44 | 40 |
| Calcium | mg | 35 | 16 | 16 | 17 | 16 |
| Potassium | mg | 260 | 325 | 305 | 407 | 325 |
| Magnesium | mg | 87 | 23 | 23 | 20 | 23 |
| Sodium | mg | 98 | 78 | 77 | 58 | 78 |
| Zinc | mg | 0.337 | 0.075 | 0.35 | 0.357 | 0.35 |
| Coblat | mg | 0.009 | — | — | 0.0016 | — |
| Chromium | mg | 0.005 | — | — | 0,003 | — |
| Manganese | mg | 0.540 | 0.329 | — | 0.244 | 0.329 |
| Nickel | mg | 0.052 | — | — | 0.011 | — |
| Copper | mg | 0.087 | 0.35 | — | 0.082 | 0.075 |
| Selenium | mg | 0.001 | — | — | 0.0006 | 0.0007 |
| Iron | mg | 0.60 | 0.80 | 0.79 | 0.890 | 0,80 |

**Table 3.**
*Mineral content of beetroot per 100 g.*

many enzymes that catalyze carbohydrate metabolism and amino acid synthesis, an antagonist to potassium, and has a synergistic relationship with phosphorus [26]. Of the trace elements, it contains the largest amount of iron, which plays a key role in the uptake, transport, and storage of oxygen [27].

CSIKKELNÉ et al. [28] found that the microelement content of different parts within the beetroot plant is significantly different. Overall, the leaf usually has the highest mineral content, much less the peel and flesh of beetroot body. While the calcium content of the leaf is 156 mg/100 g, this value is 21 mg/100 g in the peel of beetroot body and 10 mg/100 g in the flesh of beetroot body. Concentrations of potassium, sodium, and magnesium are also much higher in beetroot leaves than in the body. In contrast, in terms of phosphorus content, the peel of beetroot body (66 mg/100 g) and the flesh of beetroot body part (49 mg/100 g) contains higher concentrations than the leaf (37 mg/100 g). Among the trace elements, iron, copper, manganese, and zinc occur in the largest quantities in the leaves.

Vitamins C, B1, B2, B6, and folic acid are significantly detectable in beets (**Table 4**).

### 5.3 Reducing compounds and other bioactive components

Consumption of beetroot juice is quite advantageous because it contains large amounts of antioxidants and other bioactive components [29]. Beetroots contain betalains [3], ascorbic acid [4], carotenoids [8, 30, 31], polyphenols, flavonoids, saponins [32, 33], and high levels of nitrate [34, 35] (**Figure 2**). Some bioactive components are present in small amounts, such as glycerin, betanin [36], and folic acid [37].

#### *5.3.1 Phenolic compounds*

Beetroots have a significant content of phenolic acid and flavonoids. In a study carried out by KATHIRAVAN and coworkers [38], 50-60 μmol/g DW phenolic acid

| Vitamins | Unit | RODLER [18] | NEELWARNE [19] | YASHWANT [13] | SOUCI et al., [20] | USDA [21] |
|---|---|---|---|---|---|---|
| Ascorbic acid € | mg | 13 | 4.9 | 3.6 | 10 | 4.9 |
| Thiamine (B1) | µg | 25 | 31 | 0.31 | 22 | 31 |
| Riboflavin (B2) | mg | 0.03535 | 0.057 | 0.27 | 42 | 40 |
| Niacin | mg | 0 | — | — | — | 0.334 |
| Pantothenic acid | mg | — | 0.155 | 0.145 | 0.130 | 0.155 |
| Pyridoxine (B6) | mg | 0.07 | 0.067 | 0.067 | — | 0.067 |
| Folic acid | µg | 73 | 109 | 80 | 83 | 109 |
| Retinol (A) | µg | 0 | 33 IU | 2 | 1.8 | 2 |
| Carotene | µg | 0 | 20 | — | 0.011 | 0.020 |
| Calciferol (D) | µg | 0 | — | — | — | — |
| Tocopher€(E) | mg | — | 0.04 | — | — | — |
| Biotin | µg | 5.0 | — | — | — | — |

**Table 4.**
*Vitamin content of beetroots per 100 g.*

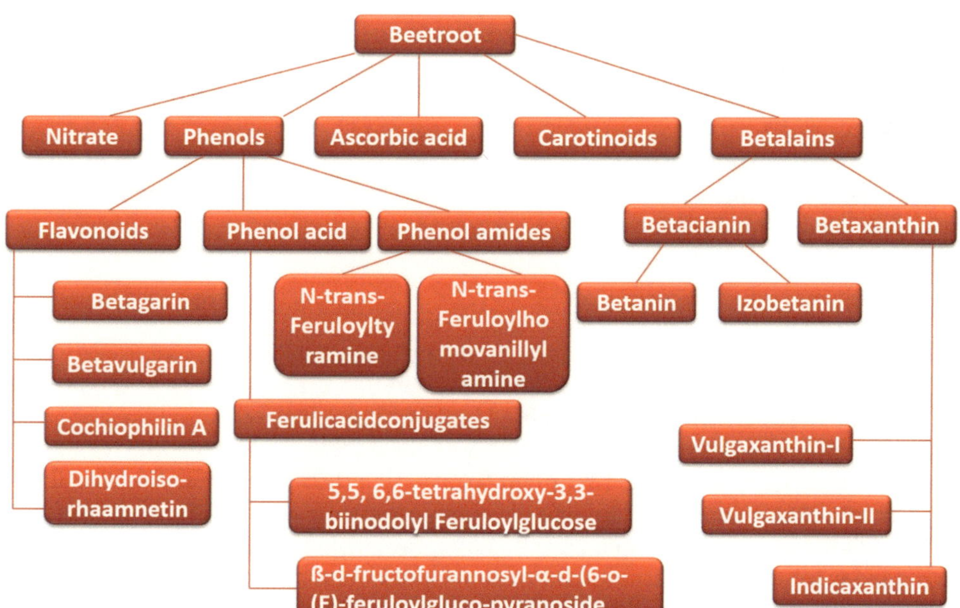


**Figure 2.**
*Bioactive components in beetroots [17].*

content was detected in the beetroot. NEMZER et al. [25] isolated highly unstable phenolic components from the beetroot peel, which were dimers of 5,5,6,6-tetra-hydroxy-3,3-biindolyl and 5,6-dihydroxyindole carboxylic acid. In addition, two phenolic amides, N-trans-feruloyltyramine and N-trans-feruloyl homovanillylamine were detected in the beet seed wall. MARAIE et al. [39] reported that beetroots contain significant amounts of hydroxybenzoic acid and hydroxyquinamic acid derivatives, including epicatechin, catechin hydrate, rutin, p-coumarin, caffeic acid, proline, and monoterpene dehydrovomololol. VASCONCELLOS et al. [40] compared the total phenol content of beetroot juice, beetroot chips, beetroot powder, and roasted beetroots. In their experiment, they found that beetroot juice (3.67 GAE mg/g) and roasted beets (2.79 GAE mg/g) had higher total polyphenol contents than beet chips (0.75 GAE mg/g) and beetroot powder, but the lowest value was detected in raw beetroot.

Flavonoids are among the biologically active components because they have excellent antioxidant potential and a number of positive health effects [17]. Flavonoids are characterized by a C6-C3-C6 backbone. The basic structure offers an extremely large number of variations, 4000 types of flavonoids with different structures are currently known. The degree of antioxidant properties of flavonoids depends fundamentally on the structure of the particular molecule, and the potency of the antioxidant is strongly and positively correlated with the degree of hydrolysis [41]. Flavonoids occur in higher plants and are secondary metabolites responsible for coloring the stems, leaves, flowers, and fruits of plants. Flavonoids can produce a variety of colors, such as yellow, orange, red, violet, and blue, but some are colorless [42]. Numerous studies have shown that beetroots contain significant amounts of flavonoids. It includes among others catechin, epicatechin, rutin, betagarin, and betavulgarin. VULIC et al. [43] reported the main group of flavonoids in beetroots.

Saponins are bioactive compounds that plants produce against pathogens and herbivores. According to previous studies, 11 triterpene saponins are found in beetroots. Each of the saponins contained oleanolic acid derivatives. Betovulgarosides I, II, III, IV, VI, VII, VIII. saponins were identified from the root, while betonulgarosides I, II, III, IV, V, IX, and X saponins were detected in leaves [33, 44]. MIKOLAJCZYK-BATOR et al. [45] also reported the presence of 26 triterpene saponins in beetroots, of which 17 triterpene saponins were not previously isolated in beets and 7 triterpene saponins were identified as new compounds.

### *5.3.2 Betalains*

Betalains are pigments of higher plants, nitrogenous and water-soluble compounds found in plants belonging to the *Caryophyllales* order [46]. Ten of the families in this order were identified as producing betalain. The *Chenopodiaceae* family, i.e. the *Beta vulgaris* family, is one of them. The name comes from the Latin name for beetroots (*B. vulgaris*), as it was the first plant from which they were identified. Betalaines are generally classified according to their characteristic structure (**Figure 3**). A total of 70 betalains are known, which have diazoheptamethine base frame. Betalains can be divided into two subgroups: betacyanin compounds, which give a reddish–violet color, and betaxanthines, which are responsible for yellowish colors [48]. The structural resonance of the parent compound of betalains, diazo-heptametin, results in its color. Accordingly, it can be distinguished the two major groups [49]. Betacyanins present in plants include betanin, isobetanine, protetanine, and neobetanin, and betaxanthines include vulgaxanthin, miraxantine, portulaxanthin, and indicaxanthin [50]. Plant physiology is uncertain about the role of betalains in plants, but KIMLER [51] reported the fungicidal properties of betalains.

Betalains were initially referred to as "nitrogen anthocyanins," but it was later found to be incorrect to assume structural similarity between colorant [52].

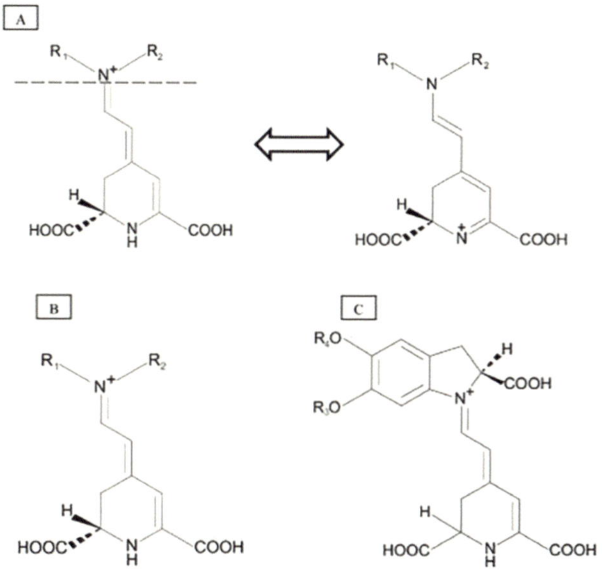


**Figure 3.**
*Resonance structure of betalain (a), general structure of betacyanin (B), and betaxanthin (C) [47].*

Both betalains and anthocyanins are water-soluble pigments found in the vacuoles of plant cells. However, betalains are structurally and chemically completely opposite to anthocyanins and no evidence has been found to occur in the same plant. Each betalain is composed of a glycoside sugar and a colored part (**Figure 3**). Their synthesis is facilitated by light [53]. Factors influencing the stability of betalains are shown in **Figure 4**.

Each colorant group is characterized by R1-N-R2 moieties. More than 50 betalains have been described with the same basic structure. Betalain has been used as a food

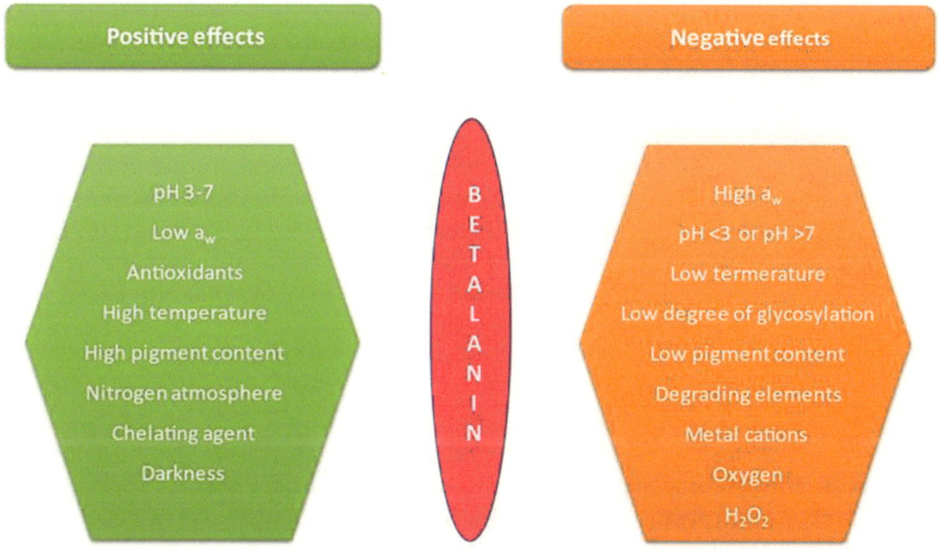


**Figure 4.**
*Effects affecting the stability of betalains [17].*

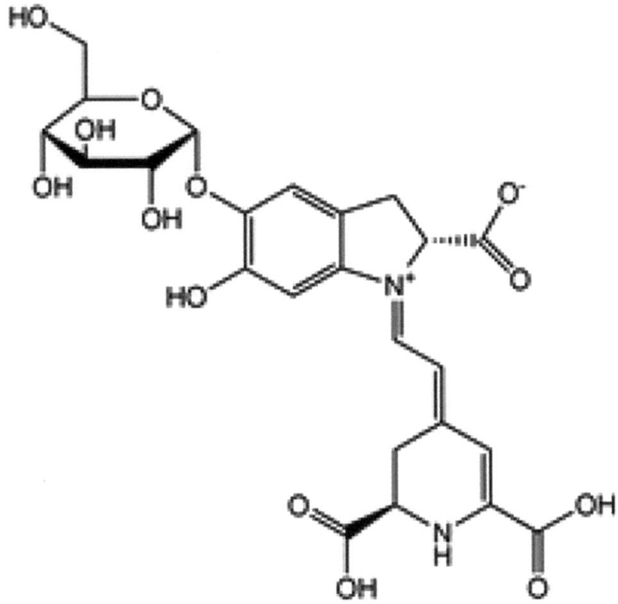


**Figure 5.**
*Structural diagram of betanin [17].*

coloring since the 20th century. Initially, a betalain-containing Phytolacca esculenta was used to imitate the color of red wine [48].

The most abundant betaline compound in beetroots is betanin (**Figure 5**), a secondary metabolite of betacyanins. The storage is located in the root, where it can reach concentrations of up to 0.5 g/kg [54]. In addition to betanin, its isomer is also found in the colorants of the plant. This compound is isobetanin, and the two substances together can account for up to 88–93% of colorants [55, 56]. For the four beet cultivars analyzed by HPLC by KUJALA et al. [57], the detected vulgaxanthin I and II were between 1.4 ± 0.3 mg / g and 4.3 ± 0.4 mg / g DW; betanin 7.6 ± 0.1 mg / g and 2.9 ± 0.2 mg /g DW, isobetanin 0.02 ± 0.01 mg /g and 3.1 ± 0.1 mg/g DW.

Beetroot colorants are available in the form of a concentrate as a natural colorant, which is typically prepared by vacuum evaporation or spray drying. The chemical degradation of colorant is affected by the duration and temperature of the heat treatment as well as the pH and water activity of the product [25]. By fermentation, approximately 75% of betacyanin content can be maintained by lowering the pH to about 4. With this acidic medium, the negative effect of the heat treatment process on reducing antioxidant capacity can be avoided [58].

### *5.3.3 Carotenoids*

Carotenoids are a group of phytochemicals that are responsible for the color of various fruits and vegetables. The carotenoids present in beetroots also function as antioxidants, anticarcinogens, and immune enhancers, in addition, their protective role and attribute mutagenesis inhibitory activity can reduce the risk of developing cancer [59]. Beetroot leaves contain β-carotene and xanthophylls such as lutein [17]. REBECCA et al. [31] detected 1.9 mg/100 g carotene in beetroots.

## 6. Effect of nutritional physiology

The results of pharmacological studies performed by several researchers confirm that beetroots can be used effectively and advantageously in the treatment of various diseases [17]. Betacyanins found in large quantities in beetroots counteract the harmful effects of oxidative stress and free radicals, have antibacterial and antiviral properties, inhibit the proliferation of cancer cells, and are involved in the prevention of cardiovascular disease [29, 60]. Because beetroots have many properties that have a positive effect on the human body, it has also become known as an herb. In addition, beetroots also have anti-inflammatory and hepatoprotective effects [61].

Its role in folk medicine is of great importance. A decoction made from the seeds of a vegetable has been used to treat intestinal and genital tumors. Beetroot juice has been thought to help fight tumors, leukemia, or other types of cancer. Among some of the components of the juice, betacyanins play an important role in inhibiting the metabolism of cancer cells. In addition, two very important components of amines are still present: choline and its oxidized form, betaine, in the absence of which experimental results have shown that tumors in the mice body have developed more likely [62].

VÁLI et al. [35] investigated the hepatoprotective properties of beetroot bioactive substances in a rat model of ischemia-reperfusion injury. As a result of feeding, global liver parameters and enzymatic antioxidants (glutathione peroxidase and superoxide dismutase) were significantly increased, indicating a positive effect of treatment. The results show that a diet rich in natural antioxidants has a positive effect on redox homeostasis during hepatic ischemia-reperfusion injury.

A colorant called betanin is known to have primarily lymphatic tumor inhibitor and antihypertensive effects. The proof of its antitumor effect is due to Sándor Ferenczi, who discovered in 1961 in animal experiments that the betanin colorant acts against cancer cells. He advised his patients to consume 1 liter of squeezed beetroot juice daily for 3 months, which not only works against cancer cells but also improves blood counts. In addition, next to Ferenczi, Rudolf Breuss the naturopath also recommended consuming 2.5–3 liters of beetroot, potato, celery, radish, and carrot juice. He discovered that this juice fasting cure helps to suppress tumor cells [63]. NYIRÁDY et al. [64] administered a commercially available natural beetroot preparation in a dose of 2 × 10 g to 24 patients with hormone-resistant and metastatic prostate cancer receiving taxane chemotherapy for 1 month to improve the quality of life. Their results showed that in the vast majority of patients, beetroots had a beneficial effect, and significantly high levels of Zn and free protoporphyrin in tumor patients were reduced, and transmethylation processes were accelerated.

Consumption of beetroots due to their high betanin content can cause beeturia (red urine) and red feces in people who are unable to degrade [65, 66]. The interest of the food industry towards betalaines has increased as they may provide protection against oxidation of low-density lipoproteins [67].

Like many other colorful vegetables, beetroots are considered an antioxidant gold mine [3, 68]. FIDELIS et al. [69] showed that beetroot juice (pH 5.45, 9°Brix) has higher total phenol (1169 mg GAE/l), flavonoid (925 mg catechin equivalent/l), and pigment content (854 mg/l) than citrus fruits, yellow passion fruit, apples and blueberries, which also results in a better antioxidant profile (325 mg AAE/l). WOOTTON-BEARD and RYAN [70] found that betanin and aglycone betanidine have extremely high antioxidant activity, which has been shown to be effective in preventing lipid peroxidation [38].

Beetroot contains a number of bioactive compounds, which result natural anti-anemic, anti-inflammatory, antihypertensive, anti-cancer, antipyretic, antibacterial, detoxifying and diuretic properties [34, 71], as well as stimulating the immune system, and liver protection [72]. SLAVOV et al. [61] demonstrated that betalain pigments play a role in the chemoprevention of lung and skin cancer and inhibit cell proliferation of various human tumor cells. IGLESIAS et al. [73] have also proved their anti-cancer effects and slightly reduce the inflammatory response and modulate the immune response.

Nitrates present in beetroots are able to lower blood pressure, protect against ischemic reperfusion injury and modulation of mitochondrial function [74], and reduce bad cholesterol [29]. NINFALI and ANGELINO [8] also report the antihypertensive effect and hypoglycemic activity of beetroot extracts. A study by MONTEIRO and AZEVEDO [75] found that regular consumption of beetroots reduces the risk of inflammation (instinctive reaction, including infection, erythema, edema, trauma, fever, and cell damage caused by pain).

Beetroots are healthy food for the entire digestive system. The water, in which the beetroots are cooked, can be used to treat skin infections, acne, and ulcers [76]. Beetroot juice helps clean the blood, regenerate and reactivate red blood cells, and provide the body with fresh oxygen [77]. The copper content of beetroots promotes the absorption of iron. Beetroot can also be used to treat fever and constipation [13].

## 7. Processing of beetroot and effects of processing on its biologically active components

The use of beetroots as food has been studied by many researchers and the food industry alike due to the specific effect of color, taste and nutrients. Beetroots are consumed worldwide, even in Eastern Europe, beetroot soup is popular, while in South America, pickled beets are a traditional dish [13].

Utilization and processing possibilities of beetroots:

- Fresh beetroots: immediate use, storage

- Semi-finished products: aseptic pulp, filtered concentrate (60–68 Bx$^{o}$), semi-concentrate (40–45 Bx$^{o}$)

- Heat-preserved products: 100% clarified filtered or fibrous juice, nectar, baby food, baby drinks, pickled beet products, beetroot jelly

- Fermented products: fermented juice, fermented lump preparations

- Dried preparations: conventional, vacuum-dried, microwave-vacuum-dried pieces (rings, cubes, cloves), spray-dried powder, flakes, instant powder, candied beetroots, lyophilized beetroots

- Non-thermal technologies: PEF, HHP, irradiation

- Quick-frozen products: cubes, slices, strips, puree, cream

- Other: colorant (in yoghurts, ice creams, cheeses, sauces, juices, jams), ingredient in food supplements

Cylindrical beetroot is made into sliced, while round beetroot is made into a risky product. As a baby beetroot, the round varieties are preserved. In several European countries, soups and salads are made from the young leaves of beetroots [6]. Due to its high antioxidant content, it is used in the manufacture of several preparations in addition to canned products. Beetroot juice is appearing as a product of more and more manufacturing companies, which is also available in a version made from organic vegetables [78].

In order to preserve the bioactive active ingredients of beetroot as efficiently as possible, producers shall endeavor to give priority to humane technologies, e.g. vacuum technologies, PEF or HHP treatment.

High hydrostatic pressure (HHP) technology allows microorganisms to be gently removed from beet juice, thus prolonging their shelf life by inactivating pathogenic microorganisms. Betaxanthins in beetroot juice were more stable than betacyanins under high pressure. Losses of betalain pigments under high hydrostatic pressure at ambient temperature were small compared to heat treatment. The significant reduction in the number of spoilage and pathogenic microorganisms without the recovery of sublethal-damaged cells and the slight degradation of pigments indicate the possibility of industrial application of high pressure to preserve beetroot juice [79].

One of the best ways to preserve fruits and vegetables is drying. In the case of vegetables, conventional convective drying was the most common; due to its simplicity, economical design and operation costs, however, this method of drying can greatly reduce the content values of vegetables [68].

In an experiment conducted by MALAKAR et al. [80], betalain pigment retention was 63.98% higher for drying with an evacuated tubular solar dryer (ETSD) than for day drying. The color changes were higher when dried in the sun. The mean phenol content and antioxidant capacities were 31.07% and 21.87% higher in ETSD, respectively than in the sun-dried. Therefore, ETSD can efficiently dry other foods with reduced drying time and significant preservation of quality characteristics.

According to a study by MELLA et al. [81], vacuum drying (VD) can be a suitable alternative to freeze-drying (FD). The effect of both drying techniques on the physicochemical properties, betalain pigment, antioxidant potential, and individual phenolic compounds of beetroots have been studied. The results showed that the increase in temperature promotes growth in the drying rate and effectively shortens the drying time. In general, VD samples retained the approximate composition of beets better than FD. Nevertheless, VD (50° C) excels in FD in terms of total polyphenol content (TPC) and oxygen radical absorption capacity (ORAC). In addition, syringic acid was identified in the VD samples at 50° C but not in the FD samples.

Kazimierczak et al. [29] found that beetroots and fermented beetroot juices from organic farming contained more vitamin C than conventional ones. The results reveal that organic and conventionally produced beetroots and fermented beetroot juice have different chemical properties and different effects on cancer cells. During the lactic fermentation of beetroot juices, 75% of betacyanins were retained compared to their original concentration. By adjusting the pH to about 4, this process also promotes antioxidant activity and avoids the negative effects of heat treatment, which reduces antioxidant content.

Beetroots are suitable as a substitute for synthetic colorants [58] and can thus become a marketing tool in the food industry [82]. This is because synthetic colorants can have negative effects on human health, cause allergies, and long-term consumption can be carcinogenic [83]. Dried and concentrated beetroot juice is also used in many foods to increase the intensity of the red color. Examples of such products are ice creams, jams, desserts, tomato concentrates, beverages, and dairy products [84]. Fresh beetroots, beetroot powder, or extracted pigments are used in soups, sauces, confectionery, ice creams, and breakfast cereals [68, 85]. The choice depends largely on the manufacturing technology, not only on the state of the excipient but also on its heat sensitivity. When exposed to heat, it changes color to brown [86], but it still occurs in heat-treated and then refrigerated beverages, desserts, ice creams, dairy products, and confectionery [87].

Processing methods have a significant effect on antioxidant activity and the availability of its phytochemicals. Some processing methods, such as microwave vacuum drying, fermentation, and irradiation, enhance antioxidant capacity and pigment stabilization, while convective drying reduces color retention [88, 89]. VASCONCELLOS et al. [40] examined the total antioxidant activity of beetroot chips (95.70%), beetroot powder (95.31%), boiled beetroots (85.79%), and beetroot juice (80.48%). According to their results, there was no significant difference between the antioxidant activity of beetroot chips and beetroot powder, and higher values were detected in them than in cooked beetroot and beetroot juice. The liquid chromatography method developed by PIETRZKOWSKI and THRESHER [90] was able to increase the betalain content of the beetroot powder by passing the beetroot juice through a silica gel column before drying. Thus, a concentration of up to 45% w/w betalain is achieved. NEMZER et al. [26] found that the higher betalain content of beetroot powder produced with this new technology is 615 mg/100 g of vitamin C, while the value of beetroot powder produced by the traditional method in an oven is only 1–7 mg/100 g.

Factors, that affect the stability of antioxidants or betalains, are storage, pH, temperature, water activity, oxygen, metals, and ion radiation [29, 91]. Optimal stability of betalains is achieved in the pH range of 3–7, suggesting that it is worth using in acidic food preparations. Thus, betalains are stable in foods with a pH 5, even in products below pH 3 the color of betanin changes to violet, and above pH 7 to blue color due to the longer wavelength [91]. Betanin is degraded in an alkaline environment, hydrolysis of aldimine to form ferulic acid with an amino group. The degradation of betanin at pH 3 is three times higher than at pH 5 under fluorescent light. Betalain was found to be more stable between pH 5.5 and 5.8 in the presence of oxygen. Under anaerobic conditions, betalain is more stable at pH 4–5 [92–94].

Water activity regulates the rate of biochemical conversion and influences the stability of betanin by regulating the water-dependent hydrolytic reaction of aldimine bond cleavage. A decrease in water activity (below 0.63) during various treatment procedures, such as drying and evaporation, enhances the stability of betalains [95]. An increase in water activity raises the rate of betalain degradation from 0.32 to 0.75. However, in the case of encapsulated beetroot pigment, the greatest degradation of betanin occurs at a water activity value of 0.64 [96].

Temperature also affects the stability of betalains. An increase in temperature results in degradation of betalain. However, thermal decomposition is also affected by temperature range, degree of heating, presence of oxygen, and pigment concentration [97].

Colorants are oxidized and degraded in the presence of light. There is an inverse relationship between light intensity in the range of 2200−−4400 lux and the stability of betalain. Absorption of ultraviolet and visible light excites the chromophore electrons of betalain, which induces higher reactivity or lower molecular activation energy. However, the effect of light under anaerobic conditions is negligible [92–94].

Some metal cations have been identified that promote or accelerate the degradation of betanin, such as iron, copper, tin, aluminum, and so on. According to a study, beetroot juice is less sensitive to metal ions because it contains metal complexing agents. Chelating agents (citric acid and EDTA) have been shown to stabilize betanin against metal-catalyzed degradation [92–94].

## 8. Conclusion

Numerous studies show that beetroots contain large amounts of valuable vitamins and minerals, as well as antioxidants, especially potassium, magnesium, iron, vitamins A, C, and B6. Not only the root but also the tender leaves can be eaten, for example, mixed into a salad. Beetroots greatly improve blood circulation, including in our brains, and the compounds in the beetroot are extremely helpful in maintaining the health of the nervous system. By consuming beetroots regularly, the risk of various types of dementia, such as Alzheimer's can be reduced, and keep the memory sharp. In addition to improving its blood circulation, it also contains substances that, when transformed in our body, also lower blood pressure.

## Acknowledgements

The authors acknowledge EFOP-3.6.3-VEKOP-16-2017-00005 for the support in this study.

## Author details

Dóra Székely* and Mónika Máté
Department of Fruit and Vegetable Processing Technology, Institute of Food Science and Technology, Hungarian University of Agriculture and Life Science, Budapest, Hungary

*Address all correspondence to: szekely.dora@uni-mate.hu

## References

[1] Ceclu L, Nistor O. Red beetroot: Composition and health effects - a review. J Nutri Med Diet Care. 2020;**6**:43. DOI: 10.23937/2572-3278.1510043

[2] Koubaier HBH, Snoussi A, Essaidi I, Chaabouni MM, Thonart P, Bouzouita N. Betalain and phenolic compositions, antioxidant activity of Tunisian red beet (Beta vulgharis L. conditiva) roots and sterm extracts. International Journal of Food Properties. 2014;**17**:1934-1945

[3] Guldiken B, Toydemir G, Nur Memis K, Okur S, Boyacioglu D, Capanoglu E. Home-processed red beetroot (*Beta vulgaris* L.) products: Changes in antioxidant properties and bioaccassibility. International Journal of Molecular Sciences. 2016;**17**:858. DOI: 10.3390/ijms17060858

[4] Clifford T, Howatson G, West DJ, Stevenson EJ. The potential benefits of red beetroot supplementation in health and disease. Nutrients. 2015;7(4):2801-2822. DOI: 10.3390/nu7042801

[5] Bálint I, Höhn M. Növényrendszertan. 3th ed. SZIE KTK; Budapest. 2001. pp. 48-56

[6] Kovács A. Gyökérzöldségek termesztése. Budapest: Mezőgazda Kiadó és Kereskedő Kft; 2011. pp. 87-91

[7] Takácsné HM. In: Kovács A, editor. Gyökérzöldségek termesztése. Budapest: Mezőgazdasági Kiadó; 2011. pp. 87-96

[8] Ninfali P, Angelino D. Nutritional and functional potential of Beta vulgaris cicla and rubra. Fitoterapia. 2013;**89**:88-199. DOI: 10.1016/j.fitote.2013.06.004

[9] Nottingham S. Cooking Beetroot. London: Times; 2004. Available from: http://www.stephennottingham.co.uk/beetroot.htm

[10] Hodossi S, Kovács A, Terbe I. Zöldségtermesztés szabadföldön. Mezőgazda Kiadó: Budapest; 2010. pp. 128-135

[11] Kezi J, Sumathy JH. Betalain - a boon to the food industry. Discovery. 2014;**20**:51-58

[12] Balázs S. Zöldségtermesztők kézikönyve. 2nd ed. Budapest: Mezőgazdasági kiadó; 2004. pp. 578-583

[13] Yashwant K. Beetroot: A super food. International Journal of Engineering Studies and Technical Approach. 2015;**1**(3):20-26

[14] Hájas M. Gyökérzöldségek Termesztése. Budapest: Mezőgazdasági Kiadó; 1976. pp. 157-167

[15] Gyaraky Z, Kovács J. Az élelmiszer-előállítás és forgalmazás gyakorlati kézikönyve. Budapest: Novorg; 1996. pp. 49-53

[16] Marscher AP. The compact jets of TeV blazars. Astroparticle Physics. 1999;**11**:19-25. DOI: 10.1016/S0927-6505(99)00019-5

[17] Chhikara N, Kushwaha K, Sharma P, Gat Y, Panghal A. Bioactive compounds of beetroot and utilization in food processing industry: A critical review. Food Chemistry. 2019;**272**:192-200. DOI: 10.1016/j.foodchem.2018.08.022

[18] Rodler I. Új tápanyagtáblázat. Budapest: Medicina Könyvkiadó Rt.; 2005. pp. 308-309

[19] Neelwarne B. Red Beet Biotechnology: Food and Pharmacentical

Applications. New York: Springer Science and Business; 2013. pp. 199-243

[20] Souci SW, Fachmann W, Kraut H. Food Composition and Nutrition Tables. 7th edidtion ed. South Africa: MedPharm Scientific Publishers; 2008. p. 788

[21] USDA 2021: Available from: https://fdc.nal.usda.gov/fdc-app.html#/food-details/169145/nutrients

[22] Bavec M, Turinek M, Grobelnik-Mlakar S, Slatnar A, Bavec F. Influence of industrial and alternative farming systems on contents of sugars, organic acids, total phenolic content, and the antioxidant activity of red beet (Beta vulgaris L. ssp. vulgaris rote kugel). Journal of Agricultural and Food Chemistry. 2010;**58**(22):11825-11831. DOI: 10.1021/jf103085p

[23] Murray R, Paul GL, Seifert JG, Eddy DE, Halaby GA. The effects of glucose, fructose, and sucrose ingestion during exercise. Medicine & Science in Sports & Exercise. 1989, 1989;**21**(3):275-282

[24] Wruss J, Waldenberger G, Huemer S, Uygun P, Lanzerstorfer P, Müller U, et al. Compositional characteristics of commercial beetroot products and beetroot juice prepared from seven beetroot varieties grown in Upper Austria. Journal of Food Composition and Analysis. 2015;**42**:46-55. DOI: 10.1016/j.jfca.2015.03.005

[25] Nemzer BZ, Pietrzkowski A, Spóma PS, Thresher T, Michalowski S, Wybranie C. Betalainic and nutritional profiles of pigment-enriched red beet root (Beta vulgaris L.) dried extracts. Food Chemistry. 2011;**127**(1):42-53. DOI: 10.1016/j.foodchem.2010.12.081

[26] Venkatesan S, Jayaganesh S. Characterisation of magnesium toxicity, its influence on amino acid synthesis pathway and biochemical parameters of tea. Research journal of. Phytochemistry. 2010;**4**(2):67-77. DOI: 10.3923/rjphytoo.2010.67.77

[27] Pais I. A mikroelemek jelentősége az életben. Budapest: Mezőgazda Kiadó; 2000. pp. 40-42

[28] Csikkelné SA, Takácsné HM, Kiss AS. Céklanövény különböző részeinek makro- és mikroelem-tartalma a fajták függvényében. 44th ed. Baja: Magyar Spektrokémiai Vándorgyűlés; 2001. pp. 25-27

[29] Kazimierczak R, Hallmann E, Lipowski J, Drela N, Kowalik A, Püssa T, et al. Beetroot (*Beta vulgaris* L.) and naturally fermented beetroot juices from organic and conventional production: Metabolomics, antioxidant levels and anticancer activity. Journal of the Science of Food and Agriculture. 2004;**94**(13):2618-2629. DOI: 10.1002/jsfa.6722

[30] Dias MG, Camoes MFGFC, Oliveira L. Carotenoids in traditional Portuguese fruits and vegetable. Food Chemistry. 2009;**113**(3):808-815. DOI: 10.1016/j.foodchem.2008.08.002

[31] Rebecca J, Sharmila S, Das MP, Seshiah C. Extraction and purification of carotenoids from vegetables. Journal of Chemical and Pharmaceutical Research. 2014;**6**(4):594-598

[32] Atamanova SA, Brezhneva TA, Slivkin AI, Nikolaevskii VA, Selemenev VF, Mironenko NV. Isolation of saponins from table beetroot and primary evaluation of their pharmacological activity. Pharmaceutical Chemistry Journal. 2005;**39**(12):650-652. DOI: 10.1007/s11094-006-0038-2

[33] Mroczek A, Kapusta I, Janda B, Janiszowska W. Triterpene saponin

content in the roots of red beet (*Beta vulgaris* L.) cultivars. Journal of Agricultural and Food Chemistry. 2012;**60**(50):12397-12402. DOI: 10.1021/jf303952x

[34] Lidder S, Webb AJ. Vascular effects of dietary nitrate (as found in green leafy vegetables and beetroot) via the nitrate-nitrite-nitric oxide pathway. British Journal of Clinical Pharmacology. 2013;**75**(3):677-696. DOI: 10.1111/j.1365-2125.2012.04420.x

[35] Váli L, Stefanovits-Bányai É, Szentmihályi K, Fébel H, Sárdi É, Lugasi A, et al. Liver-protecting effects of table beet (Beta vulgaris var. rubra) during ischemia-reperfusion. Nutrition. 2007;**23**(2):172-178. DOI: 10.1016/j.nut.2006.11.004

[36] Zwart FJ, Slow S, Payne RJ, Lever M, George PM, Gerrard JA, et al. Glycine betaine and glycine betaine analogues in common foods. Food Chemistry. 2003;**83**(2):197-204. DOI: 10.1016/S0308-8146(03)00063-3

[37] Jastrebova J, Witthoft C, Grahn A, Svensson U, Jagerstad M. HPLC determination of folates in raw and processed beetroots. Food Chemistry. 2003;**80**(4):579-588. DOI: 10.1016/S0308-8146(02)00506-X

[38] Kathiravan T, Nadanasabapathi S, Kumar R. Standardization of process condition in batch thermal pasteurization and its effect on antioxidant, pigment and microbial inactivation of ready to drink (RTD) beetroot (Beta vulgaris L.) juice. International Food Research Journal. 2014;**21**(4):1305

[39] Maraie NK, Abdul-Jalil TZ, Alhamdany AT, Janabi HA. Phytochemical study of the Iraqi Beta vulgaris leaves and its clinical application for the treatment of different dermatological diseases. World Journal of Pharmacy and Pharmaceutical Science. 2014;**3**:05-19

[40] Vasconcellos J, Conte-Junior C, Silva D, Pierucci AP, Paschoalin V, Alvares TS. Comparison of total antioxidant potential, and total phenolic, nitrate, sugar, and organic acid contents in beetroot juice, chips, powder, and cooked beetroot. Food Science and Biotechnology. 2016;**25**(1):79-84. DOI: 10.1007/s10068-016-0011-0

[41] Hegedűs A, Stefanovitsné BE. Természetes antioxidáns forrásunk: a gyümölcs. Debrecen: Debreceni Egyetem, AGTC, Kertészettudományi Intézet; 2012. p. 48

[42] Gasztonyi K. Az élelmiszerkémia alapja. Budapest: Mezőgazdasági Kiadó; 1979. pp. 21-26

[43] Vulic JJ, Cebovic TN, Canadanovic-Brunet JM, Cetkovic GS, Canadanovic VM, Djilas SM, et al. In vivo and in vitro antioxidant effects of beetroot pomace extracts. Journal of Functional Foods. 2014;**6**:168-175. DOI: 10.1016/j.jff.2013.10.003

[44] Lim TK. Beta vulgaris. In: Edible Medicinal and Non-Medicinal Plants. Switzerland: Springer; 2016. pp. 26-68

[45] Mikolajczyk-Bator K, Blaszczyk A, Czyżniejewski M, Kachlicki P. Identification of saponins from sugar beet (Beta vulgaris) by low and high-resolution HPLC–MS/MS. Journal of Chromatography B. 2016;**1029**:36-47. DOI: 10.1016/j.jchromb.2016.06.038

[46] Tarnavölgyi G. Az élelmiszer-színezékek technológiai és humánegészségügyi vonatkozásai. 2. rész. In: Élelmezési ipar. Nagykőrös: MÉTE Kiadó. 2009;93(5):133-136

[47] Francisco DV, Paredes-Lopez O. 8. Chapter. In: Natural Colorants for Food

and Nutraceutical Uses. Boca Raton: CRC Press; 2003. pp. 166-220

[48] Delgado-Vargas F, Jiménez AR, Paredes-López O. Natural pigments: Carotenoids, anthocyanins, and betalains - characteristic, biosynthesis, processing, and stability. Critical Reviews in Food Science and Nutrition. 2000;**40**(3):173-289. DOI: 10.1080/10408690091189257

[49] Csapó J, Csapóné KZ. Élelmiszerkémia. Budapest: Mezőgazda Kiadó; 2003. pp. 409-410

[50] Salisbury FB, Ross CW. Plant Physiology. 4th ed. Belmont: Wadsworth Publishing; 1991. pp. 325-326

[51] Kimler LM. Betanin, the Red Beet Pigment, as an Antifungal Agent. St. Louis, Missouri, United State: Botanical Society of America, Abstracts of papers; 1975. p. 36. DOI: 10.1080/02652030600986032

[52] Lawrence W, Price J, Robinson G, Robinson R. The distribution of anthocyanins in flowers, fruits and leaves. Philosophical Transactions of the Royal Society of London. Series B, Biological Sciences. 1939;**230**:149-178. DOI: 10.1098/rstb.1939.0006

[53] Stafford HA. Anthocyanins and betalains: Evolution of the mutually exclusive pathways. Plant Science. 1994;**101**(2):91-98. DOI: 10.1016/0168-9452(94)90244-5

[54] Strack D, Vogt T, Schliemann W. Recent advances in betalain research. Phytochemistry. 2003;**62**(3):247-269. DOI: 10.1016/S0031-9422(02)00564-2

[55] Gasztonyi MN, Daood H, Hájos MT, Biacs P. Comparison of red beet (Beta vulgaris var conditiva) varieties on the basis of their pigment components. Journal of the Science of Food and Agriculture. 2001;**81**(9):932-933

[56] Liu X, Gao Y, Xu H, Wang Q, Yang B. Impact of high-pressure carbon dioxide combined with thermal treatment on degradation of red beet (Beta vulgaris L.) pigments. Journal of Agricultural and Food Chemistry. 2008;**56**:6480-6487. DOI: 10.1021/jf800727q

[57] Kujala TS, Vienola MS, Klika KD, Loponen JM, Phihlaja K. Betalain and phenolic compositions of four beetroot (Beta vulgaris) cultivars. European Food Research and Technology. 2002, 2002;**214**(4):505-510. DOI: 10.1007/s00217-001-0478-6

[58] Kusznierewicz B, Smiechowska A, Bartoszek A, Namie'Snik, J. The effect of heating and fermenting on antioxidant properties of white cabbage. Food Chemistry. 2008;**108**(3):853-861. DOI: 10.1016/j.foodchem.2007.11.049

[59] Sardana RK, Chhikara N, Tanwar B, Panghal A. Dietary impact on esophageal cancer in humans: A review. Food & Function. 2018;**4**:1967-1977. DOI: 10.1039/c7fo01908d

[60] Kapadia GJ, Rao G. Anticancer Effects of Red Beet Pigments. New York: Springer Science+Business Media; 2013. pp. 125-154

[61] Slavov A, Karagyozov V, Denev P, Kratchanova M, Kratchanov C. Antioxidant activity of red beet juices obtained after microwave and thermal pretreatments. Czech Journal of Food Sciences. 2013;**31**(2):139-147. DOI: 10.17221/61/2012-CJFS

[62] Duke JA. 1983. The handbook of energy crops. Available from: http://www.hort.purdue.edu/newcrop/duke_energy/dukeindex.html

[63] Veres Zs. A zöldséglevek világa. In: Hajtatás, korai termesztés. Debrecen: Debreceni Egyetemi Kiadó; 2005;36(3):16-18

[64] Nyirádi P, Sárdi É, Bekő G, Szűcs M, Horváth A, Székely E, et al. A Beta vulgaris L. ssp. esculenta var. rubra bioaktiv vegyületeinek hatása metasztatikus prosztatarákban. Orvosi Hetilap. 2010;**151**(37):1495-1503

[65] Escribano J, Pedreño MA, García-Carmona F, Muñoz R. Characterization of the antiradical activity of betalains from Beta vulgaris L. roots. Phytochemical Analysis. 1998;**9**(3):124-127

[66] Frank T, Stintzing FC, Carle R, Bitsch I, Quaas D, Strass G, et al. Urinary pharmacokinetics of betalains following consumption of red beet juice in healthy humans. Pharmacological Research. 2005;**52**(4):290-297. DOI: 10.1016/j.phrs.2005.04.005

[67] Tesoriere L, Allegra M, Butera D, Livrea MA. Absorption, excretion, And distribution of dietary antioxidant betalains in LDLs: Potential health effects of betalains in humans. The American Journal of Clinical Nutrition. 2004;**80**(4):941-945. DOI:10.1093/ajcn/80.4.941

[68] Singh B, Hathan BS. Chemical composition, functional properties, and processing of beetroot-a review. International Journal of Scientific & Engineering Research. 2014;**5**(1):679-684

[69] Fidelis M, Santos JS, Coelho ALK, Rodionova OY, Pomerantsev A, Granato D. Authentication of juices from antioxidant and chemical perspectives: A feasibility quality control study using chemometrics. Food Control. 2017;**73**:796-805. DOI: 10.1016/j.foodcont.2016.09.043

[70] Wootton-Beard PC, Ryan L. A beetroot juice shot is a significant and convenient source of bioaccessible antioxidants. Journal of Functional Foods. 2011;**3**(4):329-334. DOI: 10.1016/j.jff.2011.05.007

[71] Hobbs DA, George TW, Lovegrove JA. The effects of dietary nitrate on blood pressure and endothelial function: A review of human intervention studies. Nutrition Research reviews. 2013;**26**:210-222. DOI: 10.1017/S0954422-413000188

[72] Miraj S. Chemistry and pharmacological effect of beta vulgaris: A systematic review. Der Pharmacia Lettre. 2016;**8**:404-409

[73] Iglesias R, Citores L, Di Maro A, Ferreras JM. Biological activities of the antiviral protein BE27 from sugar beet (Beta vulgaris L.). Planta. 2015;**241**:421-433. DOI: 10.1007/s00425-014-2191-2

[74] Satyanand V, Vali SM, Krishna BP. A study of beet root derived dietary nitrate efficacy on performance of runners. Indian Journal of Basic and Applied Medical Research. 2014;**3**(2):690-695

[75] Monteiro R, Azevedo I. Chronic inflammation in obesity and the metabolic syndrome. Mediators of Inflammation. 2010;**2010**:289645. DOI: 10.1155/2010/289645

[76] Gamal AAE, Alsaid MS, Raish M, Al-Sohaibani M, Al-Massarani SM, Ahmad A. Beetroot (*Beta vulgaris* L.) extract ameliorates gentamicin-induced nephrotoxicity associated oxidative stress, inflammation, and apoptosis in a rodent model. Mediators of Inflammation. 2014;**2014**:983952. DOI: 10.1155/2014/983952

[77] Coles LT, Clifton PM. Effect of beetroot juice on lowering blood pressure

in free-living, disease-free adults: A randomized, placebo-controlled trial. Journal of Nutrition. 2012;**11**:106. DOI: 10.1186/1475-2891-11-106

[78] Kähkönen MP, Hopia AI, Vuorela HJ, Rauha JP, Pihlaja K, Kujala TS, et al. Antioxidant activity of plant extracts containing phenolic compounds. Journal of Agricultural and Food Chemistry. 1999;**47**:3954-3962. DOI: 10.1021/jf990146l

[79] Sokołowska B, Nasiłowska J. Controlling spoilage and pathogenic microorganisms in beetroot (*Beta vulgaris*) juice by high hydrostatic pressure. In: Safety Issues in Beverage Production. Cambridge, Massachusetts: Academic Press; 2020. pp. 79-104

[80] Malakar S, Alam M, Arora VK. Evacuated tube solar and sun drying of beetroot slices: Comparative assessment of thermal performance, drying kinetics, and quality analysis. Solra Energy. 2022;**233**:246-258

[81] Mella C, Vega-Gálvez A, Uribea E, Pasten A, Mejias N, Quispe-Fuentesa I. Impact of vacuum drying on drying characteristics and functional properties of beetroot (*Beta vulgaris*). Applied Food Research. 2022;**2**(1):100-120

[82] Yadav DN, Sharma M, Chikara N, Anand T, Bansal S. Quality characteristics of vegetable-blended wheat–pearl millet composite pasta. Agricultural Research. 2014;**3**(3):263-270. DOI: 10.1007/s40003-014-0117-7

[83] Panghal A, Janghu S, Virkar K, Gat Y, Kumar V, Chhikara N. Potential non-dairy probiotic products - a healthy approach. Food Bioscience. 2018;**21**:80-89. DOI: 10.1016/j.fbio.2017.12.003

[84] Kaleta A, Górnicki K. Some remarks on evaluation of drying models of red beet particles. Energy Conversion and Management. 2010;**51**(12):2967-2978. DOI: 10.1016/j.enconman.2010.06.040

[85] Sruthi PD, Anoohya PN, Vasu AT, Latha BS, Chavali M. Portrayal of red pigments extracted from red beet (*Beta vulgaris*, L.) and its potential uses as antioxidant and natural food colourants. VFSTR Journal of STEM. 2014;**2**(1):24-32

[86] Agrawal A. Scope of betalains as a food colorant. International journal of advanced scientific and technical research. 2013;**3**(3):22-36

[87] Latorre ME, Narvaiz P, Rojas AM, Gerschenson LN. Effects of gamma irradiation on bio-chemical and physico-chemical parameters of fresh-cut red beet (Beta vulgaris L. var. conditiva) root. Journal of Food Engineering. 2010;**98**:178-191. DOI: 10.1016/j.jfoodeng.2009.12.024

[88] Chapman S. Guidelines on Approaches to the Replacement of Tartrazine, Allura Red, Ponceau 4R, Quinoline Yellow, Sunset Yellow and Carmoisine in Food and Beverages. Scotland: Food Standards Agency in Scotland; 2011. Available from: https://www.reading.ac.uk/foodlaw/pdf/uk-11026-removing-colours-guidance.pdf

[89] Gokhale SV, Lele SS. Dehydration of red beetroot (Beta vulgaris) by hot air drying: Process optimization and mathematical modeling. Food Sciences and Biotechnology. 2011;**20**:955-964. DOI: 10.1007/s10068-011-0132-4

[90] Pietrzkowski Z, Thresher W. Solid Betalain Compositions and Methods. 35 U.S.C: Patent Application Publication. 2008

[91] Wootton-Beard PC, Moran A, Ryan L. Stability of the antioxidant capacity and total polyphenol content of commercially available vegetable juices

before and after in vitro digestion as measured by FRAP, DPPH, ABTS and Folin Ciocalteu methods. Food Research International. 2011;**44**(1):217-224. DOI: 10.1016/j.foodres.2010.10.033

[92] Manchali S, Murthy KNC, Nagaraju S, Neelwarne B. Stability of betalain pigments of red beet in Red Beet Biotechnology. Boston: Springer. 2013. pp. 55-74. DOI: 10.1007/978-1-4614-3458-0_3

[93] Ravichandran K, Palaniraj R, Saw NMMT, Gabr AM, Ahmed AR, Knorr D, et al. Effects of different encapsulation agents and drying process on stability of betalains extract. Journal of Food Science and Technology. 2014;**51**:2216-2221. DOI: 10.1007/s13197-012-0728-6

[94] Paciulli M, Medina-Meza IG, Chiavaro E, Barbosa-Cánovas GV. Impact of thermal and high pressure processing on quality parameters of beetroot (*Beta vulgaris* L.). LWT-Food Science and Technology. 2016;**68**:98-104. DOI: 10.1016/j.lwt.2015.12.029

[95] Kearsley MW, Katsaboxakis KZ. Stability and use of natural colours in foods red beet powder, copper chlorophyll powder and cochineal. International Journal of Food Science & Technology. 1980;**15**:501-514. DOI: 10.1111/j.1365-2621.1980.tb00969.x

[96] Serris GS, Biliaderis CG. Degradation kinetics of beetroot pigment encapsulated in polymeric matrices. Journal of the Science of Food and Agriculture. 2001;**81**:691-700. DOI: 10.1002/jsfa.864

[97] Herbach KM, Stintzing FC, Carle R. Betalain stability and degradation structural and chromatic aspects. Journal of Food Science. 2006;**71**(4):R41-R50. DOI: 10.1111/j.1750-3841.2006.000222.x

Chapter 9

# Prospects of Cassava Development in Indonesia in Supporting Global Food Availability in Future

*Fachrur Rozi, Dian Adi Anggraeni Elisabeth, Ruly Krisdiana, Adri Adri, Yardha Yardha and Yanti Rina*

## Abstract

Climate change is a major factor endangering sustainable food production. Various efforts have been made to prevent potential food shortages in future. Meanwhile, access to adequate food is an important part of human rights. In Indonesia, the opportunity for cassava development is still widely open and has potential to provide for the world's needs in future. Cassava is well-known by farmers in Indonesia and can be easily cultivated in all areas of Indonesia, even though the soil fertility is low. The current problems are that cassava is still considered as an inferior commodity and is only used for direct consumption. Indonesia is able to meet the world's cassava needs only by utilizing 54% of the total available land, which is suitable for planting cassava. Cassava utilization is actually large and has potential as raw material for many strategic industries. The map of cassava development in Indonesia is in the phase of growth and product expansion (diversification). Thus, the efforts in preparing cassava in Indonesia to meet the world's needs in future, including (1) increasing cassava productivity, (2) improving cassava added value by-product diversification, and (3) enhancing cassava bio-economy by implementing a bio-industry system integrating cassava and livestock farming.

**Keywords:** cassava development, cassava-livestock integrated farming, climate change, global food, sustainable food production

## 1. Introduction

Climate change is a major factor endangering sustainable food production. Various efforts have been made to prevent potential food shortages in future. Cassava and sago are not considered as main staple foods, and their production is not significantly influenced by climate. Access to adequate food is an important part of human rights, in addition to the rights of being free from hunger, obtaining safe drinking water, and accessing to any resources, including fuel. Food sovereignty is the right to sufficient food, which means that every people both individually and collectively in their community, must have access to food at all times physically and economically [1–3].

One of the agricultural commodities, which is a focus for development in Indonesia is cassava. Cassava has a variety of highly prospective and sustainable derivative products, both food and non-food. In general, cassava is processed into tapioca. Cassava starch can be further processed into modified cassava flour (mocaf) as an alternative to wheat flour and hydrolyzed starch can be further processed into glucose syrup and its derivatives. Meanwhile, for non-food purposes, cassava is utilized as raw material in cosmetics, bioethanol, chemicals, and textile industries. The benefits of cassava are divided into local staple foods, agricultural industrial products, and industrial raw materials, so cassava has the potential to be developed [4–6].

In Indonesia, cassava is also an important food crop commodity after rice, corn, soybean, peanut, and mung bean. Cassava is utilized for food, feed, bioethanol, and industrial raw materials. In terms of food utilization, cassava is not only for meeting the need of carbohydrates as a rice substitute but it is also developed for food diversification. In addition, cassava has wide adaptability, easy to store, and has good taste, so that by diversifying cassava products it is expected that it can create new business opportunities and increase farmers' income [7–9].

Regarding food security improvement, the Government of Indonesia through the Ministry of Agriculture continuously makes efforts in reducing rice consumption by looking for food substitutes, such as cassava. In several areas in Indonesia, cassava has been used as a food ingredient, such as in form of blocks, chips, and traditional sun dried-slice cassava or *gaplek,* which have longer shelf-life. However, cassava is still considered as an inferior commodity, so it is not in demand by community. Therefore, the strategy for increasing the community interest in cassava consumption begins with processing cassava into various products that have added value and high-selling value [10].

The potential and opportunities for cassava development are still widely open in line with the development of processed food products, livestock industry, and other industries, such as alcohol, sorbitol, fructose, and many others, and also be supported by research and innovation. Currently, the plastic industries start using tubers, including cassava, as their raw material for biodegradable plastics that are more environmentally friendly [11, 12].

Cassava as raw material for food has not been able to compete with rice or wheat flour. It can be seen from food business actors who use rice or wheat flour as raw materials more than the local ones, such as tapioca, mocaf, arrowroot flour, and so on. Businesses with local raw materials are not nonexistent, but they are few in number, and their products are not widely known by the community. In addition, there are still few business actors who specifically process local food products. Most of the existing industries are still labor intensive and not supported yet by good infrastructures, so their productivity is still low. In fact, local food ingredients are usually consumed in form of their derivative products, such as flour, which is further processed into noodles, cakes, and so on. This problem causes the distribution of local food products is not as wide as imported food products. The improvement in efficiency and effectiveness of local food postharvest and processing, both into intermediate and end-products that have added value should be carried out to succeed in food diversification.

## 2. Potentials and performances of cassava in Indonesia

The performance of cassava production in Indonesia is continuously increasing since 2018 at 1.51%. The five provinces with the highest cassava production are

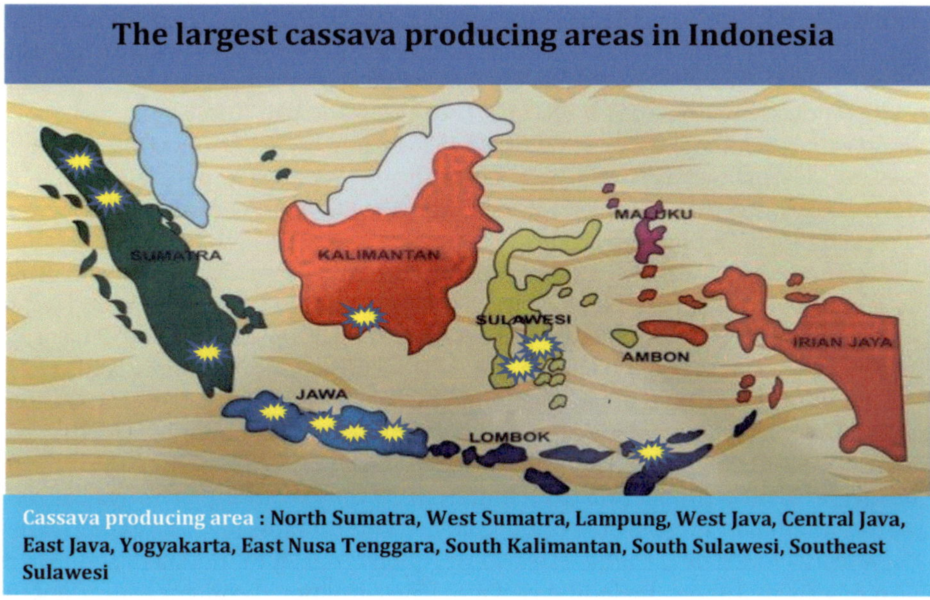


**Figure 1.**
*The largest cassava-producing areas in Indonesia (Source: [13]).*

Lampung, Central Java, East Java, West Java, and East Nusa Tenggara (**Figure 1**). The domestic cassava productivity fluctuates where in the last five years the average production is 23.99 tons/ha [13].

The realization and prediction of cassava production in Indonesia are described in **Table 1**. At the farming level, cassava productivity is very low, only around 20–26 tons, and still far below the yield of the research result, which is an average of 40 tons/ha. The low productivity of cassava farming causes by conventional cultivation, which is still dependent on existing agro-climatic conditions, and the low mastery of farmers on cassava technology, particularly for those who live in marginal areas (dry land and forest edges). As a result, the supply of cassava to various industries is not continuous. It can be overcome by applying the recommended technology for both improved varieties and cultivation, as well as cropping patterns setting [14].

Indonesia is the fourth largest cassava-producing country in the world with a total production of 19–20 million tons after Nigeria (57 million tons), Thailand (30 million tons),

| Year | Planting (ha) | Harvesting (ha) | Productivity (ku/ha) | Production (tons) | Production rate (%) |
|---|---|---|---|---|---|
| 2020 | 663,137 | 631,559 | 262.25 | 16,587,900 | — |
| 2021 | 669,488 | 637,608 | 262.89 | 16,762,073 | 1.05 |
| 2022 | 677,110 | 644,866 | 263.44 | 16,988,361 | 1.35 |
| 2023 | 686,146 | 653,473 | 264.26 | 17,268,669 | 1.65 |
| 2024 | 697,375 | 664,167 | 265.58 | 17,638,945 | 2.14 |

*Source: [13]*

**Table 1.**
*Realization and prediction of planting area, harvesting area, and cassava productivity in Indonesia.*

and Brazil (23 million tons). The area of cassava planting in 2019 was 628,305 ha with a production of 16.35 million tons spread across 13 provinces. Over the last five years, cassava productivity tends to increase *c.a.* 2.85%. The average growth rate increased by 2.64% per year, with productivity from 97.51 ku/ha in 2011 to 239.13 ku/ha in 2016. The average growth of cassava export volume and value in 2000–2015 increased by 96.21% and 118.22% per year, respectively. Indonesia exports cassava in fresh and processed forms (flour, dried-shredded cassava, and pellets), especially to Taiwan, the Philippines, Australia, Malaysia, England, and Brunei Darussalam [15, 16].

The opportunity for cassava development is very large, considering the land availability is relatively wide. Based on data from BPS in 2005, the potential for dry land in Indonesia was 25,955,901 ha consisting of 10,775,051 ha of upland, 3,839,093 ha of idle land or *ladang*, and 11,341,757 ha of temporarily uncultivated land. These lands have the potential for agricultural areas, including the development of cassava cultivation. The minimum world demand for cassava is 271.6 million tons (**Table 2**) [17–20]. With the large potential of land in Indonesia for cassava cultivation, the world's needs can be met by cassava production from Indonesia by only requiring a cassava harvest area of 13.9 million ha or 54% of the potential dry land suitable for cassava cultivation.

There are 77 kinds of carbohydrates-source food crops in Indonesia in addition to rice. Among them, tubers, including cassava, have nutritional content equivalent to rice or wheat. As an alternative to non-rice food, cassava can be served in daily menu, as long as it is enriched with high protein food [21, 22]. Besides being processed directly from fresh roots, cassava can also be processed into an intermediate product in the form of flour, which can be further processed into food products with a longer shelf-life and higher-selling value.

Cassava is a rice substitute, which has an important role in supporting the food security of regions in Indonesia. However, there are still many obstacles faced in changing the existing consumption patterns in the community. Therefore, in regard

| | **Growth rate for utilization 1993–2020 (% per year)** | | | **Utilization in 2020 (million tons)** | **Production in 2020 (million tons)** |
|---|---|---|---|---|---|
| | **Food** | **Feed** | **Total** | | |
| Southeast Asia | 1.4 | 0.13 | 1.25 | 27.0 | 51.1 |
| China | −1.27 | 2.08 | 1.19 | 3.9 | 4.2 |
| Other East Asia | −0.95 | 1.09 | 0.63 | 3.5 | 0.0 |
| India | 1.00 | 0.00 | 1.00 | 7.6 | 7.8 |
| Other South Asia | 1.00 | 0.00 | 0.83 | 0.6 | 0.6 |
| Latin America | 0.26 | 1.26 | 0.78 | 39.3 | 40.5 |
| Sub-Saharan Africa | 2.51 | 0.29 | 2.47 | 166.0 | 166.0 |
| Developing | 2.01 | 1.18 | 1.88 | 28.8 | 271.1 |
| Developed | 0.03 | 0.01 | 0.02 | 22.7 | 0.4 |
| World | 2.01 | 0.59 | 1.68 | 271.6 | 271.6 |

*Source: [18, 19]*

**Table 2.**
*Production and utilization of cassava in the world.*

to food security in regions, it is necessary to disseminate cassava-based food diversification as an alternative to rice or corn.

In line with the increasing demand for food and industrial materials, the availability of raw materials with quantity and quality which fulfill each demand requirement is indispensable. For example, as raw material for flour, cassava should have dry matter and starch content of >20%; and for foodstuffs and food industry material, in addition to high starch content, hydrogen cyanide (HCN) content must also be <50 mg/kg [23]. The strategies on it include providing suitable improved varieties and increasing productivity through cassava cultivation technology improvement, especially fertilization and pest control.

## 3. Indonesia's performance in world cassava trade flows

Indonesia ranks fourth in the world as a cassava producer with a production share of 9.26% and an average production of 22.819 million tons (**Table 3**), Cassava in the world is traded in form of fresh cassava and processed dried cassava.

The average growth of cassava export volume from 2011–2016 increased by 96.21% per year. Exports of cassava from Indonesia are in the form of fresh and processed, such as cassava flour, shredded cassava, and cassava pellets. Indonesia's cassava exports are mainly to Taiwan, the Philippines, Australia, Malaysia, England, and Brunei Darussalam. Meanwhile, the volume of cassava imports in the same period also fluctuated with a tendency to increase by 76.32% per year. Indonesia's cassava imports are generally in the form of cassava flour, shredded cassava, and cassava pellets, mainly from Thailand, Vietnam, and Myanmar [24–26].

As a cassava-producing country with an average production of 22.819 million tons, Indonesia is only able to export dried cassava on average 41,241 tons or only 1.90% share of world cassava exports. Meanwhile, the volume of dried cassava imports in Indonesia is quite small, with an average of 406 tons (0.01%) or ranks 31st in the world. Most of the cassava production is still absorbed by the domestic market for consumption and industry. The problem of "large production but small exports" is not only experienced by Indonesia but also by several countries, such as Nigeria and Brazil [27].

The low export of cassava from Indonesia causes the low price of imported cassava starch. As a result, cassava farmers cannot compete and affect the selling price of

| Cassava commodity | Quantity (tons) |
|---|---|
| Production | 22,819,353 |
| Consumption | 13,112,440 |
| Export: | |
| • Fresh cassava | 128,587 |
| • Processed-dried cassava | 41,241 |
| Import: | |
| • Fresh cassava | 1230 |
| • Processed-dried cassava | 406 |
| *Source: Processed data [24–26]* | |

**Table 3.**
*Cassava trade in Indonesia (in average units), 2011–2016.*

cassava in Indonesia. In addition, as food, feed, and industrial raw materials, cassava does not yet have an agreement on the basic price. In regards to a strategic approach to developing competitive cassava exports, the efficiency of farming and the improvement of cassava added value should be carried out considering the higher level of global competition [28, 29].

## 4. Cassava agro-industries from upstream to downstream

The developed agricultural product processing industries based on local resources ranging from home industries to large industries, as well as regional core competencies is one of the ideals of the Indonesian industry, with the expectation that the potential of each region can be optimally utilized and does not depend on imports. Thus, there will be no more inequality because each region is able to develop its industries. Well-managed industries in each region will further strengthen the structure of national manufacturing industries [30].

Cassava is very well-known by farmers in Indonesia and can be planted easily in all regions in Indonesia even though the soil fertility is low. Cassava is also very flexible in farming and harvesting age, resistant to biotic and abiotic stresses, and can produce well in a sub-optimal environment compared to other food crops. Furthermore, with the advances in agricultural technology, cassava productivity can be increased to 100% of the average farmer productivity. Thus, it becomes great opportunity for the development of a bioindustry based on cassava.

Cassava is the daily staple food consumed by households and most of it is obtained from their farm themselves. Cassava has proven to be suitable for local agricultural and food systems and becoming a major food crop in some areas. Therefore, cassava should be promoted to maintain and even increase its productivity and to ensure that households can maintain their dietary pattern and livelihoods. Policies to support cassava production and processing, as well as to promote the availability of cassava products in the market can contribute to improving rural food security, especially during climate change [31].

The food diversification program aims to utilize various local food sources, such as cassava, corn, sweet potatoes, sago, and others. However, currently what has happened is wheat consumption experiencing a significant increasing trend and there is a diversification of wheat flour-based food products, which results in the increase in wheat flour imports to Indonesia. Wheat flour contributes 20% of total food consumption in Indonesia. The value of wheat flour imports reached more than IDR 30 trillion, even higher than the budget value of the Ministry of Agriculture of IDR 27 trillion. Moreover, wheat flour is the only agricultural commodity with 0% imported tax. Therefore, for reducing wheat flour and its derivatives consumption, it is necessary to develop local food diversification which has a higher substitution value [32].

The quality of cassava from Indonesia in terms of moisture content and starch color intensity is better than that from Thailand and Vietnam, so that when converted into modified cassava starch, the moisture content of starch is lower and the starch is brighter [26]. Cassava has the potential to be developed as a raw material for the carbohydrate-based food industry. Efforts for cassava utilization as a buffer for food security include the development of cassava flour. In addition to extending the shelf life of the product, the purpose of flour production is also that the product is more preferred by consumers, and with cassava flour derivatives into modified cassava flour (mocaf), the physicochemical properties of cassava flour will increase, so that it is suitable for a wheat flour substitute in processed food products, such as cakes, bread, and noodles [33].

Starch in cassava can be used as raw material and adhesive in textile, paper, and certain confectionery industries [34, 35]. The use of α-amylase and amyloglucosidase in the hydrolysis of starch in cassava peels allow the fermentation process for the production of alcoholic beverages, vinegar, and bioethanol, and becomes an added value for the utilization of cassava wastes mainly from starch production [36, 37].

The development of cassava production provides a competitive industrial system, such as cheap raw material, easy to plant, and wide growth ability (both on fertile and marginal land), so that cassava products will create new business opportunities and increase farmers' welfare. However, supporting institutions for cassava farming systems for industrial purposes have not been well organized. The development of cassava-based agro-industries can be carried out from home-scale to large-scale industries. Several cassava-based industries of various scales, including chips, dried-slice cassava (*gaplek*), and dried-shredded cassava (*sawut*) can be established in the upstream industrial sector and create partnership activities with cassava farmers. Besides food and feed utilization, cassava can be developed as a raw material for bioethanol. Bioethanol is an alternative energy source for fuel and is carried out on large-scale industries [26].

Therefore, in developing a food diversification program to support food security, cassava is one of the food crop commodities that has an important role to support the program. Cassava production has great potential to be increased and the tubers can be processed into various products that can encourage the development of the agro-industry.

Based on the cassava industry tree (**Figure 2**), there are 28 cassava products and about 80% of them are for non-food purposes [38], and the rest of 10–20% are for food purposes (staple food, intermediate, and end-products). Preferences of cassava as industrial raw material has not well-known yet by farmers; while many cassava varieties have characteristics and specifications, which are suitable for industrial purposes. For instance, cassava with good taste for food industries, high biomass for

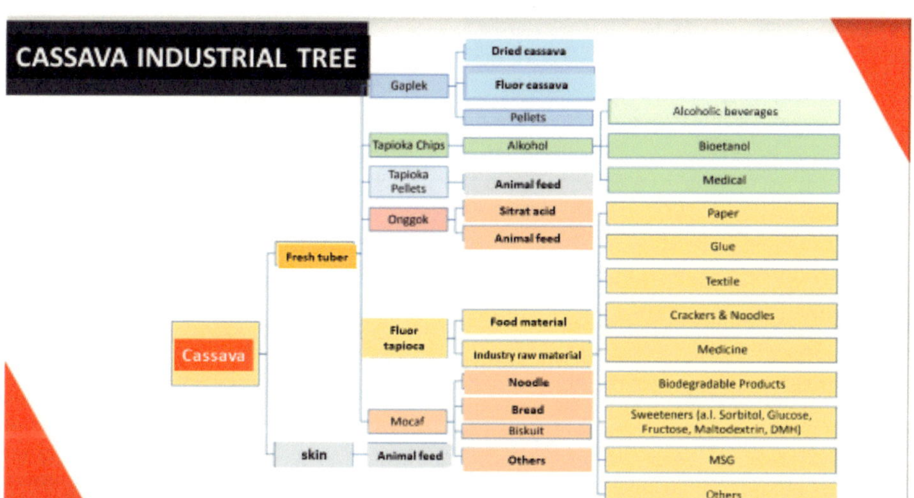


**Figure 2.**
*Cassava industrial tree (Source: [38]).*

| Variety | Year of release | Harvested age (months) | Productivity (tons/ha) | HCN content (mg/kg) | Starch content (%wb) | Flour yield (%) | Other characteristics |
|---|---|---|---|---|---|---|---|
| **Improved cassava varieties suitable for food.** | | | | | | | |
| Adira 1 | 1978 | 7–10 | 22 | 27.5 | — | 45 | Not bitter, yellow-fleshed tubers |
| Malang 1 | 1992 | 9–10 | 36 | <40 | — | 32–36 | Not bitter, yellowish white-fleshed tubers |
| Malang 2 | 1992 | 8–10 | 31 | <40 | — | 32–36 | Not bitter, light yellow-fleshed tubers |
| Darul Hidayah | 1998 | 8–12 | 100 | <40 | 25–31.5 | 35–45 | Fairly bitter, white-fleshed tubers |
| Improved cassava varieties suitable for industry. | | | | | | | |
| Adira 2 | 1978 | 8–12 | 22 | 124 | — | 41 | Bitter, white-fleshed tubers |
| Adira 4 | 1978 | 10 | 35 | 68 | 18–22 | 39 | Bitter, white-fleshed tubers, ethanol conversion 4.5–4.7 kg peeled tubers/l |
| Malang 4 | 2001 | 9 | 40 | 100 | 25–32 | — | Bitter, white-fleshed tubers |
| Malang 6 | 2001 | 9 | 36 | 100 | 25–32 | 43 | Bitter, white-fleshed tubers, ethanol conversion 4.7–5.1 kg peeled tubers/l |
| UJ 3 | 2000 | 8–10 | 27 | >100 | 20–27 | 41 | Bitter, yellowish-fleshed tubers, ethanol conversion 4.9 kg peeled tubers/l |
| UJ 5 | 2000 | 9–10 | 31 | >100 | 19–30 | 46 | Bitter, white-fleshed tubers, ethanol conversion 4.5 kg peeled tubers/l |
| Litbang UK 2 | 2012 | 9–10 | 42 | 31 | 18–31 | 43 | Bitter, white-fleshed tubers, ethanol conversion 4.3 kg peeled tubers/l |

*Note: % wb = percentage in wet basis; Source: [41]*

**Table 4.**
*Improved cassava varieties suitable for food and industrial raw materials.*

feed industries, high starch content for ethanol industries, and many others have not been identified yet. To increase cassava productivity, it is necessary to apply improved varieties and cultivation technologies that are adaptive for each cassava-producing area, and suitable for product utilization.

From 1978–2015, about 12 improved varieties of cassava have been released by the Indonesian Agency for Agricultural Research and Development (IAARD), including Adira 1, Adira 2, Adira 4, Malang 1, Malang 2, Darul Hidayah, UJ 3, UJ 5, Malang 4, Malang 6, Litbang UK 2, and UK 1 Agritan with the yield >30 tons/ha [14]. Therefore, to suppress the yield gaps in farming level can be done by planting those IAARD's high-yielding varieties.

Among production technology components, improved varieties have important and strategic roles due to they are essential for increasing crop productivity. As mentioned before that for food purposes, the improved cassava varieties with good taste, fluffier, and low HCN content, such as Adira 1, Malang 1, Malang 2, and Darul Hidayah are suitable. While the improved cassava varieties are suitable for industrial raw materials producing flour and starch should have high yield, high dry matter, and starch content. The HCN content is not a requirement of cassava for industrial raw material because most of it will be lost in washing, heating, and drying processes. Several improved varieties that are suitable for industrial raw materials, including Adira 4, Malang 6, UJ 3, UJ 5, and Malang 4 varieties, which have been well-known and planted by farmers [39]. While Litbang UK 2 variety has potential as a biofuel due to its high ethanol content. The potential yield of 96% bioethanol of Litbang UK 2 is 144,72 l/ha [40]. **Table 4** describes the improved cassava varieties released by the IAARD for the utilization of food and industrial raw materials.

In the downstream industrial sector, cassava-based agro-industry aims to increase the added value of cassava by processing the commodity into various high-value products. Various cassava-based products (intermediate and end-products) have been produced, both in small-scale industries with simple equipment and large-scale with modern machinery [42]. Tapioca as a cassava intermediate product, has been growing rapidly in Indonesia. In recent years, agro-industry modified cassava flour (mocaf) has also been started [43]. Several agro-industries produce cassava end-products, such as cakes, chips, brownies, traditional sweets (*dodol*), fermented cassava (*tape* or tapai), and so on. In addition, cassava processing wastes or by-products can also be processed into fertilizer, especially for plantation crops and the cassava peels can be processed into animal feed [43].

## 5. Cassava value chain for agro-industry

The importance of agricultural sectors strengthened by the integration among related sectors from upstream to downstream can increase regional economy, absorb labor, and equalize regional development, which leads to an increase in community welfare, as well as strengthen the national economy. It can be realized by increasing the role in the value chain, whereby adding activities and the ability to increase product value will provide independence for the regions producing agricultural commodities. The goal is that the region is not only an object of development but is able to become a subject due to the ability to process and market agricultural commodities independently. According to Kaplinsky and Morris (in [44]), a value chain consisting of various actors (main producers, processors, traders, and service providers) can be established if all actors at the chain work in such a way to maximize the value along the chain.

The structure of the cassava value chain ideally includes five elements, namely, end market (consumer) opportunities, supportive business environment, vertical relationships, horizontal relationships, and supporting markets. If these five elements work properly, marketing costs can be streamlined and can improve coordination [45]. Meanwhile, in terms of marketing, the lack of access to information on prices and goods, which are mostly controlled by brokers and wholesalers, uneven road access, and the inability of farmers to diversify cassava commodities have caused farmers to not have a competitive advantage. In addition, too many actors are involved from cassava production to marketing, including traders from inside and outside region, brokers or middlemen, and wholesalers, causing the unstable price of cassava. Therefore, strengthening the cassava value chain and trading system are very important, so that it will be able to improve the bargaining position of farmers.

The impact of cassava economy improvement will be widespread and involve many stakeholders. The biggest stakeholders are cassava farmers and small-scale processors in rural areas. Middle-upper industries and middlemen have an important role to play in enhancing future linkages. The demand for cassava-based processed products plays an important and effective role as an economy driver. Unfortunately, not all rural farming communities can develop toward a value-added improvement orientation. The rural agro-industry or bio-industry is still lagging behind, causing the added value to flow out of the region [46].

Study of value chain for cassava based on observations and desk study to its actors in Indonesia resulted in three models of cassava supply chain [47, 48], that is, (I) direct sale of cassava fresh tubers, (II) cassava tubers for food, and (III) tapioca starch. **Figure 3** depicts the schematic diagrams for the three models. The simplest supply chain model is the direct sale of cassava fresh tuber to consumers (Model I). This model involves three actors, namely, farmers, retailers or small traders, and consumers. Small traders act as middlemen between farmers and consumers. In Model I, cassava has low economic value due to there is no further product transformation, and fresh tubers are only used for direct consumption or for animal feed. Model II

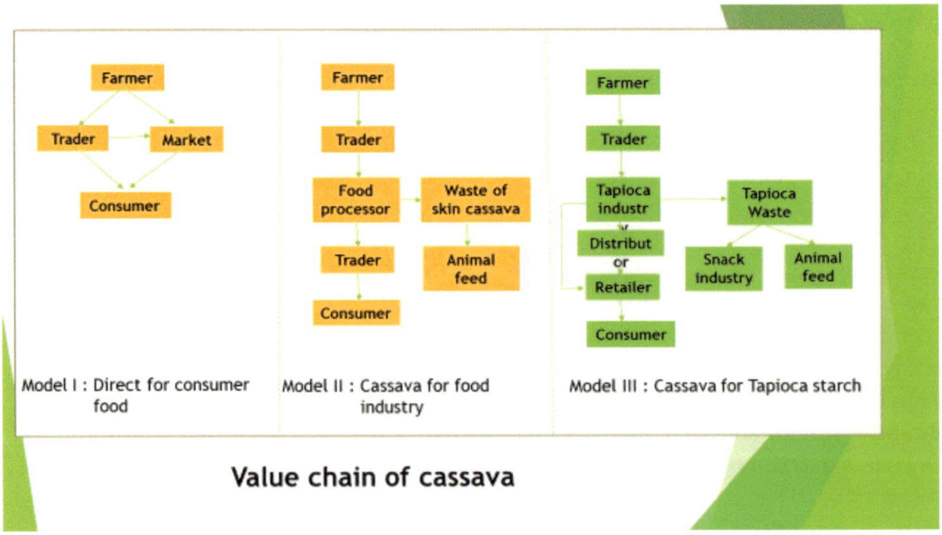


**Figure 3.**
*Cassava value chain (Source : [47, 48]).*

involves four actors, namely, farmers as cassava suppliers, middlemen as collectors of cassava from farmers, processors who sometimes also act as food vendors for their cassava products to consumers, and consumers as the users of the end-product. In Model II, there is a slight increase in cassava's economic value due to it has undergone further processing. The last existing model is Model III, which is the most complex supply chain of all existing supply chains. There are eight actors involved in Model III, namely, farmers, traders (middlemen), tapioca industries, distributors, retailers, consumers, snack industries, and cattlemen. In this supply chain model, cassava undergoes the highest product transformation and the highest economic value.

The reality of the current yield of cassava in Indonesia is cultivated by farmers in marginal land with small land ownership area, poor transportation facilities, and the majority of the yield is used for direct consumption or snack (70–80%). Meanwhile, based on the utilization of industrial trees, cassava is widely used for non-food raw materials for strategic industries (**Figure 2**). So, there is a lack of raw materials supply for non-food products. The non-food industries are only supplied by large industries engaged in the cassava business and about 10–15% by cassava farmers around the industrial locations with easy transportation facilities. Thus, in general, the cassava value chain is quite short, so it is necessary to increase the cassava added value to extend the cassava value chain from farmers.

Realizing that added value of cassava from the processing activities in the downstream sector is much higher than the primary products in the upstream, the future agricultural development approach is directed toward product development. The added value development is carried out through agro-industrial development, which processes primary products into competitive intermediate and end-products [49].

## 6. Map of cassava business development

The development of cassava is carried out from upstream at the farm level to downstream for food, feed, and industrial businesses. These four cassava businesses are described in four quadrants (**Figure 4**). The development of cassava at upstream or farm level is in quadrant I where this position is between the strength (S) factor and the opportunity (O) factor. In this area, the business is at a growth rate, which is further development is carried out by managing or optimizing the business' strengths to seize opportunities. The direction (slope) of cassava farming development tends toward the opportunity (O) factor, so that the implementation includes meeting high market demand by increasing cassava productivity (yield) [50].

Position of cassava development as raw material for food and feed is also in quadrant I with the directions (slopes) of those development positions tend towards strength (S). Businesses of cassava as food and feed raw materials have strength factors larger than the weakness. Meanwhile, by looking at the influence of the environment on cassava farming development on food, the opportunity for cassava development is easier to achieve, due to the opportunity (O) value being higher than the threat (T) value. The threats for cassava development at the farm level, as well as for food and feed businesses, can be anticipated by seizing the best opportunities supported by the use of great strength.

The development of cassava farming for industrial raw materials is in quadrant II, which means that the direction of development is still leaning toward threat (T) factor than to strength (S) factor. Thus, the right strategy is exploring the strength owned by the business to overcome the existing threat, namely, product

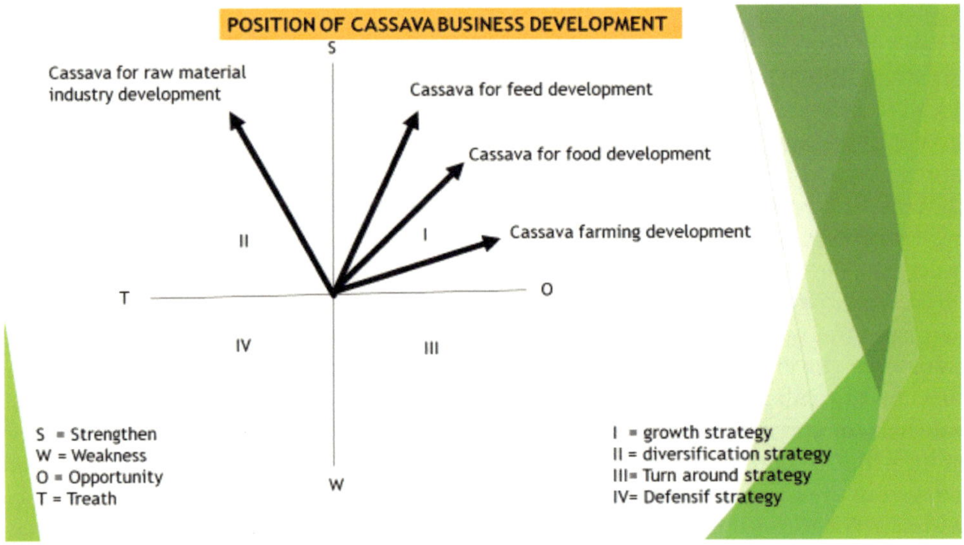


**Figure 4.**
*Map of cassava business development (Source: [50]).*

diversification strategy. Product diversification will enhance the longer and more numerous value chain, so that it will reduce the threat in the cassava business.

There is a lesson learned from the small-scale cassava industry, namely, UD. Riang in Malang Regency, East Java Province produces three kinds of processed cassava in form of instant food, namely, instant *gatot*, instant sweet *tiwul*, and instant plain *tiwul*, would show us how this small business is trying to manage its strengths to seize market opportunities [51].

*Gatot* and *tiwul* are traditional specific Javanese food and usually consume in some barren areas in Java, such as in Gunung Kidul (Yogyakarta), Wonogiri (Central Java), and Trenggalek (East Java). Javanese people in that area tend to consume cassava as their staple food as a rice substitute during famine times. Firstly, cassava was processed into *gaplek* by drying the cassava under the sun, so that it can be stored longer. For consumption purposes, *gaplek* can be further processed into *gatot* and *tiwul*. *Gatot* is made from sliced black brown-color *gaplek* that is overnight-soaked then steamed, while *tiwul* is made from chopped white-color *gaplek* into a coarse powder that is mixed with water and then steamed. *Gatot* and *tiwul* have unique texture affected by the gelatinization of starch from *gaplek*. The steamed *gatot* is usually consumed as a sweet snack and served with some sugar, salt, and grated coconut flesh, while *tiwul* is usually served as a rice substitute by adding side dishes with plain or salty taste or consumed as a sweet snack by further processed using with brown sugar and sprinkled with grated coconut flesh [52].

UD. Riang has a production capacity of 100 kg per product per one production time, where the one-time production process of instant sweet and plain *tiwul* is 3 days and instant *gatot* is 7 days. UD. Riang is trying to continue to improve the image of processed cassava as a prestigious food by doing innovations in processing cassava from traditional Javanese food with local wisdom into the current trend of people's eating patterns that demand fast and practical things, in form of instant food.

UD. Riang also improved product packaging to keep up with the current times and included the information of expiration date and the Ministry of Health's permit in the packaging. Product quality is always maintained and no preservatives are used in this instant traditional food processing. UD. Riang actively participates in traditional food festivals to expand the marketing of their products and introduce them to the community. The management of UD. Riang continues to innovate the products because they realize that the sales of these three instant cassava products are very volatile in the market, so they need to be maintained in order to do not fall and disappear from the market. UD. Riang is also trying to expand the marketing segment by reaching potential young consumers, so that the unique and attractive products should be displayed where they can compete with other processed food products that are already in demand by young consumers [51].

## 7. Bio-economic potential of cassava

The potential and opportunities for developing cassava as a bio-economy in the future are widely open. This potential is in line with the development of the livestock industry, processed food, and other industries, such as alcohol, sorbitol, fructose, and many others. In the future, the plastic industry will also use tubers, including cassava, as raw material. The added value of cassava commodities obtained from the development of processed products in the downstream sector is much higher than the upstream at the farm level. Therefore, the approach for future agricultural development should be more directed at developing products postharvesting.

Agro-industry is a sector that is able to provide added value to cassava commodities. Agro-industry has a direct relationship with primary agriculture, where the industry processes primary agricultural commodities into intermediate products, such as flours and direct consumption products. Thus, the cassava agro-industry needs to be supported by the availability of research technology starting from the production (cultivation and varieties) up to postharvest and processing for food, feed, and other industries.

So far, cassava farming has not implemented an efficient business concept considering that there are many potentials and opportunities that have not been utilized optimally, for instance, the waste (biomass) from cassava crops, which can be used for unconventional feed [53]. Based on the definition of the Food and Agricultural Organization (FAO), the characteristics of unconventional feed are as follows: (1) it is the end result of production, which can no longer be used or recycled, (2) it is a solid or liquid organic material, (3) it has low economic value compared to the cost of collection and processing, (4) it is a source of fermentable carbohydrates, and (5) it is a bulky material containing high crude fiber and low nitrogen [54]. Cassava peel is an excellent raw material for feed. With cassava production of 18.9 million tons per year, white colored-inner peel waste can reach 1.5–2.8 million tons, while brown colored-outer peel waste reaches 0.04–0.09 million tons [55].

The potential economic value of cassava farming is IDR 71,790,000 at farmer productivity level of 42.5 tons/ha with the price of cassava is IDR 1,200/kg. Indirect economic benefits are IDR 20,698,000 or 29.7% of total economic benefits or 40% of direct use economic value received by farmers. Besides, there are intrinsic values that have not been detected and will be known in further cassava development (**Table 5**) [56].

The bio-economic approach is needed to create business sustainability in the form of a bio-industrial system integrating cassava crops and livestock. Besides tubers as the main yield of cassava, there are cassava by-products (biomass) that can be used for livestock feed. While livestock manure can be utilized as fertilizer or processed

| Category | Use value (IDR 000) | | | Intrinsic value | | Total (IDR 000) |
|---|---|---|---|---|---|---|
| | Direct | Indirect | Optional | Existence | Pride/ heritage | |
| Farm level: | 51,000 | | | | | 51,000 |
| • Leaves | | 1,028 | | | | 1,028 |
| • Stem | | 6,670 | | | | 6,670 |
| Feed: | | | N.d. | N.d. | N.d. | |
| • Cassava peel | | 500 | | | | 500 |
| • Stem stump (*bonggol*) | | 5,200 | | | | 5,200 |
| Industry: | | | | | | |
| • Cassava peel | | 300 | | | | 300 |
| • Solid waste (*gamblong*) | | 7,000 | | | | 7,000 |
| • Fluid waste | | worthless | | | | |
| Total | 51,000 | 20,698 | | | | 71,798 |

*Note: N.d. = not detected; Source: [56]*

**Table 5.**
*Potential economic value of cassava for agro-industry from the upstream industrial sector.*

into biogas for energy or fuel in the industry. Thus, the interdependence and beneficial integration creating biological and economic circulation can enhance the bioeconomy of cassava farming [57].

Cassava biomass processing can produce various products with high value-added and facilitate the essential nutrients recycling is needed to maintain the sustainability of land productivity. Cassava is also one of the most efficient crops for harvesting

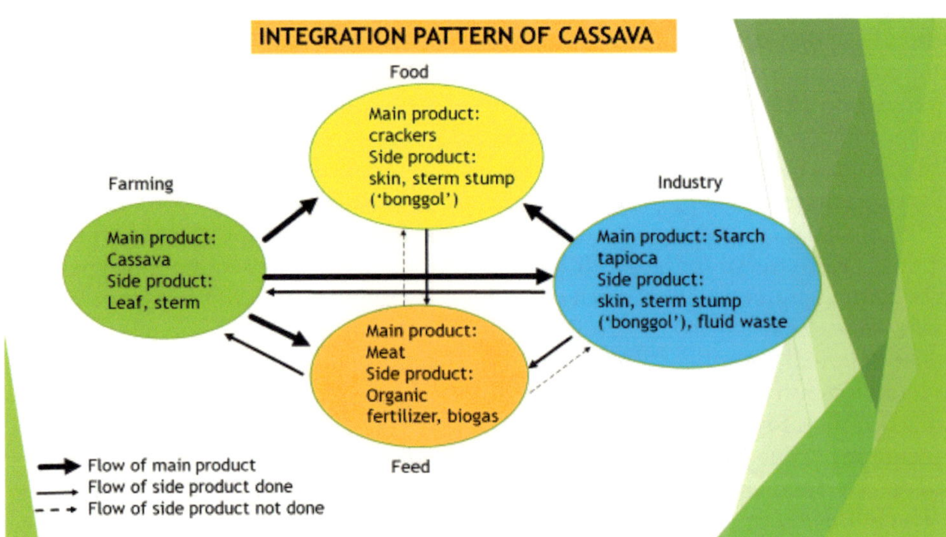


**Figure 5.**
*Integration pattern of cassava (Source: [56]).*

solar energy. The integration of cassava cultivation with certain processing methods will form an industrial symbiotic structure, such as in the pattern of symbiotic interaction in biological community life (**Figure 5**) [56].

## 8. Upcoming strategy in cassava development

Based on the cassava business development map (**Figure 4**), policies and operational strategies for cassava development can be carried out. **Table 6** formulates policies and operational strategies for the four cassava business sectors. The operational strategies are concrete steps in implementing policies that must be carried out to develop cassava toward certain targets.

*First*, the strategy carried out at the farm level to increase cassava production in the form of intensive cultivation with the application of new technology at the production and post-production stages, as well as expansion of the planting area. The expansion of the planting area includes: (a) new land clearing, (b) double planting, and (c) increasing harvest index. Meanwhile, increasing production through intensification can be done by applying several alternative cultivation technologies, including (1) improved varieties, (2) seed preparation, (3) land preparation, (4) planting, (5) fertilization, (6) plant maintenance, and (7) harvest.

*Second*, the policy for the food business sector is to regulate the production system of cassava commodities from upstream to downstream. The continuity of the supply

| **Business level** | **Strategy** | | **Target** |
|---|---|---|---|
| | **Policy** | **Operational** | |
| Farm | S-O Strategy:<br>Intensive cultivation with the use of new technology in the production and post-production stages and planting area expansion | 1. The assistance on new improved cassava seeds.<br>2. Demonstration of cassava cultivation technology.<br>3. Training on cassava cultivation techniques for farmer groups.<br>4. Expansion of cassava planting area. | Farmers. local government |
| Food business | S-O Strategy:<br>Regulate cassava production system from upstream to downstream sector | 1. Arrange cassava planting system and planting area zoning.<br>2. The assistance on capital from financial institutions.<br>3. Training on production to improve food product quality.<br>4. Support of local government for marketing and business partnership.<br>5. Promotion of functional food products to improve the image of cassava in the community. | Farmers. community, processors, industries, financial institutions, and local government |

| Business level | Strategy | | Target |
|---|---|---|---|
| | Policy | Operational | |
| Feed business | S-O Strategy: Disseminate the processing technology on cassava biomass for livestock feed | 1. Introduction and promotion of affordable cassava feed technology.<br>2. Training on cassava biomass technology for farmers. | Farmers, processors |
| Industrial raw material business | S-T Strategy: Improve business scale by doing product development or diversification and business partnership and producer capacity improvement | 1. Counseling on liquid waste from the production process.<br>2. Training on various products from cassava.<br>3. Entrepreneurship training for processors.<br>4. Capital assistance<br>5. Marketing and partnership supporting. | Processors, industries, financial institutions, and local government |

*Source: Processed data [49]*

**Table 6.**
*Policy and operational strategies for cassava development.*

of cassava raw materials is important because the cassava harvest period is relatively long and cassava planting is mostly done during the rainy season. Therefore, it is necessary to arrange the planting system and the zoning of the cassava planted area in order to provide sustainable yields for the supply of raw materials, such as for tapioca production.

The inappropriate planning in cassava procurement causes a significant loss of starch content. This inappropriate occurred on farmers' land by suspending the harvest time to anticipate higher-selling prices, and in the warehouse of the cassava processing industry where do the overstock due to concerns that the supply of raw material would be inconsistent during the low season so that it was unable to meet the factory's demand for a year. Therefore, appropriate planning in the cassava procurement will increase production efficiency and reduce production costs, and at the same time will maintain a consistent supply of raw material to starch industries [58].

*Third*, the policy for the feed business sector will be implemented is the socialization of livestock ration technology using cassava biomass (tuber peels, stems, and leaves), which is quite potential and needs to be explored. The optimal utilization of local resources is a strategic step in achieving business efficiency in ruminant livestock production. This efficiency will be more significant if these resources are not directly needed by competitors, namely, humans and livestock, other than ruminants. Feed is closely related to productivity and production costs; therefore, the efficient use of local resources will greatly affect the development of ruminants [59].

Cassava is an important multi-purpose crop. Cassava is an affordable alternative feed for livestock and has great potential in the future due to the increasing demand for cassava products. Cassava tubers are low in protein but a good source of carbohydrates that can be used as a supplement in poultry feed. The leaves have a moderate protein content, which can replace part of conventional protein sources in livestock feed. Anti-nutritional substances, especially cyanide, reduce the feeding value of cassava commodities, but the appropriate processing can reduce the level and make the product safe for the feed. Cassava commodity has entered the industrial market and has great potential in the feed industry [60].

*Fourth,* the policy of cassava for the industrial raw material sector is to increase business scale through product development or diversification, business partnership, and producer capacity improvement. The policy is expected to overcome the obstacles of cassava as industrial raw material because the position of cassava development in this sector is facing opportunity (O) and threat (T) factors.

For instance, some cassava enterprises in the Central Java and Yogyakarta Provinces had already performed business-oriented processing and quality assurance methods, which require the best quality of cassava as industrial raw material. Some business units that diversify and intensify their products perform more effective business and result in a better selling price. However, most of the business units have not kept accounting records, resulting in the lack of data about the financial position. Business partnership, business diversification and intensification, and business record keeping can guarantee the sustainability of business [61].

## 9. Conclusions

Indonesia has a great opportunity in supplying the world's food needs in the future—only by utilizing 54% of the total available land, which is suitable for planting cassava, by preparing the potential of cassava. The potential of this land area that is also supported by human resources (farmers) who have long experience in cassava farming can be realized through some efforts as follows:

1. Increase cassava productivity by the availability of cassava planting system technology in form of new improved varieties and cultivation technology. This potential will be easily realized because the cassava development business map at the farm level is in the growth phase.

2. Increase the added value of cassava, so that the cassava value chain will be longer. This increase in cassava added value is also supported by the position of the cassava development business for industrial raw materials in the diversification phase. Product diversification will provide added value for cassava to support global needs.

3. Increase farmers' income and welfare, as well as meet cassava raw materials supply in the future by creating a cassava bio-economy. The bio-economic approach will enhance business continuity or optimum sustainable yield by obtaining maximum profit from the business. The integration of cassava crops and livestock farming is a form to realize cassava economic circulation.

## Conflict of interest

The authors declare no conflict of interest.

## Author details

Fachrur Rozi*, Dian Adi Anggraeni Elisabeth, Ruly Krisdiana, Adri Adri, Yardha Yardha and Yanti Rina
National Research and Innovation Agency, Jakarta, Indonesia

*Address all correspondence to: f_rozi13@gmail.com

## References

[1] Bantacut T. Indonesian staple food adaptations for sustainability in continuously changing climates. Journal of Environment and Earth Science. 2014;**4**:202-216

[2] Syahyuti S. Paradigma Kedaulatan Pangan dan Keterlibatan Swasta: Ancaman terhadap Pendekatan Ketahanan Pangan (?). Analisis Kebijakan Pertanian. 2016;**9**:1-18

[3] FAO. The Future of Food and Agriculture – Trends and Challenges. Rome: FAO; 2017

[4] Pabon-Pereira C, Slingerland M, Hogervorst S, van Lier J, Rabbinge R. A sustainability assessment of bioethanol (EtOH) production: The case of cassava in Colombia. Sustainability. 2019;**11**:3968

[5] Dinata FS, Kartawiria IS. Bioethanol potential from whole parts of cassava plant in Indonesia. Jurnal Teknologi dan Industri Pertanian. 2021;**31**:20-33

[6] Ajala EO, Ajala MA, Tijani IA, Adebisi AA, Suru I. Kinetics modelling of acid hydrolysis of cassava (*Manihot esculanta* Cranz) peel and its hydrolysate chemical characterisation. Journal of King Saud University. 2020;**32**:2284-2292

[7] Suharko S, Hudayana B. Rural woman and food security: Diversification of cassava-based foods in Gunungkidul District, Yogyakarta. Sodality Journal of Sosiol Pedesaan. 2020;**8**:1-14

[8] Mangku IGP. Food diversification to support food scurity. Wicaksana Journal of Lingkung dan Pembang. 2017;**1**:1-10

[9] Arief RW, Novitasari E, Asnawi R. Food diversification of cassava as non-rice based functional food in Lampung. Journal of Agro Science. 2018;**6**:62-69

[10] Ministry of Agriculture. Draft Strategic Plan of the Ministry of Agriculture for 2010-2014. 2009.

[11] Wahyuningtiyas NE, Suryanto H. Analysis of biodegradation of bioplastics made of cassava starch. Journal of Mechanical Engineering and Science Technology. 2017;**1**:24-31

[12] Akhmad R. Comparing the Use of Bioplastic Cassava and Plastics Ethylene as Packaging Cayenne Pepper. Ecology, Environment, and Conservation Journal. 2021;**21**:122-127

[13] Ministry of Agriculture. Ungkit Produktivitas Ubikayu Kementan Dorong Pengembangan Benih Bersertifikat. Dirjen Tanaman Pangan; 2021. [Internet]. Available from: https://tanamanpangan.pertanian.go.id/detil-konten/berita/378

[14] Balitkabi. Deskripsi Varietas Unggul Aneka Kacang dan Umbi. Badan Litbang Pertanian; Malang, Indonesia; 2016

[15] FAO F. Food Outlook: Biannual Report on Global Food Markets. Rome: FAO; 2018

[16] Adam P. Cassava: International Market Profile. Trade Mark Div Food Agric Organ United Nations; Rome, Italy; 2005

[17] Howeler RH, Hershey CH. Cassava in Asia: Research and Development to Increase Its Potential Use in Food, Feed and Industry: A Thai Example. International Center for Tropical Agriculture; Rome, Italy; 2002

[18] Rosegrant MW, Gerpacio RV. Roots and Tubers in the 21st Century: Their Role and Importance in the Global Food Market. IFPRI Discuss Doc. Washington, DC: IFPRI; 1997

[19] Scott GJ, Rosegrant MW, Ringler C. Roots and Tubers for the 21st Century:

Trends, Projections, and Policy Options. IFPRI; Washington DC, USA; 2000

[20] FAOSTAT FAO. Statistics. Rome: Food and Agriculture Organization of the United Nations; 2020

[21] Sediaoetama AD. Ilmu gizi II. Dian Rakyat Jakarta; Jakarta, Indonesia; 2001

[22] Almatsier S. Prinsip dasar ilmu gizi. Gramedia; Jakarta, Indonesia; 2002

[23] Ginting E. Cyanide reduction in cassava root products through processing and selection of cultivars in relation to food safety. Buletin Palawija. 2013;**2013**:26-36

[24] INDEXBOX. Cassava—Indonesia. IndexBox, Inc. 2022. [Internet]. Available from: https://app.indexbox.io/table/071410/360/

[25] WITS. Indonesia Starch; Manioc (Cassava) Exports by Country in 2019. World Trade Organization (WTO). 2019. [Internet]. Available from: https://wits.worldbank.org/trade/comtrade/en/country/IDN/year/2019/tradeflow/Exports/partner/ALL/product/110814

[26] Pertanian K. Outlook Komoditas Pertanian Tanaman Pangan Ubi Kayu. Jakarta: Pus Data Dan Sist Inf Pertan Kementrian Pertan; 2016

[27] Dirgantoro K. Bodo Amat, Pokoknya Indonesia Sampai Impor Singkong Itu Salah Pemerintah. MOJOK. CO. 2017. [Internet]. Available from: https://mojok.co/esai/impor-singkong/

[28] Asriani PS. Perdagangan Ubikayu Indonesia Di Pasar Dunia (Indonesia Cassava Trade in World Market). Jurnal AGRISEP: Kajian Masalah Sosial Ekonomi Pertanian dan Agribisnis. 2010;**9**:184-196

[29] Hutabarat NAP. The export performance of Indonesian dried cassava in the world market. Agric Socio-Economics Journal. 2017;**17**:134-139

[30] Kuncoro M. Masalah, kebijakan, dan politik ekonomika pembangunan. Erlangga; Jakarta, Indonesia; 2010

[31] Saediman H, Limi MA, Rosmawaty AP, Indarsyih Y. Cassava consumption and food security status among cassava growing households in Southeast Sulawesi. Pakistan Journal of Nutrition. 2016;**15**:1008-1016

[32] Hipotesa FEM IPB. Diversifikasi Pangan: Solusi Terwujudnya Kemandirian Pangan Berkelanjutan 2019. [Internet]. Available from: https://www.kompasiana.com/hipotesafemipb/5d7b45ad097f36749874a413/diversivikasi-pangan-solusi-terwujudnya-kemandirian-pangan-berkelanjutan?page=all#section3

[33] Hadistio A, Fitri S. Tepung Mocaf (Modified Cassava Flour) untuk Ketahanan Pangan Indonesia. Jurnal Ilmu Pangan Halal. 2019;**1**:20

[34] Tonukari NJ. Cassava and the future of starch. Electron Journal of Biotechnology. 2004;**7**:5-8

[35] Mtunguja MK, Beckles DM, Laswai HS, Ndunguru JC, Sinha NJ. Opportunities to commercialize cassava production for poverty alleviation and improved food security in Tanzania. African Journal of Food Agriculture Nutrition and Development. 2019;**19**: 13928-13946

[36] Claire. Modified Starch Market by Raw Material, Function, Application, and Region—Global Forecast to 2022. 2017. [Internet]. Available from: https://www.prnewswire.com/news-releases/modified-starch-market-by-raw-material-function-application-and-region---global-forecastto-2022-300412218.html. Cision US Inc

[37] Souto LRF, Caliari M, SOARES MS, Fiorda FA, Garcia MC. Utilization of residue from cassava starch processing for production of fermentable sugar by enzymatic hydrolysis. Food Science and Technology. 2016;**37**:19-24

[38] Julian. Serba Manfaat, Singkong bukan lagi Pangan Inferior. Tabloid Sinar Tani. 2021. [Internet]. Available from: https://www.tabloidsinartani.com/detail/indeks/pangan/17311-Serba-Manfaat-Singkong-bukan-lagi-Pangan-Inferior

[39] Puslitbangtan. Varietas Unggul Ubikayu untuk Bahan Pangan dan Bahan Industri: Litbang Pertanian. 2014. [Internet]. Available from: https://www.litbang.pertanian.go.id/info-teknologi/1648/

[40] Sholihin. LITBANG UK-2, Varietas Ubi Kayu Hasil Umbi dan Ethanol Tinggi: Balitkabi. 2015. [Internet]. Available from: https://balitkabi.litbang.pertanian.go.id/infotek/litbang-uk-2-varietas-ubi-kayu-hasil-umbi-dan-ethanol-tinggi/

[41] Saleh N, Taufiq A, Widodo Y, Sundari T, Gusyana D, Rajagukguk RP, et al. Pedoman Budidaya Ubi Kayu di Indonesia. Jakarta (diakses tanggal 3 July 2018): IAARD Pres; 2016

[42] Banowati E, Sari Y, Mustofa MS, Rahmawati SN. Cassava farming based on rural agribusiness development. International Journal of Indonesian Social Culture. 2020;**12**:58-68

[43] Wibowo Y, Purnomo BH, Wicaksono EP. Rancang bangun sistem informasi potensi agroindustri berbasis singkong di Kabupaten Jember. Agrointek J Teknol Ind Pertan. 2015;**9**:50-62

[44] Kusumawardani MH. Membuat rantai nilai lebih, berpihak pada kaum miskin: buku pegangan bagi praktisi analisis rantai nilai. Australian Centre for International Agricultural Research (ACIAR); Canberra, Australia; 2012

[45] Sewando PT. Urban markets-linked cassava value chain in Morogoro rural district, Tanzania. Journal of Sustainable Development. 2012;**14**:283-300

[46] Fitriani F. Pertanian Perdesaan Lampung: Peluang dan Tantangan Lampung rural agriculture: Opportunities and challenges F. Journal of Food System Agribus. 2017;**1**(2):43-52

[47] Suryaningrat IB, Amilia W, Choiron M. Current condition of agroindustrial supply chain of cassava products: A case survey of East Java, Indonesia. Agricultural Science. 2015;**3**:137-142

[48] Tambun R, Manik Y. Value chain analysis of cassava (*Manihot esculeta*) in Toba Samosir Regency, North Sumatera, Indonesia. Journal of Engineering Science and Research. 2020;**2**:39-43

[49] Simatupang P. Meningkatkan daya saing ubikayu, kedelai dan kacang tanah untuk meningkatkan pendapatan petani, ketahanan pangan, nilai tambah, dan penerimaan devisa. Prosiding Seminar Hasil Penelitian Tanaman Aneka Kacang dan Umbi. 2012:1-12

[50] Rozi F, Harnowo D. Pilihan Ubikayu sebagai Komoditas Unggulan untuk Mewujudkan Kedaulatan Pangan. Konf Nas Kedaulatan Pangan 2016 "Pertan Berkelanjutan dan Agroindustri Halal dalam Mendukung Kedaulatan Pangan untuk Menghadapi Masy Ekon Asean." Universitas Darussalam Gontor; Ponorogo, Indonesia; 2016

[51] Dina Rahmawati F. Analisis keuntungan berbagai produk olahan ubi kayu (Studi Kasus pada UD. Riang di Desa Telogorejo Kecamatan Pagak Kabupaten Malang). University of Muhammadiyah Malang; Malang, Indonesia; 2007

[52] Jesica C, Cansa E, Fidelia J. Gaplek, Tiwul and Gatot as staple food in Javanese barren area. International Journal of History and Research. 2017;**7**:1-6

[53] Chuzaemi S. Arah dan Sasaran Penelitian Nutrisi Sapi Potong Lokalit Sapi Potong. Puslitbang Peternakan Bogor; Bogor, Indonesia; 2002

[54] Anggraeny YN, Umiyasih U, Krishna NH. Potensi limbah jagung siap rilis sebagai sumber hijauan sapi potong. Pontianak: Pros Lokakarya Nas Jejaring Pengemb Sist Integr Jagung-Sapi Puslitbangnak; 2006. pp. 9-10

[55] Widyastuti P. Pengolahan limbah kulit singkong sebagai bahan bakar bioetanol melaui proses fermentasi. Jurnal Kompetensi Teknik. 2019;**11**:41-46

[56] Rozi F, Pudjiastuti Q. Produk Samping Tanaman Ubikayu Sebagai Potensi Bioekonomi Untuk Pertanian Masa Depan. SOCA Jurnal Sosial Ekonomi Pertanian. 2019;**13**:433

[57] JBM R, Indrasti R, Bakrie B. Sustainable agricultural bioindustry development: Integration of cassava cultivation with beef cattle husbandry in North Sulawesi Province. International Journal of Environmental and Agricultural Biotechnology. 2018;**3**:264410

[58] Suvittawat A, Soni P, Igel B, Khang DB. Cassava root supply chain in Nakhornratchasima Province: Challenges and opportunities for local starch mills. Journal of Food and Agricultural. 2014;**12**:332-338

[59] Simanihuruk K, Sirait J, Syawal M. Penggunaan silase biomassa tanaman ubikayu (kulit umbi, batang dan daun) sebagai pakan kambing peranakan Etawah (PE). Pastura. 2013;**2**:79-83

[60] Devi A, Diarra SS. Factors affecting the utilisation of cassava products for poultry feeding. Egyptian Journal of Veterinary Science. 2021;**52**:387-403

[61] Maulida YF, Hardyastuti S. The urgency of institutional development of Cassava industry in Daerah Istimewa Yogyakarta and Jawa Tengah. Sodality Journal of Sosiol Pedesaan. 2021;**9**:e33369

Chapter 10

# Production of Potato Quality Seeds in Mountainous Region of Central Africa

*Nyamangyoku Ishibwela Obedi*


## Abstract

Potato production in the mountainous region of Central Africa (CA) remains very low (10–15 T/ha), while the commonly cultivated varieties have a genetic potential to reach more than 40 T/ha. This low productivity is due, among other reasons, to a low quality of propagation material (seed tubers). Certified potato seed tubers are not only expensive but also not sufficient on the market to cover the needs of potato producers, leading them to use year-to-year seeds potatoes already infected by diseases and totally degenerated. Due to constraints of getting appropriate facilities for good Potato seeds production among producers, various cutting methods to obtain quality potato seeds have been developed. Indeed, in several experiments carried out in Rwanda and in other countries, the use of good healthy propagation material has increased yields by more than 35%. Since this mountainous zone of Central Africa is very suitable for the cultivation of potatoes for its climatic conditions, the methods developed in this chapter for producing quality potato seeds and appropriate agricultural practices (crop rotation, good fertilization, disease control and pests) will make it possible to significantly increase yields, and thus allow the varieties grown in Central Africa to express their genetic potential of more than 40 T/ha.


**Keywords:** potato tuber seeds, Central Africa, cuttings methods, yield, potato varieties

## 1. Introduction

The mountainous region of Central Africa where the potato crop is cultivated at more than 1800 m altitude and covered by this chapter concerns four countries, namely Burundi, DRC, Rwanda and Uganda. **Figure 1** shows the mountainous zone suitable for potato crops in Central Africa.

In Burundi, potatoes are grown everywhere except in areas of low altitudes (Bujumbura and Rumonge) [1]. While in the Democratic Republic of the Congo, the potato is grown mainly in the North Kivu and South Kivu provinces. Potato is cultivated across Rwanda and growing in popularity. But the majority of the crop is produced in the northwestern region of the country in the districts of Burera, Musanze, Nyabihu and Rubavu [2]. The Main potato producing districts in Uganda

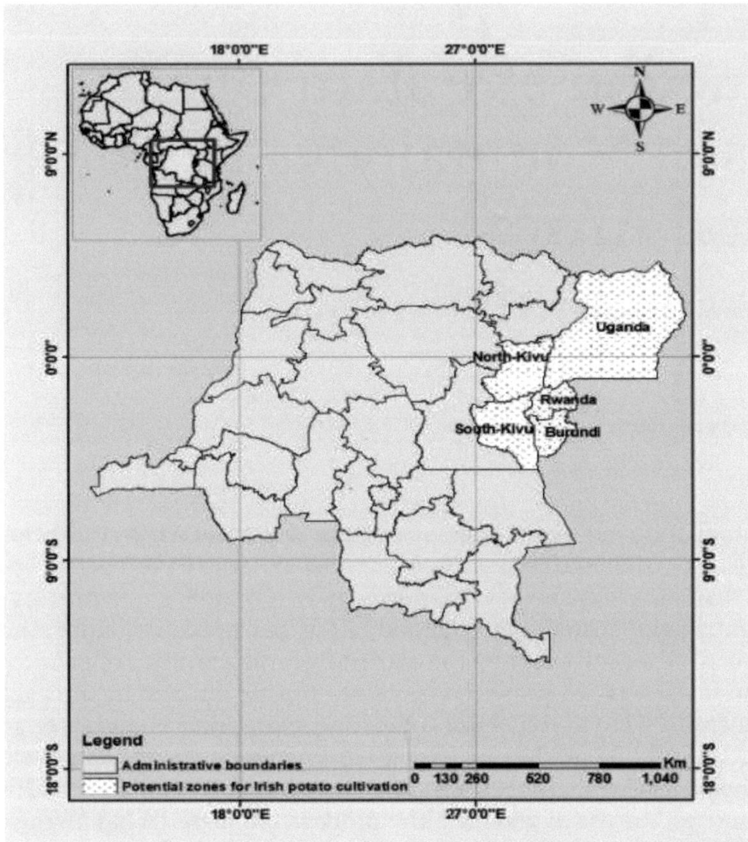


**Figure 1.**
*The mountainous area of Central Africa where potatoes are cultivated, comprising Burundi, Democratic republic of Congo, Uganda and Rwanda countries.*

are Bushenyi, Isingiro, Kabale, Kabarole, Kapchorwa, Kisoro, Kyenjojo, Masaka, Mbale, Mbarara, Mubende, Nebbi and Sironko [3].

The potato is a staple food source for many people around the world, particularly in Central Africa, it is cultivated for the following reason: It is an important crop, ranked 4th after wheat, rice and corn; takes less time to mature; a source of income in high altitude regions in Africa; job offer; source of raw material if industrialized; takes less cooking time compared to cereals; competes well in nutritional value compared to cereals; has a higher yield per unit area and in a given time; crop assuring food security in many rural areas and can be used in rotation with cereals. The nutritional value of the potato is given in **Table 1** [4].

Despite its nutritional importance and the favourable climate for its production in the mountainous zone of Central Africa, potato productivity is still low (**Table 2**), notably for the following reasons: (1) Low soil fertility, (2) Seed degeneration and low potential of existing varieties, (3) Inadequate cultivation techniques, (4) Short rotations and no intensive crop system, (5) Crop attacks by diseases (mainly late blight, bacteria and viruses) and pests, (6) Shortage and poor seed quality, (7) post-harvest losses, (8) Non-structuring of the potato sector, (9) Insufficient technology transfer, and (10) Insufficient training and technical information [1–3].

| Component | Content (%) | Component | Content (mg/100g) | Component | Content (mg/100g) |
|---|---|---|---|---|---|
| Dry matter | 15–28 | Asparagine (free) | 110–529 | Vitamin C | 8–54 |
| Starch | 12.6–18.2 | Glutamine (free) | 23–409 | Vitamin E | 0.1 |
| Glucose | 0.01–0.6 | Proline (free) | 2–209 | Folic acid | 0.01–0.03 |
| Fructose | 0.01–0.6 | Other amino acids | 0.2–117 | Potassium | 280–564 |
| Sucrose | 0.13–0.68 | Polyphenols | 123–441 | Phosphorus | 30–60 |
| Fiber | 1–2 | Carotenoids | 0.05–2 | Calcium | 5–18 |
| Lipid (fat) | 0.075–0.2 | Tocopherols | Up to 0.3 | Magnesium | 14–18 |
| Protein | 0.6–2.1 | Thiamin B | 0.02–0.2 | Iron | 0.4–1.6 |
| Nitrogen (total) | 0.2–0.4 | Riboflavin | 0.01–0.07 | Zinc | 0.3 |
| | | Vitamin B6 | 0.13–0.44 | Glycoalkaloids | < 20 |

**Table 1.**
*Potato chemical composition based on fresh weight.*

However, with good quality seed, and the use of fertilizers, the yield can reach more than 40 tons per hectare. Reason why it is important to focus more attention on good quality potato seed production and develop multiplication methods that could be applicable and affordable by potato seed multipliers.

## 2. Organization of the potato seed chain in mountainous region of central Africa

### 2.1 Potato seed production in Central Africa

There are three recognized potato seed systems in Central Africa: formal, informal, and semi-formal, as elaborated below [6, 7]:

#### *2.1.1 Formal seed system*

The formal seed system involves a chain of activities leading to certified seeds of officially released varieties. This is guided by scientific methodologies for plant breeding. Multiplication is controlled and operated by public or private sector specialists, with significant investments having been made throughout the process. In the formal system, production of basic seeds is mainly a responsibility of public research institutions [8, 9]. The basic seed is then passed on to public and private sector seed multipliers for bulking and distribution as certified seed. The regulator is responsible for the inspection and certification function.

#### *2.1.2 Informal seed system*

The informal seed system in CA context is defined as seed production and distribution practices where there is no legal seed certification. The system constitutes many individual small-scale farmers, who save or exchange seeds at the local level (**Figure 2**).

| Country | Area × 1000ha | | | | | Production × 1000tons | | | | | Yield (tons/ha) × 1000 | | | | |
|---|---|---|---|---|---|---|---|---|---|---|---|---|---|---|---|
| | 2016 | 2017 | 2018 | 2019 | 2020 | 2016 | 2017 | 2018 | 2019 | 2020 | 2016 | 2017 | 2018 | 2019 | 2020 |
| Burundi | 14 | 20 | 27 | 54 | 30 | 146 | 205 | 303 | 376 | 295 | 11 | 10 | 11 | 7 | 10 |
| DRCongo | 22 | 22 | 22 | 22 | 23 | 100 | 100 | 102 | 103 | 109 | 05 | 05 | 05 | 05 | 05 |
| Rwanda | 106 | 94 | 119 | 108 | 104 | 751 | 846 | 916 | 973 | 859 | 07 | 09 | 08 | 09 | 08 |
| Uganda | 32 | 39 | 48 | 55 | 51 | 171 | 299 | 327 | 326 | 309 | 05 | 08 | 07 | 06 | 06 |

*(FAOStat 2022 [5])*

**Table 2.**
*Potato data in Central African countries (2016–2020).*

**Figure 2.**
*Potato seeds grading under informal potato seeds system; Kavumu farm, Bukavu, South Kivu, DRC.*

It also includes development agencies and projects supporting community seed production with no regulatory oversight. It is considered the most flexible system and it involves the use of both local and improved varieties. The seed production and distribution processes are not monitored or controlled by government policies and regulations but rather by local standards, social structures and norms.

### *2.1.3 Semi-formal seed system:*

Semi-formal seed system has overlapping features with both the formal and informal seed systems. The major actors in this system are groups (of farmers) engaged in community-based seed production and marketing. Seed producers do not necessarily go through formal channels to get planting materials or through the formal certification process. The intermediary seed system also includes the production and marketing of seed by local farmers under financial and technical support from NGOs and breeding centres. Seed potato producers in the semi-formal system produce two different types of seed: (1) clean seed and (2) positively selected seed.

1. Clean Seed or QDS: This comprises seed multiplied at the farm level which originates from certified seed. It is produced using Good Agricultural Practices (GAPs). Most guidelines used in production of certified seed are also used in clean seed production. However, sampling, testing and certification by the

regulator are omitted or where involved, less rigorous as for the case of QDS. Quality declared seed is officially recognized in Uganda and Rwanda by law and can be legally sold through formal market channels but for localized areas only.

2. Positively Selected Seed: These are seed potatoes produced from farmer seed through a process of selecting the best-looking plants during vegetative growth by farmers trained in seed selection and management. Although the process of production lacks stringent procedure and inspection by the regulator it offers an opportunity for farmers to control diseases and improve their yields by an average of 30% per season.

### 2.2 Formal seed potato multiplication

Formal seed potato sources include public institutions, private seed companies, and registered individual seed growers. There are three main types of formal seed production systems [2]:

- *Public formal seed system*: Here, the public sector undertakes all activities involved in variety development, seed production and marketing.

- *Public-private formal seed system*: This involves the partnership of the public and private sector from variety development and seed production up to seed marketing. The public institution conducts research and breeding while the private sector multiplies the seed under the supervision of the regulator and distributes the seed to farmers.

- *Private formal seed system*: These are systems that are entirely performed by the private sector from variety development to seed multiplication and distribution. There is minimal government involvement except in seed quality control and certification.

There are three main business models used to produce seeds as described below.

- *Multiplication from breeder seed*: Licensed local or international, public or private seed producers grow seed from mini-tubers to the second stage of certified seed (C2).

- *Multiplication of imported basic seed:* In this case, basic or certified seed potato tubers are imported (under specified conditions) for further multiplication as certified seed by local seed companies and multipliers.

- *Multiplication of Clean Seed*: Seed potatoes are sourced from certified seed producers which are then multiplied by farmer groups and cooperatives with the support of the extension services.

### 2.3 Sources of seed potato

The sources of formal seed in CA are public institutions and private seed companies and registered individual seed growers. Such seeds must undergo certification by the regulator.

- *Sources of basic seed*: The official source of basic seed for public bred varieties are research institutions and private seed producers. Basic seeds can only be produced upon the assurance that the breeder materials are free from all major diseases and pests. Basic seed production begins with tissue culture where meristem tissues are multiplied in a controlled environment to produce in vitro plantlets. These plantlets are transferred to the glass house for hardening. Afterwards, they are then planted in pots, aeroponics or hydroponics to produce mini-tubers. The mini-tubers are planted in the field through a number of generations to produce pre- basic seeds and basic seeds. The basic seed is then supplied to authorized seed multipliers for production of further classes of certified seed

- *Sources of certified seed*: These are mainly the authorized seed multipliers and seed companies that source basic seed either from public institutions or private basic seed producers to produce certified seed. Provided the seed meets the quality standards, the certified seed can be further multiplied up to two cycles for production of ware potato by growers.

## 3. Production and management of quality potato seeds

The usual way to spread potatoes all over the world is by vegetative way through whole or fragmented tubers. The latter is only practised in a few European countries and in North America [3, 10].

Through this vegetative production from generation to generation, seed tubers can be infected with a high number of viral, bacterial and fungal diseases, and thus seriously affect the growth and production of the plantation. This is the reason why the production of quality seeds, storage and related legislation are discussed in the following pages

At the harvesting time, the seed tubers are usually dormant and do not present growing buds even if the ecological conditions are favourable.

### 3.1 Characteristics of potato quality seed

Many factors determining yield are affected by seed quality. Good quality seed is obtained when the seed tubers are harvested on a seed field in good condition, dry before the formation of the tubers has been naturally completed and harvested shortly after dehaulming. Quality tuber seed should be able to produce healthy and vigorous plants within the growing season. Seed quality is determined by the following parameters [6, 10]:

a. Variety: seed must be true variety and should not have different varieties or mutants;

b. Physiological maturity: At planting, seeds should be physiologically mature with multiple vigorous sprouts. This kind of seed tubers germinate rapidly and develop several stems.

c. Size: The size of a seed lot will need to be homogeneous to ensure consistent germination and development.

d. Healthy tuber: tuber seeds must be free of disease and the percentage of infected tubers in the seed lot must be below locally accepted standards. Seeds devoid of diseases and parasites are obtained from tubers that themselves were devoid of them.

e. Physical defect: Seeds must be free from external damage that could occur during harvesting or during post-harvest treatment that could make tubers vulnerable to bacteria, fungi and parasites, and thus increase the risk of low germination and growth. Seeds should also be free from any internal damage caused by poor growing and storage conditions. This can increase vulnerability to diseases.

### 3.2 Dormancy

Like other plant organs, the tuber goes through a phase during which its buds do not show significant growth. It is a period of any vegetative growth, during which the tuber is unable to germinate, even when it is placed in optimal conditions for germination, such as optimal temperature and humidity. Influence of different factors on dormancy:

#### *3.2.1 Tuber-related factors*

- *The size of the tuber*: The sprouts of small tubers take longer than the sprouts of large tubers to reach a certain length (3 mm).

- *Genotype*: the dormancy time depends in particular on the genotype.

In general, most early varieties have a rather short vegetative dormancy.

#### *3.2.2 External factors*

In general, environmental factors would have a very limited impact on dormancy. However, the effects of nitrogen, temperature and light were reported in the experiments. Nitrogen fertilization and relatively high temperatures would shorten the dormancy time. Whereas the light effect would have no significant impact.

The agricultural calendar for potato seed production should take into account the calendar for potato consumption production, the dormancy of tuber seed and the fact that tuber seed is generally earlier harvested when the tuber is still small. Therefore, three examples of calendars are given considering the dormancy duration of Potato varieties in CA countries (**Figures 3–5**), and where very early varieties, early and late ones cannot be scheduled in the same ways.

### 3.3 Production of quality seed plants/tubers

The production of potato plants requires considerable agricultural practices. This involves the vegetative production of tubers which do not deviate in any way from the original characteristics of the variety and which free diseases.

Although "True potato seed" regeneration was initially the only means available to producers a clonal breeding model was then developed at the beginning of the 20th century in order to preserve the intrinsic qualities of the cultivars produced [11–13].

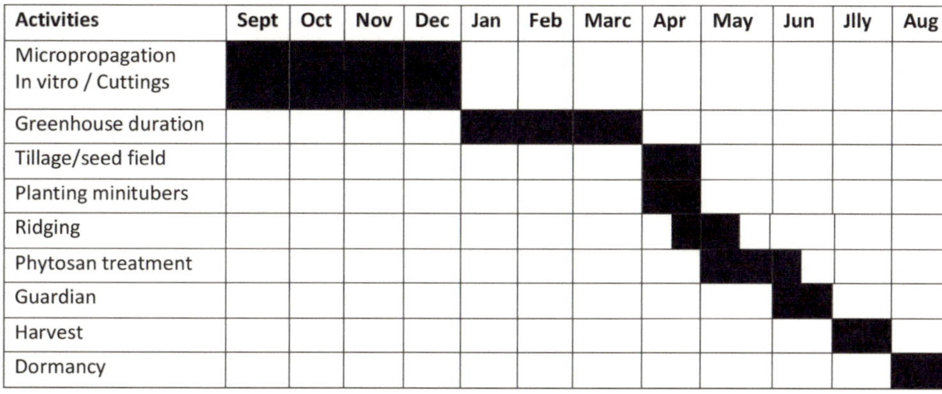


**Figure 3.**
*Agricultural calendar for very early varieties (CIP 720118, CIP 381381.20 and Rw 8201-19).*

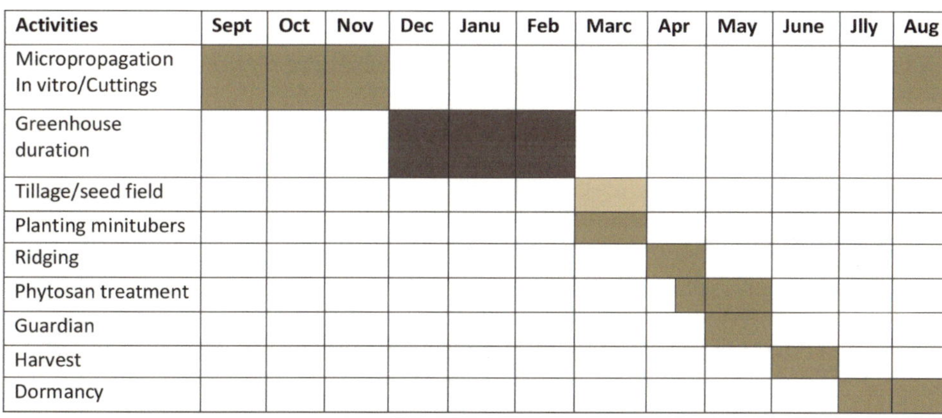


**Figure 4.**
*Agricultural calendar for Early potato Seed Varieties (CIP 386003-2, CIP 393077-54, CIP 393371-50).*

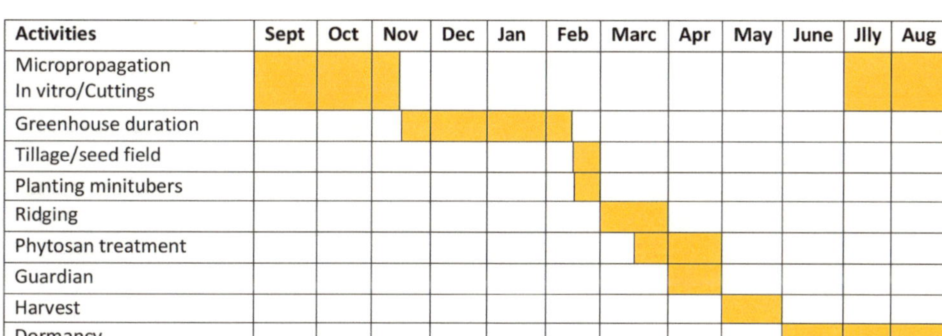


**Figure 5.**
*Agricultural calendar for Late potato Seed Varieties (CIP 800949, CIP 381391-13, CIP 381395-1, CIP 383120.14, CIP 387233.24).*

The development in the recent past of in vitro potato micropropagation techniques combined with new quality control techniques (ELISA, PCR, etc.) has led to a significant increase in the overall quality of production. Micropropagation has also provided more flexibility and speed in plant production processes. These two methods will be developed below.

#### *3.3.1 Traditional Inbreeding and clonal selection*

This method includes the application of techniques that maintain the preservation of good health over time, namely [4, 6, 13]:

1. Selection of mother tubers on apparently healthy plants,

2. Tuber growth in areas of unfavourable climate for viral infections;

3. Isolation of propagating crops from infected consumer crops;

4. Severe mass selection in propagating crops;

5. Early dehaulming to remove leaves from plants and prevent them from infections.

##### *3.3.1.1 Selection of mother tubers on apparently healthy plants*

It is essential, before multiplying a potato clone, to ensure that the seed tuber is free from any disease, including virus-borne diseases.

The tuber is chosen from the healthy batch (varietal collection, family field, etc.) and then undergoes a series of various tests (ELISA, visual checks, etc.) to verify the absence of any infection.

Sometimes there may be no healthy tubers (old varieties chronically infected with one or more viruses). Since viruses are mostly absent from seeds and meristem, regeneration from meristem and thermotherapy can help to cure these diseases and thus obtain a healthy starting material.

These very fine techniques have proven their effectiveness for a long time.

Meristem culture consists of growing in vitro a very small fragment of the apical shoot which is generally virus-free. This technique gives very good results, provided that only the strict meristematic zone is sampled (of the order of 0.3 to 0.5 mm).

##### *3.3.1.2 Tuber multiplication in unfavourable climates for viral infections*

Except for the generation F0 production or subsequent in vitro propagation and micro-cuttings, the majority of subsequent multiplication is done in the field and the impact of the environment is then essential to producing good quality plants. Indeed, zones with a climate unfavourable to the propagation and dissemination of aphids (cool and humid regions, with frequent winds) make it easier to maintain a good sanitary condition of the crops. Following methods can be applicable:

###### *3.3.1.2.1 Removing diseased plants*

In addition to these environmental precautions, it is also necessary to eliminate the sources of inoculum that are infected virus plants themselves inside the crops.

This cannot be done before planting, as the tubers infected by virus generally do not show symptoms, with the exception sometimes of some discolouration on the sprout.

On the other hand, the foliage of contaminated plants expresses various specific symptoms of the virus notably, leaf roll, mosaics, curly, stunted, etc. These symptoms may be more or less easy to observe depending on the varieties, vegetative stage and climatic conditions and laboratory techniques may be required in addition to visual examination. The knowledge of these visual symptoms helps to eliminate diseased plants (purification) in a continuous way during the period of growth of the foliage.

##### *3.3.1.2.2 Removal of regrowth and other sources of viruses*

Another source of contamination is regrowth, plants from small-scale tubers left on the ground after harvest. The number of these tubers can exceed one hundred thousand per hectare, there is no effective means against this regrowth except the respect of a certain rotation time.

##### *3.3.1.2.3 Limiting passages*

In order to limit the spread of contact-transmissible viruses (PVX and PVS), it may be useful to limit the passage of tools, machines and humans.

The cultivation operations will also be done always starting with the healthiest plots, to avoid contaminating them hard to pass into the affected plots.

##### *3.3.1.2.4 Protective treatments*

Control of vector aphids is first carried out by chemical means. It is especially effective in preventing the spread of so-called persistent viruses (PLRV) whose particles are infectious only after a certain period of residence inside aphids. So, we can destroy the aphids first. The spread of non-persistent viruses, among which PVY is particularly important, is much better contained by the application of mineral oils. These nevertheless have some disadvantages: by maintaining moisture on the foliage, they may promote the development of Late Blight and make it more difficult to purify certain varieties whose leaves may deform as a result of phytotoxicity.

#### *3.3.1.3 Isolation of propagating crops from infected consumer crops*

An important factor is a distance from external sources of contamination represented by consumer potato plots and regrowth in other crops. It may be interesting to exclude some crops from plant production areas as they may be a source of vector aphids. The standards for isolation are:

- Plots with pre-basic material of one or more varieties must be isolated at least 50 meters from any other potato crop;
- Plots for the production of basic plants are separated by at least 10 m from any other potato crop;
- Plots intended for the production of certified plants are isolated by at least 10 m from any consumption potato crop;

- Breeding plots of different varieties are separated by at least two empty rows;
- When the presence of regrowth is observed in the separation interval, these are considered as not isolated;
- In cases where the production plots of basic or certified plants are adjacent to another plot which, during the vegetative stage, presents a danger of contamination, this one shall be cleaned up on a contiguous strip with a minimum width of 10 m.

*3.3.1.4 Severe mass selection or varietal treatment in propagating crops*

Selection is mandatory from the beginning of the vegetation until the beginning of yellowing of the leaves (early maturity).

It consists of the removal of foreign and nonconforming plants, regrowth and plant with virus diseases as soon as symptoms appear, severe rhizoctonia and verticillium.

*3.3.1.5 Early dehaulming to remove leaves from plants and prevent them to infections*

Potato plots are systematically removed early, meaning that the foliage is reduced before ripening. This practice removes vegetation from aphid flight and prevents virus migration to the tubers (in the case of late contamination). The follow-up of aphids allows knowing whether or not to advance the dates of dehaulming.

By limiting the life of the foliage, removing old leaves also helps to limit the growth of the tubers. This is one of the ways to obtain a suitable proportion of small and medium-size tubers corresponding to the regulation of plant trade.

*3.3.2 Pre-base seed production*

*3.3.2.1 Production of quality plants by in vitro micropropagation*

In the in vitro system, parts of plants are propagated and regenerated into whole plants or tubers under sterile artificial conditions. For rapid multiplication, three types of material can be used: (a) Node cuttings, (b) Apical cuttings and (c) Micro-tubers [14]

Apical cuttings are not frequently used because they are not widely available. To do this, we will focus our attention on node cuttings and micro-tubers.

*3.3.2.1.1 Node cuttings*

When a large quantity of seedlings with high genetic quality and a maximum sanitary condition is desired, rapid multiplication is done by node cutting. The various steps are as follows:

**1. Source plant selection and treatment**

Three types of plant material are used:

- Tubers; where the tubers are used as starting material, they must first be pre-germinated, after which the buds are harvested. This step can take 2 months.

- Stems; stems can be harvested directly from the field or from greenhouse plants.
- Seedlings in vitro

These materials must conform to the cultivar and be viral tested (mainly PVX, PVS, PVY, PVA, PVM, and PLRV), bacteria (stem rot, brown rot and those caused by Erwinia spp) and other diseases.

**2. Disinfection of stems**

The leaves must be removed from the stems, after which the stems are disinfected for a few seconds into alcohol 70–96%, followed by a few minutes in a solution of sodium hyperchlorite 10%. After disinfection, rinse with sterile water (distilled water).

**3. The first in vitro culture**

The cuttings are cut into pieces including an axillary bud. They are then placed in the culture medium, with the base pushed into the medium. The test tube must be closed. After 4 days, the bud could start to grow into a stem.

Throughout the in vitro development stages (Steps 3 and 4), only 1 medium type is used. The Murashige and Skoog growing medium contain a wide range of nutrients including the auxin hormone that induces root development. Consequently, the use of this growing medium during stages 3 and 4 leads to the production of seedlings [15]. The composition of the culture medium

| Composition of the medium | Growing conditions |
|---|---|
| Agar 80.0 g/l | Light duration 16 h |
| Murashige & Skoog 4.2 g/l | Light intensity 800–500 lux |
| Sucrose 25.0 g/l | Temperature 20–23°C |
| Alar 85% 0.001 g/l | |

**4. Multiplication**

Seedlings produced in Step 3 are cut into pieces with an axillary bud and leaf. These individual pieces are placed in a new tube medium, with the bud just above the nutrient medium. This multiplication step can be repeated every 4 weeks, resulting, on average, in 5 new vitro-plants from a single initial vitro-plant.

**5. Acclimatization**

After the desired number of vitro-plants is produced, they must be fortified for later uses in greenhouses, under shelters or in fields. For acclimatization, the vitro-plants are planted in small pots containing the soil, then the temperature is reduced from 20–23 to 18°C in the greenhouse.

If the micro-tubers are produced directly in a greenhouse, the vitro-plants have no interest in being fortified (physiological maturity), they can be used as such.

#### 3.3.2.2 Production in a protected system or in a well-controlled environment

The semi-in vivo environment refers to sterile (aseptic), artificial (in vitro) and natural (in vivo) conditions. An example of a semi-in vivo environment is the greenhouse or under-shelter. The semi-in vivo propagation systems are: (a) Bud cuttings, (b) Stem cuttings, (c) Leaf bud cuttings, (d) Single node cuttings and (e) Micro-tubers

##### 3.3.2.2.1 Bud cuttings

The bud cutting is a rapid propagation method [16, 17] which requires following steps (**Figure 6**).

**1. Tuber selection**

The tubers selected are those that have passed the dormant stage and are healthy systemic pathogens.

**2. Treatment of tubers**

After breaking of dormancy, a vigorous bud is stimulated by the transfer of tubers every 7 to 10 days from the dark medium to the light (indirect) medium and back. Darkness promotes the growth and development of the internodes, while indirect light improves the vigour of the buds and induces short internodes. After treatment with dark and indirect light, when the buds are 3 cm long, the apical part is sectioned by a very sharp knife. This stimulates the formation of lateral buds and consequently the increase in the number of cuttings. After removal of the apical part, the tubers with their buds are submerged for a maximum of 10 minutes in a solution of

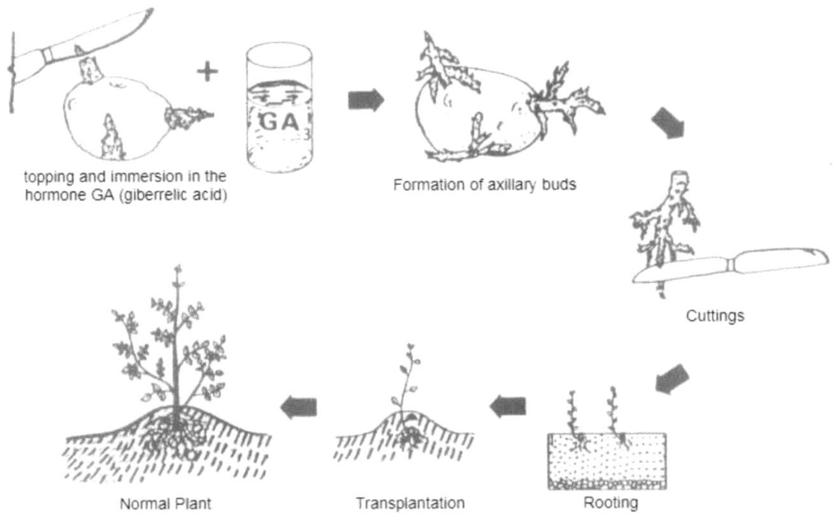


**Figure 6.**
*Bud cutting technique (Stapes to follow).*

gibberellic acid of 1–2 ppm in order to improve the growth of the buds. After which the proper distance of the internodes is regulated by placing the tubers in darkness or under indirect light. Root formation can be stimulated by high relative humidity.

**3. Cuttings**

A small portion of the buds will have to remain on the tubers if another harvest of the bud cuttings is expected. Two to three bud harvests can be taken from each tuber if they are physiologically young.

If the tuber is to be planted, at least a harvest can be made. After removal of the tubers, the buds are sectioned into pieces of one or more nodes. Cuttings should have at least one apical bud and two small roots to ensure good growth of the new seedling. A single tuber can produce up to 40 buds, depending on the size of the tuber, the number of eyes and the management of the buds.

**4. Planting Bud Cuttings**

The bud cuttings are planted on a well-drained substrate (fine sand 1 mm) in tray. The apical bud should be slightly higher above the sand after first watering. Terminal cuttings grow very quickly and should therefore be planted separately from those from the base.

**5. Transplantation**

After about 15 days, the cuttings will have formed roots and are thus ready for transplantation. Two days before transplant a foliar application of fertilizer is desired. The transplant can be done directly in the field or in pots in the greenhouse.

- *In the field:*

The soil-cuttings contact must be optimal. At least one leaf node must be underground; this is best done by watering after transplantation. Good results are obtained when the fertilizer solution is made of a concentrated mixture of $P_2O_5$ and water.

Two to three weeks after transplant, cuttings are treated as normal potato plants. Early earthing-up must take place to maximize tuber production. The average yield of 500 gr per plant can be obtained.

- *In the pots:*

Cuttings planted in pots can be considered mother plants for the production of node cuttings. If 3–4 cuttings are transplanted into a large pot, they can be used for future propagation of stem cuttings, leaves or for tuber production.

*3.3.2.2.2 Stem cuttings*

Stem cuttings used as rapid propagation can produce 20 to 60 cuttings from each parent plant. The advantage of stem cuttings is that non-systemic diseases and nematodes can actually be eliminated because only the top part is used for propagation [18, 19]. The procedure is as follows (**Figure 7**).

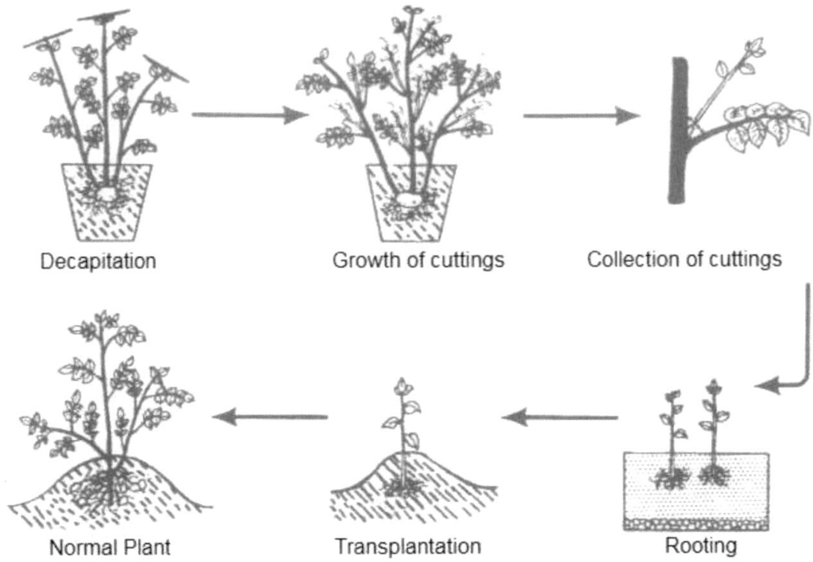


**Figure 7.**
*Stem Cutting Technique (Stapes to follow).*

## 1. Management of mother plants

Mother plants are grown in pots in a greenhouse from the best buds of tubers or cuttings with disease free. For optimal use of the greenhouse space, plants with 3 or 4 stems are preferable, depending on the size of the pot and the cultivar.

When tubers are used to produce mother plants two methods of planting exist:

- The tubers are planted in the pot and lightly covered with the soil. The shoot emerges together with the aerial stolons.

- When tuber production is desired in pot, the earthing-up must take place early.

Aerial stolons are harvested as cuttings. When mother plants are 25–30 cm tall, each stem is decapitated. This eliminates apical dominance and stimulates the development of lateral branches from the axillary buds of each leaf. After 15–20 days, the cuttings are ready for harvest.

Mother plants must be fertilized with nitrogen for rapid growth and phosphorus for rapid root development. Liquid fertilizer is desired after beheading and after each harvest of cuttings. A suitable solution is obtained when 5 gr of 12-14-12 fertilizer is dissolved in 1 litre of water.

## 2. Cutting

When the lateral branches (cuttings) are 12–15 cm long, they are cut with a sharp knife. The section must be made near the new axillary bud that will produce the branch, which will produce cuttings. After the first harvest, additional harvests are made at intervals of 12–15 days. These have cuttings yield of 30–60% more than that of the first harvest. In total, between 20 and 60 cuttings are obtained per mother plant.

Cuttings should preferably have a stem 4–5 cm below the node of the first leaf. If cuttings cannot be planted immediately, they can be stored in the refrigerator at 4–6°C for up to 2 days.

**3. Culture of cuttings**

The cuttings are planted at 5 × 5 cm spacing, on tables (bins) containing a well-drained substrate, of washed sand (size 1–2 mm). If necessary, cuttings can be dipped in a hormone-containing solution (auxin) to activate root formation for 2–3 days. When cuttings are planted, the lowest leaf node should be below the sand and the roots should have good contact with the soil. If the hormone has been used, irrigation should be done at least in 2 hours to allow its impregnation. Plants must be grown under shade.

**4. Transplantation**

Fourteen days after the extraction of mother plants, plants are ready for harvest if they have been well rooted. The plants can be transplanted in pots or directly in field.

- Pot transplants are made to produce new mother plants or tubers. Plants should be put under the soil as for tubers (see point 1). One or more nodes should be covered with soil.

- When transplanted in the field, the plants are placed at a uniform distance with one or more leaf nodes under the ground. Liquid fertilizer with a high phosphorus concentration will need to be applied to improve root development. This should be done early and lightly.

*3.3.2.2.3 Leaf bud cuttings*

As with stem cuttings, the use of leaf bud cuttings as a rapid propagation technique eliminates non-systemic pathogens from the soil and tubers [20, 21]. The procedure is as follows (**Figure 8**).

**1. Selection of the mother plant**

The first step in the production of leaf bud cuttings is to choose a healthy and suitable mother plant. The mother plant should be cultivated under the conditions of a long photoperiod (long days) and then kept for 10 to 15 days in short photoperiod (short days) before cutting to induce tuberization. A plant that begins senescence (when the basal leaves are mature) is ready to cut the cuttings of leaf buds.

**2. Cutting**

The best cuttings yield comes from the central part of the plant. Those coming from the lower part of the plant produce smaller micro-tubers, the cuttings of the top parts produce few micro-tubers and tend to produce aerial roots and buds.

After removing the stems from the base of the plant, they are cut (main stems) into cuttings of leaf buds of 1–3 cm, depending on the cultivar. In the centre of the

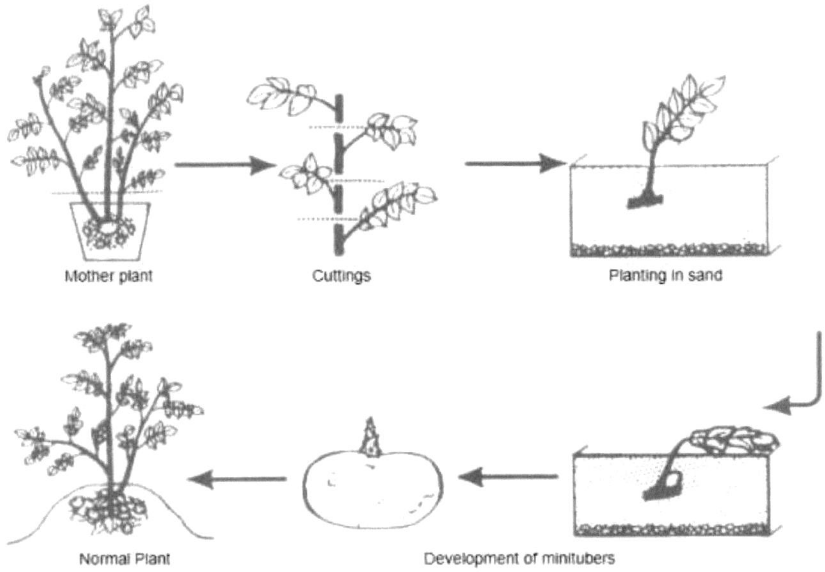


**Figure 8.**
*Cutting technique of leaf buds (Stapes to be followed).*

cuttings, a node should be present with an undifferentiated bud and a leaf. Approximately 70–100 cuttings can be produced per plant.

### 3. Cultivation of cuttings

Leaf bud cuttings are planted with part of the stem in a well-drained substrate (fine sand, 1 mm), with the bud below and the leaf above the surface.

The cuttings are planted in line 5–7 cm apart, depending on the size of the leaves, and their leaves should be covered with sand more or less completely. Cuttings and sand substrate should be well in contact. Irrigation should be done carefully with very fine water droplets. When the light intensity is high, shading can be provided. The temperature in the greenhouse should be relatively low, about an optimum 20°C.

### 4. Harvest

After one or two weeks, the germs begin to develop. When all the leaves have died (4–6 weeks after planting, depending on the cultivar and temperature), the micro-tubers are harvested. Generally, 1 micro-tuber is harvested by cutting, but sometimes 2 can be obtained. As a result, 80–120 tubers are harvested from a single-parent plant. The typical size of tubers produced in 31 days is between 0.5 and 1.0 cm (0.2–1.0 gr).

By the technique of leaf buds, more than 1000 tubers can be produced on 3 $m^2$.

### 5. Storage of Micro-tubers

Micro-tubers can be stored for 4–6 months at 4°C and relative humidity of 90% after which the dormancy period is generally over.

**6. Planting in the field**

The spacing between the lines is usually 15–20 cm, depending on the cultivar, the soil conditions and the desired tuber size. In general, micro-tubers produce only one main stem and average yield of 500 gr each (depending on cultivar, soil and climatic and management conditions).

*3.3.2.2.4 Single node cuttings*

"Single node" cuttings should not be confused with "node cuttings". Single node cuttings are produced under in vivo conditions, while node cuttings are produced under in vitro conditions (previous section) [17, 18]. The single node cuttings technique is often used to produce many seedlings in the first generation of the pre-basic seed production program.

When the produced plants are transplanted into the field, the yield of the tubers can be about 500 g/plant, while the produced tubers have an ideal size of the seed tubers. The process is as follows (**Figure 9**).

**1. Mother Plant Management**

The small mother plant of single node cuttings can come from the cuttings of buds, plants from the vitro-plants, stem cuttings, micro-tubers or true seed tubers. When the plants have 5–6 leaves, the cuttings are taken, but 2–3 days before this, foliar

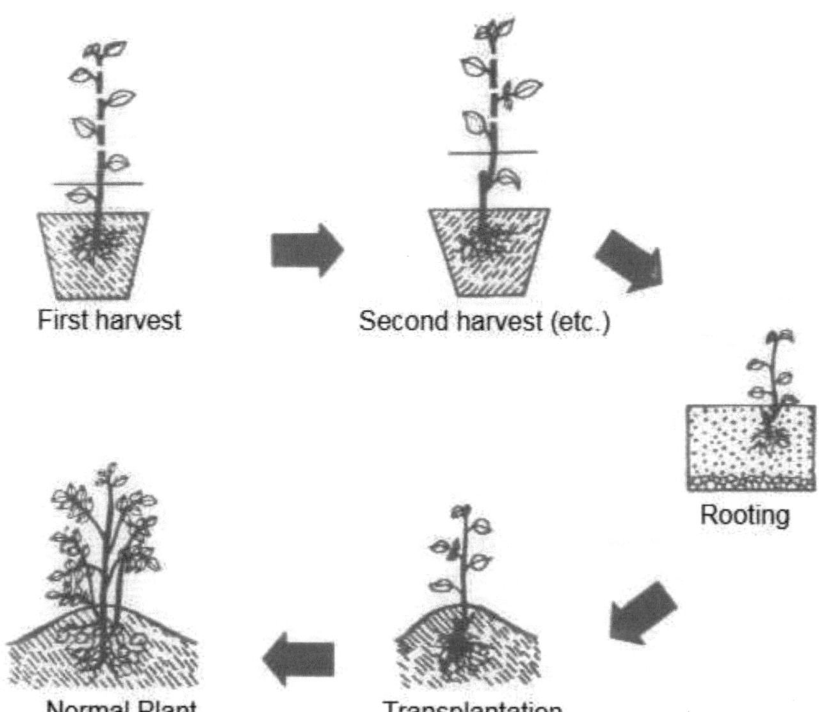


**Figure 9.**
*Stapes of Single node cutting technique.*

fertilization must be applied. The stem is cut from the mother plant leaving a strong whole leaf. After each harvest of cuttings, the mother plant is fertilized with a liquid NPK fertilizer to stimulate the growth of new cuttings. An application of 5 g of a 12-14-12 fertilizer per litre of water is appropriate for each plant. A new stem is formed at the leaf node remaining at the base of the stem.

A temperature of 23–26°C and a long photoperiod stimulate rapid growth. In addition, these conditions not only stimulate growth but also do not induce the formation of tubers. The new stem is ready for harvest in 15–29 days. Each parent plant can be harvested 2–10 times and produce between 30 and 200 cuttings.

### 2. Cutting

After removing the stems from the mother plant, they are sectioned, each with a leaf and an axillary bud in the centre.

### 3. Planting

Before planting, cuttings are brought into contact with a rooting hormone. Single node cuttings are either planted individually in a pot to produce more mother plants, or they are planted together in a well-drained rooting substrate made of 1 mm of sand. The latter cuttings are planted in the soil. The plants will be transported to the field in the next phase. Cuttings should be put deeply in the soil, so the node and stem are covered with sand.

### 4. Culture of cuttings

Temperatures of 20–23°C (rooting optimum) and 23–26°C (shooting optimum) are ideal. Irrigation must be done by a fine droplet sprayer, 2–3 hours after planting to allow the hormones to penetrate the plant tissues. Usually, 15–20 days are required to produce a seedling, which can be transplanted. The seedling thus has adequate roots and 3–4 leaves.

### 5. Field Transplant

Single node cuttings could be planted in the field in narrow, spaced lines. Yields can be around 500 gr per plant. The tubers are ideal for use as seed tubers.

## 4. Plant selection and elimination

Plant selection and elimination are done to remove off-type plants, remove sources of infection such as virotic plants, and thus prevent the spread of diseases. Viral diseases as well as those caused by bacteria and fungi are, especially controlled by this approach. These practices also have long-term effects because they reduce the number of infected tubers entering the stock, where the infections could spoil the entire seed lot. These practices also reduce the level of inoculum and spread of virus infections. Plant selection and elimination can only be effective under the following conditions:

- Timing: should be done as early as possible to avoid possible spread of disease;
- Frequency: repeat operations regularly;

- Extended throughout the field;
- Carefully remove plants to avoid dispersal of vectors by more movement.

In fact, this negative selection (removal of infected plants) does not guarantee that all remaining plants are healthy. It could only have effects if the majority of plants are healthy.

Positive selection is also possible. In this case, apparently healthy plants are harvested only if they are surrounded by other apparently healthy plants.

### 4.1 Dehaulming, harvesting and post-harvest work

Another protective measure that is often needed in potato seed production is shoot elimination (dehaulming) [22]. It allows for:

- Prevent the spread within the crop (from plant to plant) of the main fungal diseases of aerial origin and even some of the soil;
- Stop the production of fungal spores that could attack tubers (e.g. *phytophtora infestans*);
- Prevent viral infections when vector populations become very important or stop the spread of viral diseases from the air to the forming tubers;
- Stop growth and thus allow the tubers to resist adverse climatic and edaphic conditions (drought, cold, heat);
- Stop the growth of the tubers in the formation when they have reached the desired size and initiate their maturation and skin colouring.
- Removing shoot can also influence tuber dormancy.
- The time between removing shoot and final harvest is important. It seems that the long period (more or less 3 weeks) between removing shoot and harvest could lead to attacks of various diseases and parasites.
- Depending on the temperature of the soil, about 10 to 20 days are required for the tubers to acquire a beautiful skin, so the harvest can be done without too much damage, due to the spread of diseases. The harvesting of tubers must be done with great care and avoid as much as possible injuring the tubers with the tillage instruments.

After harvest, tuber seeds will need to be dried cleanly to further prevent the development and spread of diseases. Initially, storage temperatures will need to be relatively high (15°C) to allow more firmness and skin coloring. Once this process is completed, the temperature should be lowered to a temperature that allows physiological development during the storage phase, with minimal loss of dry matter and water. Good ventilation is essential. Heat and cold shocks can be applied to advance the physiological maturation of the tubers.

## 5. Potato seed certification

### 5.1 Types of seed

Following are categories of types of seeds [14, 23]

*a. Breeder Seed*

- It is a class in seed certification;
- They are produced by the breeder, owner of the variety;
- They are controlled by the person or research institutions;
- This is the source of production of pre-basic seeds;
- They are generally in very small quantities.

*b. Pre-basic seeds*

It is a seed generation that is still under the control of the breeder or the research institution.

*c. Basic seeds*

It is a seed class that constitutes the last step of multiplication of initial seed and which plans for the production of certified seed. This category is generally given to multiplier farmers.

*d. Certified seed*

It is a seed generation from basic seed that has been certified to meet the standards of genetic purity established by the certification agencies

*e. Standard seed*

It is seed of lower status that is subject only to varietal conformity, purity and laboratory tests.

### 5.2 Seed certification process

Quality control objectives are two: (a) Ensure that farmers receive quality seed in terms of input to maximize their production at harvest, and (b) Ensure farmers are not at risk of receiving low-quality seed (inputs) from fraudulent traders [24].

The Seed Certification process involves the following steps: Field inspections, Seed Treatment; The seed test, Label and seal, Post controls, and Post Certification Survey

#### *5.2.1 Field inspections*

This is the first step in the certification process. It consists of:

- Registration of the field;
- Provide proof of origin from parents;
- Observe the minimum isolation distance.

The inspection is conducted to ensure that harvested crop is the true type, it has no contamination, and culture is healthy (disease free/pest free).

#### *5.2.2 Seed treatment*

Tuber seeds are harvested and treated to remove the surrounding soil, wash and then categorize (calibration) to have homogeneous seed tubers. The categories are as follows:

Grade No. 1: Are tubers retained by a sieve with a diameter of 35 mm mesh. They correspond to the largest size of seed tubers.
Grade No. 2: Are tubers retained by a sieve with a diameter of 25 mm of mesh. They correspond to the large calibre of seed tubers.
Grade No 3: Are tubers retained by a sieve with a diameter of 15 mm of mesh. They correspond to the medium size of the seed tubers
Grade No 4: Are tubers retained by a sieve with a diameter of 10 mm of mesh. They correspond to the small calibre of seed tubers.

#### *5.2.3 The seed test*

Laboratory tests are useful for determining purity, dormancy, tuber recovery capacity, moisture content and health status of seed lot.

#### *5.2.4 Labelling and sealing*

Each seed lot is labelled with a label and a mark.

#### *5.2.5 The Post-test checking*

Post-test checking is intended to show that the measurement of the previous control was effective. They ensure that varietal traits remain unchanged during multiplication

#### *5.2.6 The post-certification surveys*

This is done by all provincial offices across the country. This ensures that all seeds are well planted and on time. Samples are taken at the selling point and from farmers and are planted along control strips for comparison. Low-quality plants are therefore easily checked.

## 6. Conclusion

Potato is ranked as the fourth most important staple food crop and the number one non-grain food commodity as per the Food and Agriculture Organization of the United Nations (FAO) (2018). It has increasingly become an international commodity cultivated across the world. In Central Africa, Potato is cultivated in mountainous regions where climatic conditions are favourable. However, its productivity is still very low (5–10 tons/ha) due to the use of degenerated potato seeds and poor agricultural practices by farmers. Limited use of quality seed potatoes, by less than 4% of farmers is due to lack of appropriate facilities, in vitro laboratory, which cannot be affordable to potato seed producers. So, in this chapter different techniques of cuttings that could be applicable to farmers are developed by using meristematic tissues which are considered free diseases and restore the genetic potentiality of cultivated varieties.

It was proved that the use of good quality seeds increased the productivity of potatoes by more than 35%. In order to make them available during cropping seasons, it would be necessary to take into account the variable dormancy of the different varieties and to establish an appropriate agricultural calendar the potato seed production, which examples have been given in this chapter.

## Author details

Nyamangyoku Ishibwela Obedi
Institut Supérieur Pédagogique de Bukavu, Université Catholique de Bukavu, Bukavu, Congo

*Address all correspondence to: obedingyoku@gmail.com

## References

[1] Bararyeny A et al. Assessment of Opportunities for Burundian Small-Scale Potato Farmers to Increase Productivity and Income. European: Association for Potato Research; 2018

[2] Placide R. Overview of Potato sector in Rwanda. Rwanda Agriculture Board Presentation. 2019

[3] Storey M. The harvested crop. In: Reugdenhil DV, editor. Potato Biology and Biotechnology Advances and Perspectives. Oxford: Elsevier; 2007

[4] Netherlands Enterprise Agency. Potato Roadmap Uganda report. 2021

[5] Food and Agriculture Organization of the United Nations (FAO). Food and Agriculture Data. 2022. Available from: http://www.fao.org/faostat/en/#data

[6] East African Community. Regional Situation Analysis of the Potato Sub-Sector in the East African Community. Arusha, Tanzania: Zenith Media Limited; 2021

[7] Scientific African. Potato Market Access, Marketing Efficiency and On-farm Value Addition in Uganda. 2016

[8] Economic Policy Research Centre. Investment Opportunities and Challenges in the Irish Potato Value Chain in Uganda. Kampala, Uganda: PASIC; 2016

[9] Research Gate. Participatory Assessment of Potato Farming Systems, Production Constraints and Cultivar Preferences in Uganda. 2017

[10] Haverkort AJ. Potato Handbook, Crop of the Future. The Netherlands: Wageningen University & Research; 2018

[11] Placide R. Overview of Potato sector in Rwanda. Rwanda Agriculture Board Presentation. 2019

[12] Astere B et al. la culture de la Pomme de terre au Burundi : Acquis sur la recherche te perspectives. Bujumbura, Burundi: ISABU; 2012

[13] https://www.cultivariable.com/instructions/potatoes/how-to-grow-true-potato-seeds-tps/

[14] Struik PC, Wiersema SG. Seed Potato Technology. Netherlands: Wageningen Press; 1999

[15] Murashige T, Skoog F. A revised medium for rapid growth and bio assays with tobacco tissue cultures. Plant Physiology. 1962;**15**:473-497

[16] Goodwin PB. Rapid propagation of potato by single cuttings. Field Crop Research. 1981;**4**:165-173

[17] Haapala T. Use of single-leaf cuttings of potato for efficient mass propagation. Potato Research. 2005;**48**:201-214

[18] Cole EF, Wright NS. Propagation of potato by stem cuttings. American Potato Journal. 1967;**44**:301-304

[19] Ahmed MA et al. Using stem tip cuttings in potato production, Minia. Journal of Agricultural Research and Development. 2018;**38**:363-389

[20] Hossain MJ. Tubers from leaf-bud and apical-bud cuttings: Methods for rapid clonal propagation of potato in Bangladesh. Field Crops Research. 1993; **35**(2):147-148

[21] Pedrosa JLF et al. Yacon potato propagation from herbaceous cuttings with different numbers of buds.

International Journal of Agricultural and Natural Resources. 2020;**47**(1):46-57

[22] Upadhyay A, Bashyal S. Effects of Dehaulming in Potato (*Solanum tuberosum*) cultivation: A review. International Journal of Environmental Agricultural and Biotechnology. 2020; **2020**:2456

[23] Thiemens-Hulscher M, Delleman J, Eising J, Lammerts van Bueren ET. Potato Breeding: A Pratical Manual for the Potato Chain. Hague: Aardappel; 2013

[24] UNECE. Guide to Operating a Seed Potato Certification Service. United Nations, Geneva, Switzerland: UNICEF; 2015

Chapter 11

# Nigeria Root Vegetables: Production, Utilization, Breeding, Biotechnology and Constraints

*Emmanuel O. Ajayi, Pamela E. Akin-Idowu, Olaide R. Aderibigbe, Dorcas O. Ibitoye, Gloria Afolayan, Oluyemisi M. Adewale, Esther A. Adesegun and Benjamin E. Ubi*

## Abstract

This chapter describes the various facets, from agronomy to marketing, of Nigerian root vegetables including garlic, onion, turmeric, ginger and carrot being the world's most significant and vital root vegetables which have high culinary and medicinal value. The chapter commences with their origin and history, universal spread, production figures, areas under cultivation and goes on to explain the botany, diversity, conservation, production practices, pests and diseases management, utilization, post-harvest technology and their uses as nutraceuticals. This chapter also presents the socio-economic, market analyses, export potential of these crops in Nigeria. It would be an important reference material for researchers, agricultural and food science students at both undergraduate and postgraduate level and policy makers; and be of great interest to experts and industries involved in root vegetables and spices trade. The in-depth information and knowledge about the genetic conservation, socio-economics, production, pests and diseases management and post-harvest technology of root vegetables in Nigeria provided in this chapter would greatly help in efforts towards improving their production and utilization for enhanced nutrition and healthy living.

**Keywords:** Nigeria, root vegetables, utilization, phytochemicals, agronomy, genetic resources, genetics and breeding, biotechnology, production constraints

## 1. Introduction

Vegetables are plants cultivated for their leaves, succulent stems, young shoots, fruits or a combination of these parts. They form an integral part of everyday diet and include diverse plant species with considerable economic and medicinal properties. Vegetables are protective foods for maintenance of health and for prevention of diseases. Nigerian Root vegetables (NRVs) such as Ginger (*Zingiber officinale*), garlic

(*Allium sativum*), turmeric (*Curcuma longa*), onion (*Allium cepa*) and carrot (*Daucus carota*) can be defined as vegetables cultivated for their edible underground parts. True roots such as taproots can be botanically differentiated from tuberous roots from non-roots such as bulbs, corms, rhizomes, and tubers, the word "root vegetable" being applied to all of them as it pertains to their agricultural and culinary usage. Root vegetables are rich in nutrients such as minerals, vitamins and fiber, and play important and valuable roles in nutritional, health, economic, social, cultural and ecological aspects of rural and urban communities in Nigeria and all over the world. Root vegetables constitute an important component of the Nigerian cultural heritage where they play vital roles in the tradition, food and income security of many households. Root vegetables (RVs) are able to make significant contributions to food security and nutrition, enhance livelihoods of marginal and smallholder farmers as well as improve the wellbeing of households. Advantages of planting NRVs include the ease of incorporating them into existing cropping systems, provide relatively higher earnings than most of the cash crops, they can be produced on small and barely productive lands, can be successfully cultivated under varying climatic conditions, and short production cycles. In addition, they require few purchased inputs, requires few resources and produces high yields with robust nutritional values. To realize higher return and the desired impact, it will be necessary to increase production of NRVs within the major areas of production as clusters of micro-enterprises. Pests and diseases can have devastating effects on vegetable crops if not well managed. According to [1], damage caused by pests on vegetable at various growth stages on the field to harvesting, storage and also during conveyance can lead to 5–40% crop loss annualy which poses a devastating effect on food and nutrition security for the rising Nigerian population. However, various approaches have been employed by researchers to tackle the menace of pests and diseases in root vegetable crops ranging from cultural practices, biological control, use of indigenous knowledge, plant extracts, pheromones, synthetic pesticides, use of improved planting materials and Integrated Pest Management (IPM) approach. This study reviews some of the efforts made so far in Nigeria on root vegetables to collect and conserve their germplasm, improve their production, utilization and trade, examine the major pests and diseases affecting them and the various approaches adopted by researchers in their management and finally post-harvest technology applied to them.

## 2. Collection and conservation of Nigeria root vegetable genetic resources

With high rate of population growth, demand for agricultural crops worldwide is expected to increase with a higher purchasing power per person which implies higher consumption and use of agricultural produce [2–5]. Genetic materials of horticultural crops and their wild types are of great importance for food and nutrition security, and also serve as good sources of fodder, fuel, shelter, as well as sources of high-value industrial products to meet the high demand of an increasing global population. Genetic materials also provide useful sources of genetic variation required by plant breeders for crop improvement, and a broad genetic base within the gene pool is necessary to expand the scope of identifying and introgessing desirable genes underlying agronomically-important traits [6]. Genetic diversity has huge value for present and future generations, and more efforts must be made for its conservation and sustainable utilization [7]. Despite the usefulness of genetic resources, available genetic variability including landraces is getting eroded at an alarming rate, causing an enormous reduction of variability. This situation thus requires fast action to conserve germplasm [6]. The conservation, likewise sustainable use of germplasm is necessary in the promotion of food and nutrition security and gives room for securing diversity to respond to future challenges including the increasing climate change [8].

The Genebank of the National Centre for Genetic Resources and Biotechnology (NACGRAB) holds one of the most extensive collections in Nigeria, with a total record of over 10,500 accessions of 40 different crops, comprising of wild relatives, landraces, as well as old and more recent cultivars of crops, with germplasms of various vegetables being well represented. Vegetables include various genera and species; and are a vital element for a balanced diet which supplies vitamins, antioxidants, minerals, fiber, amino acids, as well as other compounds that improves health, and contribute to nutrition security [9]. Roughly a million accessions of crops that are entirely or partly used as vegetables are conserved ex-situ [9]. However, only 7% of the total global ex-situ conservation are fully used as vegetables which are mostly leafy or fruit vegetables. Root vegetables are not well represented in genebanks as they are mostly conserved in the fields; thus, are potentially exposed to pest and diseases and environmental variation. To avoid the risk of losing available germplasm, such crops are however amenable to in vitro conservation which is cost and labour intensive and requires consistent material transfer to fresh growth media. Root vegetables like ginger are conserved in vitro at the NACGRAB's genebank in Nigeria.

Of the ca. 7.4 million accessions of Plant Genetic Resources for Food and Agriculture (PGRFA) that are ex-situ conserved globally, only 7% (i.e. ca. 518,000 accessions) are vegetables [10]. Of these global vegetable collections, only alliums are well-represented root vegetables in ex-situ collections [9]. As a result, the level of representation of root vegetables in both National and global collections, calls for the need to explore, collect and conserve more of the variabilities in root vegetables. Though we still have the benefit of a vast agrobiodiversity, there is the need to be conscious that two out of five plant species are endangered with losses occuring on a regular basis [11] due to the increasing climate change, extension of human settlements, and substitution of landraces with hybrid cultivars. Thus, the need for additional exploration for collection and conservation of root vegetable genetic materials in areas of vegetable diversity to maintain valuable germplasm for the present and future.

## 3. Biotechnology: tool for root vegetables production and improvement in Nigeria

### 3.1 Onions

Onion being a crop that is propagated by vegetative means with high heterozygosity, the cultivars are often low in reproductive fertility, making the breeding of this diploid species a challenging effort. Onion has a relatively long breeding cycle and genetic complexity, as well as sensitivity to inbreeding depression which determines the conventional methods adopted for onion improvement. Traditional breeding methods for onion involve mass selection for disease/pest resistance, improving or maintaining quality traits such as a bulb color, shape and increasing yields and shelf-life. This has given rise to the development of many good clones. According to [12], the yield potential of onion has remained relatively constant over centuries despite rigorous breeding efforts. Conventional breeding of diploids plants often involves screening and backcrossing of large number of plants in order to obtain the desired genotype. Selection of many desirable traits at the initial stage can be ineffective and/or time consuming. Hence, onion breeders have most times needed to screen a large number of seedlings (up to a million) to enable identification of a single clonal line that can pull through to the release of a successful cultivar. Biotechnology application has provided unparallel opportunities for plant production and quality improvement [13–15].

Possibilities for improving agronomically-important traits are limitless with the biotechnological applications like in vitro culture techniques, marker-assisted breeding technologies, genetic engineering, genome editing or a combination of all the novel gene technologies. To increase onion production and improvement through biotechnology, the following directions for research of this crop plant are suggested: (1) ploidy manipulation as an alternative technique to induce haploids in onions [16]. More haploid regenerants should be produced by ploidy manipulation to improve onion for breeding purpose; (2) embryo rescue techniques to enable successful intergeneric and interspecific hybridization as has been widely reported for other species [17]. Hybrid embryo developments have been developed through embryo rescue. These hybrids could have horticulturally- and agronomically- important traits; (3) protoplast regeneration and fusion have been used to improve a plant's agronomic and horticultural characteristics such as pests and diseases resistance [18]. This technique should be used to produce more somatic hybrids that are pests and diseases resistant. Somatic variation can lead to creation of additional genetic variation in onion. Tolerance to herbicides, environmental or chemical stresses have been developed via this technique [19]. Somatic hybridization is an alternative technique to overcome both intraspecific and interspecific cross incompatibility to a large extent [20], and this technique could be used to introduce horticulturally- and agronomically-important genes in onions.

As genetic diversity is the basic input for breeding programmes [21] an understanding of genetic diversity among Nigeria onion germplasm collection is imperative for onion breeding.

Ijeomah et al. [22] studied and revealed the genetic diversity of 10 cultivars of spring onions in Nigeria using one SSR and three ISSR markers. The four markers yielded a total of 26 polymorphic alleles with polymorphic information content (PIC) values ranging from 0.6402 to 0.7569. The resulting UPGMA dendrogram showed that the 10 cultivars studied formed two main clusters with one subgroup showing no genetic distance among them. This study indicated the efficiency of SSR and ISSR markers to estimate the extent of genetic polymorphisms of spring onion cultivars with potential utility towards the conservation and management of *Allium* species.

### 3.2 Ginger

Most ginger improvement efforts have been restricted to evaluation and selection of naturally-occurring clonal variants. Conventional crossing efforts have been largely ineffective as a result of rare flowering and poor seed setting. Efforts at evolving high yielding clones through mutation and polyploid breeding indicated lack of success [23, 24]. Furthermore, the seed stock (rhizomes for vegetative propagation) seriously suffers from fungal and bacterial diseases such as *Pseudomonas solanacearum* (bacterial wilt), *Fusarium oxysporum* (yellow leaf), *Pythium aphanidermatum* (soft rot), *Phyllosticta zingiberi* (leaf spot), leading to heavy crop losses [25, 26]. Underground rhizomes are usually used as vegetative propagules for ginger which accounts for its very low multiplication rate [25, 26].

Different explant types used for micropropagation of ginger and other related species include meristem, axilliary buds, shoot tips and aerial pseudostems, although the commonly used explants are rhizome buds and shoot tips which have been reported as responsive explants for large-scale micropropagation to generate pathogen-free propagules [27].

An optimum fragment (explant) size is required for initiating successful tissue cultures. Sathyagowri and Seran [28] reported that rhizome buds of 0.5 cm in length were best for initiating ginger in vitro culture and shoot multiplication among the different tested explant sizes of 0.5, 1.0 and 2 cm long. The establishment of clean in vitro culture ginger from rhizome explants can be a daunting task as these underground explant is laden with pathogens resulting in contamination of cultures. As a general rule, absence of browning and freedom from contamination are criteria for the explants' survival for subsequent shoot multiplication. Surface sterilization of explants is commonly carried out with disinfectants such as ethanol ($C_2H_5OH$), sodium hypochlorite (NaOCl) and mercuric chloride ($HgCl_2$). Ginger aseptic cultures have been obtained by surface sterilization of the rhizome buds explants with 0.1% $HgCl_2$ solution for 10–20 min [16], turmeric [29]. Rout et al. [30] established a surface sterilization protocol for ginger sprouting buds explants using 2% (v/v) Teepol for 15 min followed by 0.2% (w/v) $HgCl_2$ solution for 25 min and several changes of sterile distilled water. Contamination-free in vitro culture of ginger can also be optimally achieved by sterilizing the rhizome buds with 20% Clorox for 20 min [27]. Although bacterial contamination can be a common challenge with the undgerground ginger rhizome bud explants, this problem can be overcomed by incorporating antibiotics to the initial culture medium.

Maintaining optimum environmental parameters such as light and temperature should also be of utmost consideration to ensure optimal growth of cultures; the cultures being generally incubated under photoperiod regime of 16 h light/day and 8 h dark/night with cool, white fluorescent light, 60–70% relative humidity and temperature of 25 ± 2°C in culture room [26, 30]. MS meadia [31] containing 0–5 mg/l cytokinin (BAP or kinetin) eiher used alone or in combination with auxin (NAA or IAA or IBA) are commonly used for multiple shoots induction and subsequent regeneration of ginger plantlets (**Table 1**) [27, 32]. Nkere and Mbanaso [33] investigated the optimal concentrations of phytohormones for in vitro ginger culture, and found a combination of 0.05 mg/l NAA and 4.0 mg/l BAP yielding the highest mean shoot multiplication rate of 4.25 (**Table 1**). Although shoot survival is a key factor in in vitro propagation, a treatment combination of 1.0 mg/l BAP and 0.05 mg/l NAA recorded a relatively high survival rate of 80% and resulted in approximately 4-fold mean shoot multiplication rate in ginger as shown in **Table 1** [33]. Sharma and Singh [34] indicated that kinetin (Kn) outperformed BAP for vegetative bud multiplication. However, [25] reported that BAP at 17.76 μM (4.0 mg/l) yielded a 4-fold multiplication rate after the second subculture (**Table 1**). Balachandran et al. [35] also found using BAP alone resulted in better shoot multiplication than when BAP was combined with Kn in turmeric and ginger clonal propagation. Considering performance of the explants and the need to lower the cost associated with micropropagation, media containing the lowest concentrations of NAA (0.05 mg/l) and BAP (1.0 mg/l) which resulted in a very good survival rate of 80% and about 4-fold mean shoot multiplication rate [33] was recommended for in vitro ginger micropropagation. Zeatin was also reported as being more effective for microrhizome induction of 'Bentong' ginger, although its effectiveness on shoot multiplication was poor compared with BAP [32],

| Explants | Best culture media | Responses | Shoot multiplication rate | References |
|---|---|---|---|---|
| Rhizome buds | MS + Zeatin 10 uM | Multiple shoots with 100% shoot induction | 4.28-fold | [32] |
| Rhizome buds | MS + 10 um BAP | 100% shoot induction | | [32] |
| Shoot tip | MS + 4.0 mg/l + 0.05 mg/l NAA | Multiple shoots with 50% survival rate | 4.25-fold | [33] |
| Shoot tip | MS + 1.0 mg/l + 0.05 mg/l NAA | Multiple shoots with 80% survival rate | 4.0-fold | [33] |
| Rhizome buds (0.5 cm long) | MS + 3 mg/l BAP + 0.5 mg/l NAA | Shoot initiation | | [27] |
| | MS + 5 mg/l BAP + 0.05 MG/l NAA | Multiple shoots | | [27] |
| Vegetative buds (0.5–1.0 cm long) | LSBM + 17.76Um BAP | Multiple shoots plantlets | | [25] |
| Active rhizome buds (0.5 cm long) | MS + 2.0 mg/l kinetin + 2.0 mg/l NAA | Adventitious shoots and plantlets | | [34] |
| | MS + BAP | Multiple shoots | | [35] |

**Table 1.**
*Studies on in vitro propagation of ginger directly from rhizome explants.*

with zeatin's less effectiveness for shoot multiplication relative to BAP likely attributed to its high oxidative cleavage with prolonged incubation.

### 3.3 Garlic

Garlic research in Nigeria has been initiated with the breeding strategy focusing on collection, introduction, adaptation and selection of superior lines. However, there is a considerable morphological and physiological variation within and among cultivars [36]. Analysis of genetic diversity and relatedness between individuals is important for breeding purpose. Although garlic displays wide morphological differences [36], clonal propagation narrows garlic variation, given rise to a genetic bottleneck. This situation complicates garlic breeding programs geared towards improving preferred agronomic traits. In view of this, assessing the morphological and molecular polymorphisms in garlic genotypes originating from Nigeria is important for breeding programmes. Molecular characterization of 115 garlic accessions has been done in Ethiopia using 11 SSR markers [37]. There is paucity of information on genetic diversity studies of garlic germplasm in Nigeria using molecular markers.

Garlic is cultivated vegetatively due to its sexual sterility [38]. Vegetative propagation is achieved via division of the ground and aerial bulbs which results in lower multiplication rate. Many of the elite garlic cultivars succumb to diseases incited by viruses, nematodes and fungi, as well as insect pests [39]. Virus infection has been reported to reduce the bulb yield by 20–60%, and up to 80% with mixed infection, depending on cultivar and stage of infection [40]. The low propagation rate coupled with continuous accumulation of serious diseases such as viruses observed in the field warrants the development of in vitro propagation of garlic [41, 42]. There is a great influence of genotype on garlic in vitro cultures [43]. A combination of 1 mg/l 2,4-D + 5 mg/l BAP + 5 mg/l NAA resulted in 100% callus induction from root apices of garlic. The shoot has been reported to have higher callus induction frequency and higher callus fresh weight relative to root apices in some cultivars of garlic [44]. The use of root tips as explant greatly increases the regeneration potential than shoot tips as the number of regenerated shoots per explant was higher than that obtained from callus induced from shoot apices [42, 44].

### 3.4 Turmeric

Turmeric (*C. longa* L.), a cross-pollinated and triploid species ($2n = 3x = 63$), is propagated vegetatively through its yellow-fleshed rhizomes. Due to its numerous beneficial uses, the National Root Crops Research Institute (NRCRI) Umudike, Nigeria commenced research on turmeric in 1998 by pioneering collection of genetic resources and indigenous knowledge pertaining to its production and use. An active gene bank consisting of strong genetic base is a basic requirement for a sound crop improvement programme. Minor root and tuber crops nearly lacked a gene bank at the onset to enable take-off of meaningful crop breeding efforts. However, several germplasm exploration trips within the last decade resulted in the collection of a total of 76 accessions of Turmeric [45]. Ekiti State topped the list of collections (12) while Abia and Kwara were the least with one collection each. The passport data of the colections including the number collected and the local utilization at the source of collection are well documented to guide future breeding work. Following multi-locational evaluation at different locations in Nigeria—Jos, Otobi, Umudike and Igbariam, 10 genotypes were identified as promising and need to be further evaluated before their official registration and release to farmers.

Scarcity of planting materials resulting from low multiplication rate of turmeric (*C. longa*) limit their massive production. Micropropagation would help solve the challenge of limited planting materials. The phytohormone, 2,4-D, has been reported as the most effective auxin for callus induction in ginger and turmeric [46, 47]. MS media containing cytokinin (BAP or kinetin) either alone or in combination with auxin (NAA or IAA or IBA) are commonly utilized to induce multiple shoots and subsequent plantlet formation in turmeric (**Table 2**) [48, 49]. MS media containing 3.0 mg/l BAP yielded the best average vigor survival rating of [35]. This can be compared with results of other researchers as shown in **Table 2** [29, 50]. This finding is in conformity with [51] who reported that the higher BAP concentration decreased shoot multiplication rate in turmeric among the different BAP levels (1–6 mg/l) tested. In turmeric, maximum rooting to multiple shoots was notable on half strength MS medium supplemented with 0.5 mg/l NAA [52]. Rooting on MS + 0.5 mg/l IBA of microrhizomes produced multiple roots/explants [53].

| Explants | Best culture media | In vitro responses | Rate (shoot per explant) | References |
|---|---|---|---|---|
| Sprouted Rhizome buds | MS + 3.0 BAP mg/l | Multiple shoots | 2.96 | [48] |
| Sprouted rhizome buds | MS + 2.0 mg/l + 1.5 mg/l NAA | Multiple shoots | 9.00 | [49] |
| Sprouted rhizome buds | MS + 2.0 mg/l BAP | Multiple shoots | 14.50 | [29] |
| Elite clones | MS + 4.0 mg/l BAP | Multiple shoots | 6.25 | [50] |
| Elite clones | MS + 4.0 mg/l BAP + 1.0 mg/l NAA | Multiple shoots | 6.70 | [50] |
| Rhizome buds | MS + 3.0 mg/l BAP | Multiple shoots | | [35] |

**Table 2.**
*Studies on in vitro micropropagation of turmeric (Curcuma sp.) plants.*

## 4. Socio-economics of root vegetable farmers

There is an enormous variation in Nigerian agricultural production landscape in terms of age groups, social class, gender, educational attainment and background. The motivations behind farmers' enterprises, sustainability of systems are basically not the same. Therefore, the general assumption that farmers are solely motivated by the need to enhance food security does not always hold true. There are other intrinsic and extrinsic factors that propel farmers in their enterprise. Farmers have the desire for mastery of their enterprise and sense of achievement which arises when they have good harvests from their crops. As indicated in a Nigerian study, farmers of NRVs particularly onion, garlic, ginger in Kano state vary greatly in educational attainment (**Table 3**). Adewale and Oladeji [56] reported that onion, garlic, ginger farmers in Kano state are spread over all education levels. In Imo and Oyo states, farmers also have one form of education or the other [57]. This submission lends credence to the position of [58] that Nigerian farmers are not totally illiterate as they have one form of educational attainment or another. With this educational attainment, farmers are able to play a positive role in their societies as formal education provides an important foundation for knowledge, decision making and acceptance of vital innovation [59].

The gender distribution of NRV farmers is also highly variable and location-specific. The different roles men and women play in the society and in agricultural value chains are caused by gender disparity which causes differences in the distribution of resources, activities, decision-making, wealth as well as enjoyment of entitlement and rights [60]. There appears to be a dominance of male farmers involved in NRVs in Kano state. This is however due to the socio-cultural and religious inclination that places restrictions on the women gender. However, in Oyo state both male and female are involved in production and marketing of root vegetable crops such as garlic, turmeric etc. As is common with production of other vegetables, there are more females involved in the production, processing and marketing [57]. Majority 51.3% of producers and 57.0% of marketers were within the age bracket of 31–50 years indicating that both producers and marketers are still strong, mentally alert and active. These enterprises thus provide windows of opportunity for younger generations to participate. Furthermore, people engaged in root vegetable production and marketing are mostly married as the marriage institution is held in high esteem among rural

| Variable | Characteristics | Producers (%) | Marketers (%) |
|---|---|---|---|
| Gender | Male | 100.0 | 100.0 |
| | Female | 0.0 | 0.0 |
| Age | 18–30 | 20.9 | 29.0 |
| | 31–50 | 51.2 | 57.0 |
| | 51 and above | 27.9 | 14.0 |
| Marital status | Single | 2.3 | 14.0 |
| | Married | 97.7 | 86.0 |
| Educational attainment | Non formal | 38.0 | 63.0 |
| | Primary | 17.8 | 20.0 |
| | Secondary | 24.8 | 17.0 |
| | Tertiary | 3.9 | 0.0 |
| | Islamic | 15.5 | Not indicated |
| Experience in spices enterprise (years) | 1–9 | 14.6 | 17.1 |
| | 10–19 | 38.0 | 45.7 |
| | 20–29 | 29.0 | 14.3 |
| | 30–39 | 13.0 | 22.9 |
| | 40–49 | 4.6 | 0.0 |
| | < 50 | 0.8 | 0.0 |

*Source: [54, 55].*

**Table 3.**
*Comparative socio-economic characteristics of spices producers and marketers in Kano State, Nigeria.*

citizens whose occupation is mostly agrarian in nature. A large proportion (38.0%) of producers and 45.7% of marketers had more than 10 years of experience in production and marketing of root vegetables, respectively. This indicates that a large proportion of them are experienced in the production and marketing of root vegetables; they are therefore knowledgeable in the enterprise and are able to fully comprehend the intricacies and complexities of the enterprises.

Large household size usually characterizes farming households in Nigeria. According to [61], Nigeria has a national average of 5.9 persons per household. Large household is essential to traditional agriculture in Nigeria where the main source of labor is the family members. In some cases, members of the community are hired to work on the farms at pre-agreed sums of money, by barter (by giving crops or animals in return for work done on the farm) or by rendering services such as working on other farmers farms in return. In Kano state, a large proportion of these farmers have large household size which range between 10 and 20 persons [56]. This resonate the position of [62] who submitted that farmers in northern Nigeria maintain large households to meet the large labor requirements during farming seasons. This way, they make use of the children and dependents for farming activities rather than pay for hired labor. Majority of NRV are cultivated on farm sizes that ranged between <1 and 5 ha of farm land. This reveals that majority are small-holder farmers which confirm the characteristic peasant nature of agriculture in Nigeria. This is also as submitted by the National Bureau of Statistics [63] revealing the system of agriculture among farmers is on small holdings. Small holder production however is labor intensive and thus able to provide employment to a large share of rural population. Furthermore, when proper conditions are put in place to enable small farms grow and have access to markets, off-farm employment increases.

The annual income from root vegetables production is also shown in **Table 4**. It is indicative of a promising enterprise as majority (54.3%) earn an annual income between of ₦ 100,000–499,000 and 38.6% earn between ₦ 500,000 and 999,000. It reveals that root vegetables are capable of increasing the income of producing households which resound the view of [64] that root vegetables such as ginger, garlic, turmeric are crops of high value with the potentiality of enhancing household income and improving livelihoods.

The estimates of marketing margin and marketing efficiency of garlic and ginger is shown in **Table 5**. The net marketing margin for garlic is ₦3, 449.0 per bag. For

| Variable | Characteristics | Percent |
|---|---|---|
| Household size | 1–5 | 20.2 |
| | 6–10 | 17.1 |
| | 11–15 | 24.0 |
| | 16–20 | 19.4 |
| | 21–25 | 16.3 |
| | More than 25 | 3.1 |
| Farm size (in ha) | <1.0–5.0 | 90.0 |
| | 5.1–10.0 | 6.9 |
| | 10.1–15.0 | 3.1 |
| Estimated Annual Income (₦) | <99,000 | 1.6 |
| | 100,000–499,000 | 54.3 |
| | 500,000–999,000 | 38.6 |
| | N 1,000,000 and above | 5.5 |

*Source: [54].*

**Table 4.**
*Household size, landholding and income of root vegetable production in Kano state.*

| Variables | Garlic | Ginger |
|---|---|---|
| Total Marketing Cost | 10,629 | 6156 |
| Total Revenue | 14,078 | 8433 |
| Marketing margin | 3449 | 2277 |
| Marketing efficiency | 1.31 | 1.37 |

*Source: [55].*

**Table 5.**
*Marketing margin and efficiency analysis for garlic and ginger in Kano state.*

ginger, the net marketing margin is ₦2277.0. Ginger has a higher average marketing efficiency (1.37) which implies that for every ₦1.00 spent in purchasing ginger in Kano state, the marketer will accrue 37 k as returns to his/her investment. For garlic in Kano state, with an average marketing efficiency of 1.31, it implies that for every ₦1.00 spent in purchasing garlic, a return of 31 k will be gained. This further reveals root vegetable production and marketing as a promising enterprise capable of generating sufficient profit.

### 4.1 Economic values and contribution to livelihoods

Root vegetables such as ginger, garlic onion do not need expanse of land for profitable production, they can be cultivated without excessive investments as they are usually produced with little inputs (cash, labor and land). They are good crops to be included in small-scale farming systems and are also suitable for small garden production [64]. Farmers can derive a lot of benefits from root vegetables as crop with high value that requires little cash that can yield more than double the income accrued from staples and other horticultural crops [64]; to enhance household income and thus improve their livelihoods. As a production enterprise, they can be a source of additional employment opportunities for the farm family and money realized can be used as a 'safety net' during periods of need in order to improve livelihoods. Value-adding activities and sales of processed root vegetables can also be a good potential for small-scale processing (on-farm or off-farm) industries to generate higher income. Their production is especially an avenue to provide opportunities for women due to the fact that they can easily be grown in gardens in and around homes. In addition to providing opportunities for the womenfolk to start commercial enterprises and be able to participate in the local economy, root vegetables production enables women earn income for themselves, be involved in trading, create social networks, enhance their family status and their social status in the community and to provide added security to their household in case of abandonment or adversities.

## 5. Agronomy and production of root vegetables in Nigeria

### 5.1 Onion (*A. cepa*)

Onion is an important spice and one of the vegetables that is commonly consumed in Nigeria which produces underground bulb that are edible, with about two million metric tonnes of Onions produced annually in Nigeria, according to Mr. Muhammad Abubakar, the Minister of Agriculture and Rural Development, at the 4th Regional Onion annual conference held in Kano, an event organized by Regional Observatory on Onions for West and Central Africa, in conjunction with National Onion Producers, Processors and Marketers Association of Nigeria (NOPPMAN). Nigeria is part of the largest onion producers in the world with over 2,000,000 tonnes produced annually [65]; Nigeria being ranked sixth among the top 10 producers of green onion, and 11th position in dry onion production globally. Major growing ares of onion in Nigeria include Kano, Kaduna, Jigawa, Sokoto, Plateau, Bauchi and Kebbi states. Production figures in 2012 indicated about 240,000 tons of green onions and 1,350,000 tons of dry onions were produced

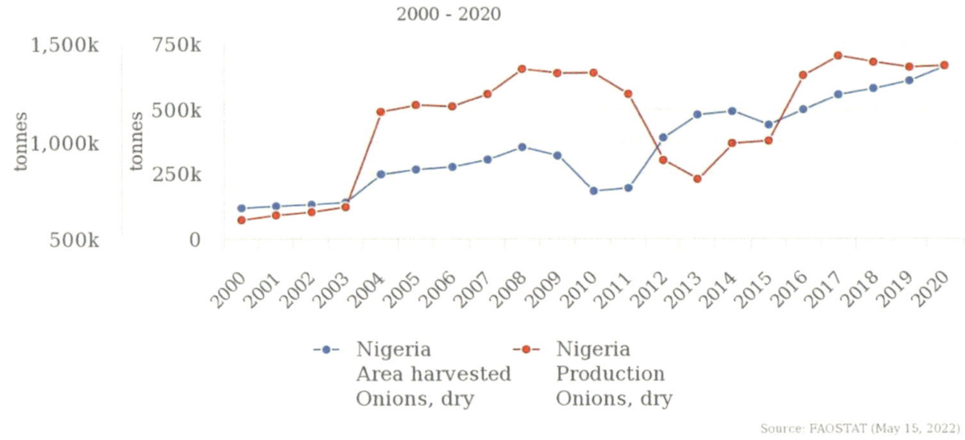


**Figure 1.**
*Production/yield quantities of onions, dry in Nigeria.*

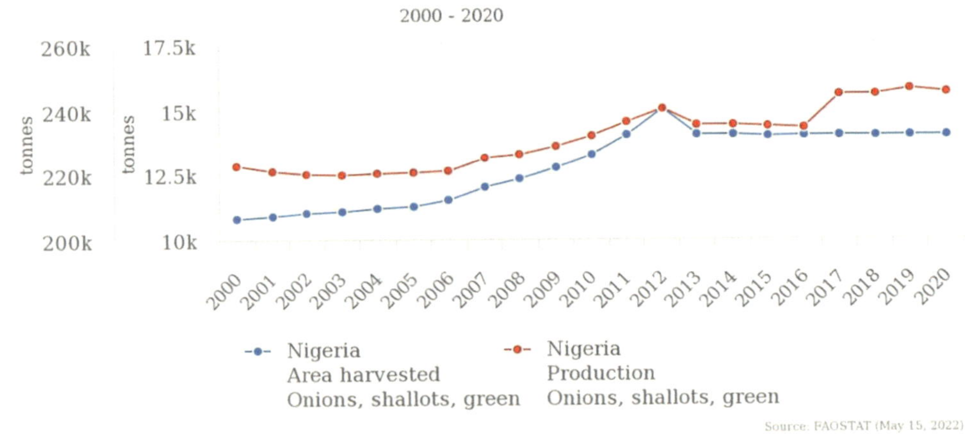


**Figure 2.**
*Production/yield quantities of onions, shallots, green in Nigeria.*

in Nigeria (**Figures 1** and **2**), with dry onion production having an increasing trend. Nigeria had a world share of 5.5% of a total 4,339,925 tons of green onions produced in 2012 and 1.6% share of a total 82,815,927 tons of dry onions produced globally in 2012 [66].

### *5.1.1 Climate and soil*

Onions can grow in different climatic conditions. However, it requires moist soil during the early growth stages, likewise hot or dry weather during the maturity and harvesting periods. It produces flowers prematurely in very cold weather conditions (resulting in smaller bulb size). Onions require a loose soil with reasonable depth. Clay soil is not suitable as it gets waterlogged and hardens when dried up. The soil needs to be slightly acidic (pH between 6.0 and 7.0), must also be fertile with humus [67].

#### *5.1.2 Land preparation*

The land should be properly plowed, farm manure could be plowed into the soil. Ridges are constructed after plowing when planting is about to start.

#### *5.1.3 Planting operations*

Onions are planted either by seeds or by setts, the seeds need to be raised in the nursery for 6–8 weeks under a conducive condition. Planting is done at a spacing of 10 cm apart around November or December. It takes about 3–4 months for the crop to reach maturity.

#### *5.1.4 Management*

The farm is kept weed-free because onion can not withstand weed competition. Weed should be removed regularly so as to enhance high yield as weed competition reduces yield. First weeding is done 14–21 days after transplanting. The plants are watered regularly because onion plants can not absorb water from deep layers of the soil, that is why the top soil is always kept moist. Irrigation system (either drip or surface) can be used in as much as the soil is not flooded, because onions do not thrive in soils that are flooded and poorly drained. After irrigation, mulching is used to conserve water and inhibit weed growth. Onion does not require much fertilizer in the case where the soil is fertile. However, 20 tons/ha of poultry manure should be incorporated into the soil a week before planting. Also, 75 kgN/ha can be applied at 4 and 6 weeks after transplanting [67].

#### *5.1.5 Harvest and storage*

Maturity period of onions is 3–4 months, and once the leaves begin to dry off naturally, it can be harvested. Onions are harvested by uprooting the bulb and cutting off the roots and leaves. In Sokoto, it was reported that onion takes between 112 and 161 days from sowing to harvest or 15 and 8 weeks for the early and late sown onions, respectively, while the time of forming bulbs was 56 days [68]. Maximum plant height of 69 cm has been recorded and number of leaves of between 10 and 13 has also been observed. Yield of up to 48 t/ha has been recorded in the early crop but yield of less than 20 t/ha has been observed for those planted after December [68]. Thorough sun-drying after harvesting is very necessary before bagging for transportation to market or storage in a well ventilated and perfectly constructed silo. This is necessary to reduce the moisture content as much as possible thereby increasing its shelf life.

### 5.2 Carrot (*Daucus carota*)

Carrots are one of the most well-known, delicious, very effective, nutritious root vegetables. Carrot is majorly grown in the Northern part of Nigeria, especially in Zaria, Sokoto, Kano and Jos [69]; and carrot production trends in Nigeria are shown in **Figure 3**. Carrot is described as a root vegetable with diverse colors like orange (most common), purple, black, red, white, and yellow. Carrot comprises mostly the taproot which is the main edible part and the green leafy part which is crispy when fresh and also eaten but not common.

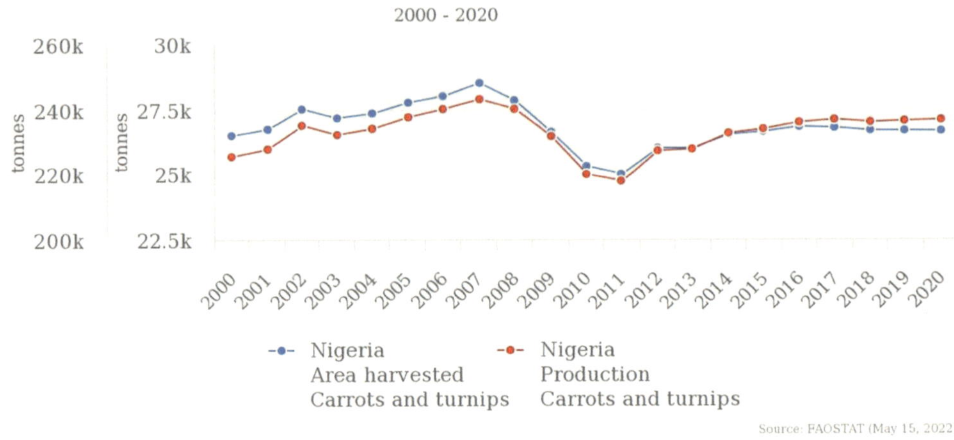


**Figure 3.**
*Production/yield quantities of carrots and turnips in Nigeria.*

### *5.2.1 Weather and soil*

Carrot is a cold-weather crop, but can also thrive well in warm climates. Optimum temperature for good growth is 16–20°C, temperatures beyond 28°C will drastically reduce its growth. Carrot grows well in sandy soil, when planted in rows of at least 10 cm apart on a raised seedbed, it starts to germinate at about 1–3 weeks and can be harvested at any size, but better when it has turned bright orange for enhanced flavor. There are several varieties of carrot that are cultivated in Nigeria with a high rate of productivity, some of which include the Danvers variety which is larger in size and requires a balanced level of soil fertility to grow well. Other varieties include Chantanay, Nantes, Armsterdam, among others. Carrots are well adapted to a wide variety of soils, although deep, loose, well-drained soils rich in humus are suitable for commercial carrot farming. In Nigeria, the best soil for cultivation of carrots is loamy or sandy loam soils rich in humus [70].

### *5.2.2 Land preparation*

Soil should be properly prepared by repeated deep plowing (at least 20–30 cm deep), harrowing, leveling, and cleaning will enhance desired yield. The soil must be loose, friable, deep, and well-drained in order to enhance effective germination of seeds. Because carrot seeds are very small and delicate, a fine seedbed of convenient size should be prepared before sowing. Carrots are taproots that penetrate and grow downwards so while cultivating the carrots, you should avoid rocky or stony areas to prevent stunted growth. It is advisable to make ridges for planting carrots, which should be higher than those for planting a crop like maize. Ridges are necessary at a considerable height level for the maximum penetration of the carrot.

### *5.2.3 Planting*

Carrots are propagated using seeds, clean, disease-free and viable seeds from reliable sources will enhance good productivity. Complete seeds germination takes approximately 7–21 days. Seeds are lightly covered with soil after planting.

Some farmers irrigate the field about 24 h prior to sowing to ensure that enough water is present in the soil at the time of sowing. The Fertilizer recommendations should be based on soil analysis.

#### *5.2.4 Irrigation*

Light irrigation should be done immediately after sowing while subsequent irrigations are applied as necessary. Excess water makes carrot to be light colored, short with large diameter.

#### *5.2.5 Harvest*

In case of special markets, carrots can be harvested early when they have not fully developed; or else, they should be allowed to reach full maturity stage in the soil. But they will become hard and unfit for consumption if they are retained in the full maturity stage. Carrots are harvested when the roots are ca. 1.8 cm or larger in diameter at the upper end. After harvesting, carrots are carefully washed, sorted by size then packaged for future endeavors.

### 5.3 Turmeric (*Curcuma longa*)

Turmeric (*Curcuma Longa* Linn) is a type of root vegetable belonging to the same family (Zingiberaceae) as ginger [71]. It is a tropical perennial plant, originated from India and Indonesia, which is widely cultivated throughout the tropical regions of the world. It is one of the most essential spices used as a culinary ingredient world wide for which it is referred to as the "golden spice of life" [72]. World production level of turmeric is 11–16 tonnes annually. Nigeria being the fourth largest world producer, produces about 3% of the world annual production [73]. About 76 cultivars of turmeric exist in the gene bank of the National Root Crop Research Institute (NRCRI), Umudike with some being evaluated in multilocational trials [74]. The prevailing favorable edaphic and climatic conditions in Nigeria place the country in a position to play a leading role in turmeric production. Turmeric is cultivated both under rain fed and irrigated conditions.

#### *5.3.1 Soil requirement*

Turmeric also thrives well in deep soil tilt with heavy manure for high yields. Fertile and friable, well-drained, loamy soil ranging from sandy loam to clay loam with high organic matter are required for turmeric cultivation [75]. Moreover, flat land with gentle or no slope is equally recommended.

#### *5.3.2 Land preparation/planting*

This commences with selection and clearing of the site followed by bed preparation because turmeric thrives well on beds, ridges or even on mounds at the onset of rains possibly around April or when the rains must have stabilized around May or June. Bed size of 3 × 2 m is advised so as to reduce human movement on the beds during farming operations. Seedbed can be prepared using tractor or manually using a spade or a hoe. Turmeric is propagated by rhizomes using mother rhizomes as planting material [76], planting distance of 30 x 50 cm [77] at a depth of 10 cm to give optimum yield [78] is recommended. The rhizomes are about 10–15 g with one or two buds [79] therefore, about 1 ton of setts are needed to plant 1 ha.

#### 5.3.3 Mulching

Mulching is important in cultivation of turmeric, the first mulching should be done immediately after planting followed by a second mulching at 8 weeks after planting. Mulching aids moisture conservation, enhances germination, controls weeds, modifies soil temperature, adds nutrients to the soil and improves soil fertility for optimum yield. Mulching can be done with elephant grass (straw) at the rate of 12 t/ha [78].

#### 5.3.4 Weeding

Pre-emergence herbicide can be used to control weeds in turmeric farm. Weed control is normally done within 4–6 weeks after planting depending on rate of weed emergence. According to [80], under Umudike condition, critical time of weed interference is between 8 and 12 weeks after planting. Yield loss in turmeric due to weed competition can range between 3 and 55%.

#### 5.3.5 Fertilizer application

Fertilizer at the rate of 60 kg N, 13 kg P and 25 kg K/ha 2 weeks after planting [81] on a sandy loam Ultisol has been reported. According to [82], this recommendation translates to application rate of 200 kg/ha N:P:K 15:15:15 and augmenting with 30 kg N/ha. Animal dung or droppings can also be used to improve soil fertility in the absence of inorganic fertilizer as application of poultry manure has been reported to influence rhizome yield of turmeric [83]. Application of lime at the rate of 2 t/ha to soil with pH of 5.91 in combination with 200 kg/ha of NPK 15:15:15 fertilizer has been reported for turmeric production on an Ultisol in South eastern Nigeria [84, 85] whereas organic manure of 30–40 t/ha plowed into the soil and inorganic fertilizer of 60 kgN, 50 kg $P_2O_2$ and 120 kg $K_2O$ per hectare in split doses is recommended by National Horticultural Research Institute (NIHORT), Ibadan [67].

#### 5.3.6 Harvesting and yield

Maturity period for turmeric is 7–9 months after planting when the leaves turn yellow and starts wilting. Harvesting is carried out by uprooting the whole rhizome with spade or hoe after which the mother and finger rhizomes are separated. Yields of turmeric ranges from 20 to 25 t/ha. Some cultivars under research-managed farms yielded 35,000 t/ha [67, 75].

### 5.4 Ginger (*Zingiber officinale*)

According to FAOSTAT (**Figure 4**), Nigeria is the third-largest producer of ginger in the world producing more than 300,000 tonnes between 2014 to 2018. The global ginger market share for Nigeria is about 11%, after only India (35%) and China (18%). Nigeria produces over 400,000 Metric tonnes of Ginger across the 36 states annually. Cultivation of ginger began in Nigeria in the year 1927 around Kwoi, Kubacha,

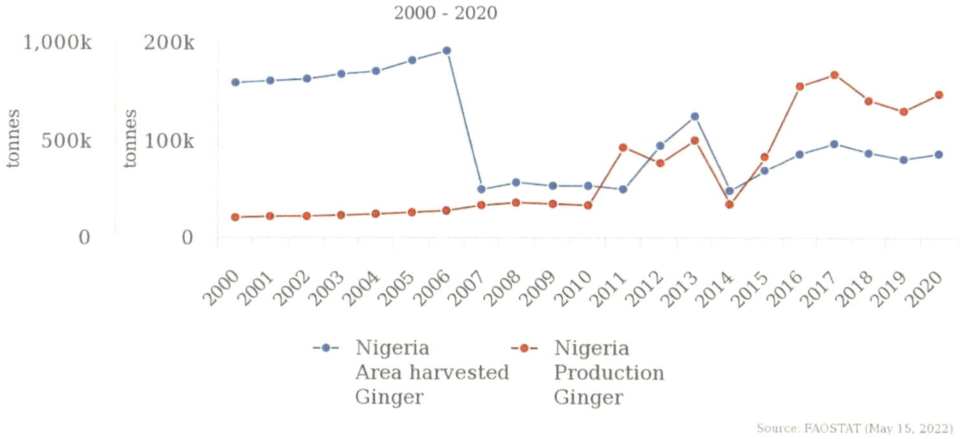


**Figure 4.**
*Production/yield quantities of ginger in Nigeria.*

Kafanchan and Kagarko areas of southern Kaduna State and some neighboring parts of Plateau State, but it is now cultivated in different parts of the country [86]. Kaduna, Bauchi, Benue, Gombe and Nasarawa are the top five producing states of ginger in Nigeria. Ginger is now cultivated in Sokoto, Osun, Anambra, Zamfara, Akwa Ibom, Oyo, Abia and Lagos states, but southern Kaduna remains the largest producer [87]. In Nigeria, two varieties (reddish and yellow) are commonly grown but different cultivars like UG1, UG2 and Maran are available in the country, UG1 produced higher yields than UG2 and it is reported to be more pungent [88]. Average yield per hectare is about 13–27 metric tonnes in Nigeria compared to the world average of about 35–40 metric tonnes.

### *5.4.1 Seed selection*

Selection of ginger seed for planting as reported by respondents in Jaba region is based on the size of the tuber irrespective of the variety, those with width of about 5–6 cm and thickness of about 2–3 cm are often selected for planting. Apart from seed size, number of budding tendencies which is about 5 and above and non-physical damage to the tuber are also part of criteria for seed selection which is usually carried out after harvest around September to October.

### *5.4.2 Planting*

As reported by respondents in the Jaba region, this is normally done in the month of April to early May. This is carried out manually after adequate moisture is ensured following successive rains. Planting entails digging with hoe, majorly by a male, while children or females drop the bud seedling in the hole and cover it. Planting distance is usually 10–13 cm side wise, this helps to reduce weed competition. After planting, the entire field is covered with dry grass to enhance quick germination and protect the seedling from the effect of sun heat. The dry grass is normally left on the field for about 1 month when the ginger is expected to have fully germinated, after which it would be removed and packed along the furrow to decompose.

#### *5.4.3 Fertilizer application*

According to Alhaji Adamu Shekari while describing the importance of of applying fertilizer to ginger stated that "the profit of ginger farming is determined by fertilizer" [89]. The general view from farmers indicated that fertilizers are applied three times during the growing period of ginger. The first is during field preparation which is mainly organic manure. Respondents gave reasons for this practice that inorganic fertilizers may cause the young ginger plants to wilt or die because it may be too hot for them when applied. This agrees with the findings of [90] that organic manure like ash has high pH of about 10.2 that neutralizes soil acidity. The next fertilizer applications involved majorly inorganic fertilizers like NPK or Urea which are applied after second and third weeding.

#### *5.4.4 Harvesting*

Maturity of ginger is usually reached when leaves of ginger plant begin to turn yellowish brown in color indicating readiness for harvest. Harvesting is done by digging out the tuber majorly by male workers while female workers or children collect the tuber into containers.

### 5.5 Garlic *(Allium sativum)*

India produces average yield of about 5.23 t/ha making it one of the world's largest garlic producers [91]. Spain, Egypt, Korean Republic, Argentina, Italy, China, and the United States are other growing countries. According to [92], garlic production is about 10 million tonnes per annum which represents only ca. 10% of bulb onions production. Garlic has been cultivated for decades in the northern states of Nigeria like Kano, Sokoto, Borno, Bauchi, Jigawa, Katsina, and Zamfara [93]; and garlic production figures in Nigeria are shown in **Figure 5**. Its wide distribution is denoted by its common "native" names in different societies. It is called "Tafarnuwa" in Hausa and "Ayu" in Yoruba [94].

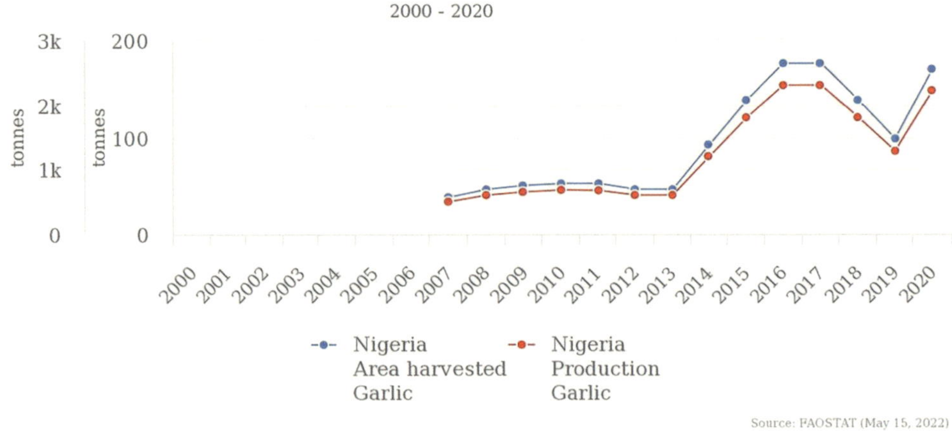


**Figure 5.**
*Production/yield quantities of garlic in Nigeria.*

### *5.5.1 Climatic and soil requirements*

Garlic can grow well in both tropical and sub-tropical environment [95] but it is a cold weather perennial crop that has a high requirement for nutrients and water [96]. It grows well in region with 600–1200 mm annual precipitation and temperatures ranging between 5–25°C and 25–40°C [97]. Fertile loamy soils free from stone and gravels and that is well-drained is suitable for garlic production. Heavy soils are unsuitable for growing garlic because the bulbs produced under such condition will be deformed and difficult to harvest [97].

### *5.5.2 Seed preparation and treatment*

Garlic cloves that are about 8–12 g in size which are detached from the bulbs and soaked in clean water for about 6 h before planting and removal of the outer skin from the bulblets and dried are used as planting materials. These cloves are mixed with fungicide and insecticide to control fungicides and seed attacking insects before planting [97].

### *5.5.3 Planting*

Garlic has fibrous root and the bulbs comprise small bulbils referred to as cloves, which are the vegetative planting materials for the crop [98]. Garlic can be planted either by dibbling, drilling or broadcasting [97]. Dibbling method is commonly practiced, it involves putting one clove in a hole 7 × 15 cm at a depth of 3–6 cm, placing the growing point upwards and covering lightly with soil. About 350–600 kg cloves are needed to plant one hectare corresponding to 400,000–500,000 plants/ha [97]. Garlic produces well when cultivated on fertile well-drained sandy or silt-loam soils with good moisture retention capacity [99]. Ahmed et al. [100] recommended large clove size and irrigation at 3-day interval for good performance under semi-arid conditions like Sokoto in Nigeria.

### *5.5.4 Fertilizer and its application*

Nitrogen is a major nutrient required for growing garlic. Bulb growth was significantly affected by applied nitrogen [101]. Longer leaves and higher number of leaves were recorded when nitrogen rate was increased to 100 kgN/ha [102]. In another experiment [96], increasing rates of nitrogen application to 150 kgN/ha increased growth and yield components of garlic but higher rates of nitrogen above 150 kg significantly reduced growth and yield. Farooqui et al. [103] reported that 200 kgN/ha significantly increased yield parameters such as neck thickness, bulb diameter, number of cloves/bulb and fresh weight of 20 cloves. However, at Samaru, Nigeria [104] growth and yield of garlic was significantly increased with application of nitrogen with 15 t/ha recorded as maximum yield when 90 kgN/ha was applied. Meanwhile, [105] reported significant increase in bulb yield with application of 75 kgN/ha but clove weight increased only at 150 kgN/ha after which a significant decrease was observed. Magaji et al. [99] recommended application of 50 kgN/ha for improved plant height (cm), number of leaves/plant, leaf area, number of bulbs, and the total yield of the garlic.

#### *5.5.5 Harvesting*

At 18 weeks after planting when leaves of garlic are partially dry and bend to the ground, it is assumed that garlic has reached optimum maturity period and the bulbs are harvested [97].

## 6. Pest and disease management in turmeric (*Curcuma longa*)

In Nigeria many pests attack turmeric in the farm. According to [106] such pests include leaf roller, shoot borer, scale insects to mention but a few. The diseases includes leaf blotch, leaf spot and rhizome rot. The symptoms appear as small oval, rectangular or irregular brown spots on any side of leaves and soon turning to dirty brown, they decrease yield and are controlled by spraying chemicals. The incidence of rhizome rot, leaf spot, leaf blotch, thrips, leaf folder and cutworm are the major cause of the yield loss in turmeric [107]. Nirmal et al. [108] reported that the rhizome rot results in 50–80% loss during storage. Jagtap et al. [109] recorded 34–57% yield loss due to the incidence of leaf spot *Colletotrichum capsici* ((Syd.) Butler and Bisbyl). The survey conducted in various turmeric growing states of South India revealed that the rhizome rot is caused predominately by *Pythium aphanidermatum* (Edson.) Fitz. [110, 111]. Fungicides like metalaxyl, mancozeb, carbendazim, alliete, propiconozle, hexoconozole etc. are widely used by the farmers for the management of disease complex in turmeric [107].

Sarathi et al. [111] has evaluated the efficacy of *C. longa* (Turmeric) as possible botanical pesticide for managing insect pests of okra on the field and they observed differences in efficacy and yield between lambda-cyhalothrin and *C. longa* which was used as the extraction solvent. Their study revealed that *C. longa* compete well with lambda-cyhalothrin in controlling pest infestation and produced yields that were significantly higher than plots without treatments. Therefore, they recommended that farmers should consider using it as a botanical pesticide. Sarathi et al. [111] also reported the antifungal activity of two varieties of turmeric (white and red) rhizome extracts on fusarium wilt pathogen (*Fusarium oxysporum* f. sp. *lycopersici*). Their findings revealed that both varieties had comparable antifungal activity on the mycelial growth of the test pathogen at 15% extract concentration and that the highest mycelial growth was 17.7 and 25.2%, respectively.

### 6.1 Pest and disease management in ginger (*Zingiber officinale*)

Ginger cultivation is affected by both biotic and abiotic factors. Biotic factors are viruses, bacteria, fungi and nematodes [112, 113] with bacteria being the most important causing wilt and soft rot. The next major pathogen after bacteria is fungi which causes rhizome rot, soft rot, sclerotium rot and yellows disease. Nematode causes root-knot disease while viruses causes mosaic and chlorotic fleck in ginger plants leading to reduction in rhizome yield of the plants. In River State of Nigeria, a survey was conducted by [114] and it was reported that the most abundant arthropods in stored ginger was mites then *Lasioderma serricorne*. Reports have it that *Aspergilus flavus, Aspergilus niger, Fusarium oxysporium* and *Rhizopus* sp. were fungi pathogens isolated from dried ginger samples with high occurrences *of A. flavus*. According to a survey in Rivers State, Nigeria, stored ginger has no immunity against arthropod pest infestation and fungal infection. It was also observed that population dynamics of arthropod pests of stored ginger was affected by seasonal variation. The four main diseaseS and pestS of ginger as reported by THE National Agricultural Advisory services (NAADS) Uganda are rhizome rot, soft rot, root-knot nematode and rhizome scale.

#### *6.1.1 Rhizome rot*

It is a partly/completely decomposition of rhizome tissue; root may dry, wet, soft or slimy and color turns black. This rot is caused by the fungus (*Fusanum Oxysporum*) both in the field and post-harvest when ginger is washed in dirty (recycled) water, poor aeration and storage in contaminated sheds where rotten piles of reject ginger is allowed to accumulate.

#### *6.1.2 Soft rot*

Soft rot is caused by fungus *Pythium gracile*, either alone or in combination with a bacterium *Erwinia* species. The fungus *Pythium* is a water-mold and develops maximally in moist conditions when temperatures are favorable. Prolonged wet weather, therefore, creates ideal conditionS for development of soft-rot disease in ginger.

#### *6.1.3 Root knot nematode*

It is caused by the nematode (*Meloidogyne incognita*) which attacks both the root and ginger rhizome, resulting in warty overgrowth on them. The nematode normally burrows themselves into the soil and form knots on the roots and rhizome. In this way, they attack the plant by feeding on the food and nutrients which should be used by the plant for growth and production. Therefore, because of starvation with nutrients and food, the plant dies.

#### *6.1.4 Rhizome scale*

The size and shape of this scale insect is about 0.5–2 mm in diameter and usually circular in shape. They normally have a shell color brown on top which act as coverage and protection. It attacks the rhizome of the plant by sucking out the juice from it thus resulting in wilting and death of the plant. It is hard or difficult to treat this pest (using insecticides or even hot-water treatment) because of the shell forming on top which protects the insect from dying.

Apart from these major pest and diseases, others include Fungi and bacteria such as Thread blight *Ceratobasidium* sp.; *Corticium* sp., Stem rot (*Athelia rolfsii*), Leaf spot (*Magnaporthe Grisea*).

#### *6.1.5 Nematodes*

Burrowing Nematode—*Radopholus similis* (associated with rhizome rot); Reniform nematode- *Rotylenchulus reniformis* (yellowing leaf drying and stunting); Pin nematode *Criconemoides onoensis*

#### *6.1.6 Yellow leafspot of ginger*

Over the years, ginger yellow leaf spot disease has posed a serious challenge to increased ginger production in Nigeria [115]. The most susceptible stage which is three- to four-leaf stage coupled with high humidity has been observed to be conducive for the disease spread. Ginger plants of up to 6–7 months old are also susceptible to this disease. Grasses have been reported as reservoir hosts while agents of dispersal are rainwater and wind [116]. This disease spread widely in ginger-growing areas with resistance to benomyl (Benlate 50wp), mancozeb (Diathane m-45), and Kocide 101 (copper hydroxide) treatments. In the rainforest agro-ecology of Nigeria, though there may be appearance of early signs of the disease but severity is usually noticeable towards the end of vegetative growth.

#### *6.1.7 Management of diseases*

The most recommended method used to control these pest and diseases is the use of hot water treatment (51°C). Spraying indoxacarb at 10 ml in 15 l of water, or novaluron at 10 ml in 15 l of water at 15 days interval has been reported to be very effective in controlling shoot borer and leaf roller. Dipping the seed rhizome in quinalphos prepared by dissolving 1 ml in 1 l of water can be used to control rhizome scale insects that destroys rhizomes. Good drainage coupled with treatment of seed rhizome by dissolving 3 g of carbendazim and mancozeb in 1 l of water for almost 30 min before storage should be done to prevent rhizome rot. Moreover, treating seed rhizome by dissolving 2 g of streptomycin in 1 l of water for 30 min canbe used to effectively control bacterial wilt that causes milky ooze when rhizomes are gently pressed.

### 6.2 Pests and disease management in onion (*Allium cepa*) and garlic (*Allium sativum*)

Nutrient-deficient and poorly irrigated soils, low technology adoption rates coupled with pests and diseases have curtailed onion and garlic cultivation in Nigeria. Ewuziem and Alleluyanatha [116] highlighted purple blotch (*Alternaria porri*), black mold diseases (*Aspergillus niger*), neck and bulb rot (*Botrytis allii*), Onion twister (*Colletotrichum cingulata*), downy mildew (*Peronospora destructor*) pink rot (*Pyrenochaeta terrestris*) and bulb rot incited by *Fusarium oxysporium* as fungal diseases of economic importance affecting onion and garlic cultivation in northern Nigeria.

The biggest threats to making profit on onion or garlic cultivation in northern Nigeria is insect pests, majorly thrips. *Thrips tabaci* (Thysanoptera: Thripidae) constitute a major threat to onion which is capable of reducing bulb yield if onion is planted late. Other minor insect pests are *Zonocerus variegatus* and *Spodoptera exigua*. Late onions can harbor about 600 thrips in a plant, with the largest population being harbored by the third youngest leaf, notwithstanding the age of the crop [117]. Thrips are also controlled with the use of Lambda-cyhalothrin, neem formulation (2–3 ml/l). Malathion (2 teaspoon/l), Diazinoon 50 WP (1 table spoon/4 l), Bayfidan (triadimenol) 20 EC (2 teaspoon/4) or Dimethoate (0.05%).

Report by Kebbi Agricultural and Rural Development Authority (KARDA) in Kebbi State of Nigeria stated that a major limitation to profitable cultivation of onions is the incidence of a disease locally called 'Danzazzalau'. The devastating effect of this disease on onion productivity necessitated the study reported and concluded that *Fusarium equiseti* was responsible for the disease. *Fusarium equiseti* living in the soil affect the roots, stem plate and fleshy leaf bases of the onion plant.

*6.2.1 Purple blotch of onion*

Pattern of variation for resistance to purple blotch (*Alternaria porri*) of onions (*A. cepa* L.) in North western Nigeria was also reported by [118]. According to the report, five cultivars of onion: Ori local, Kaharda, Sokoto, Red Creole and Koumassa were selected based on genetic backgrounds diversity in respect of resistance to *Alternaria porri (Ellis.) Cif.* In a complete diallel cross of some cultivars, 25 $F_1$s generated and their parents evaluated in a yield trial at Sokoto and Talata Mafara in Zamfara State both of Nigeria, with 31.20%, 30.58% and 5.42% disease incidence observed as phenotypic, genotypic and environmental coefficients of variability, respectively [119].

### 6.3 Garlic (*A. sativum*)

Emechebe et al. [118] evaluated the insecticidal properties of garlic aqueous extract on beans (*Phaseolus valgaris*) and maize (*Zea mays*) pests at different concentrations and concluded that there was relationship between extracts and mortality of *Sitophilus zea mays* and *Callosobruchs maculatus*. The report of the study carried out by [120] revealed that garlic (*A. sativum*) at 5% aqueous extract concentration exhibited antifungal potential when tested against the mycelial growth of southern blight pathogen (*Sclerotium rolfsii*) of tomato. Percentage inhibition was reported to be 77% at 5% concentration in vitro and disease severity was as low as 2.7 in vivo.

### 6.4 Pests and disease management in carrot

Carrot is affected by a variety of pests and diseases [121]. Reducing of plant vigor and growth are part of the effects of nematodes on crops. In nematode affected fields, some plants will be observed to be heavily infested while others will not, resulting in uneven crop maturity or reduction in the quality of the produce [122, 123]. Root knot nematodes (*Meloidogyne species*) causes general reduction in plant vigor likewise severe distortions and root swelling thereby reducing the marketability of root crops like carrots [124]. Flea beetle, white flies (*Daccus* sp.), aphids, cutworms, and horn worms are among insects that destroy carrot plants [125]. The insects can be controlled by planting resistant varieties and judicious use of short acting pesticides like malathion and carbaryl (Sevin). Calcium deficiency causes blossom end rot which can be treated by addition of lime to soil or spraying the leaves with calcium solution. This disorder can also be contolled by maintaining adequate soil moisture [122, 123]. Hill et al. [122, 123] concluded that in most locations, diseases severely restrict carrot cultivation. In Nigeria and other African countries, small holder carrot cultivation has experienced a great increase both as food and cash crops in recent years, and effective management of pests and diseases is essential for sustained carrot production. Wittwer [126] highlighted major fungal diseases of carrot as *Alternaria* and *Cercospora* leaf spots or blights, leaf mold, Fusarium wilt, target spot or early blight some of which are soil-borne. Plants are affected at any developmental stage by Septoria leaf spot [127]. Powdery mildew caused by *Oidium lycopersicum* is another important fungal disease. According to [128] scouting for disease and rogueing infected plants once they are observed is very important. Soil-borne diseases of carrot according to [129] are bacterial soft rot, cavity spot, cottony rot, crown rot, phytophthora, root die-back, root knot nematode and southern blight [126]. Carrot is considered as being capable of producing higher yields and returns to vegetable farmers in some part of the country after the rainy season [130], which calls for effective crop protection to maximize farmers'profit.

## 7. Postharvest technology of Nigeria root vegetables

Root vegetables are mainly functioning roots. Bulbs, corms, rhizomes, and tubers; they have high moisture content making them highly perishable. A moisture content of up to 52 and 72% have been reported for ginger and garlic, respectively [131], which make these root vegetables highly vulnerable to huge postharvest losses. For instance, up to 50% of onions harvested in the northern states of Nigeria were lost during storage [132]. The postharvest losses of these commodities require that they are taken through a number of postharvest operations in order to keep them wholesome for consumption and increase their utilization. There are different low-cost technologies that have been applied in Nigeria, the primary processing involves sorting, blanching, curing, splitting or peeling and drying to a moisture level of 7–12%. In some cases, pretreatments are applied to improve the quality of the final product. This step varies depending on the crop being processed. These postharvest operations will be discussed in this chapter.

### 7.1 Postharvest operations

#### *7.1.1 Curing*

This helps to release the unique aroma of root vegetables such as ginger, garlic and tumeric that are used as spices, curing is usually done before storage by spreading the bulbs to a thickness of about 5 cm in a shaded and well-ventilated place for 3–4 days. This allows the bulbs to receive dry air which helps to maintain the quality in storage. It is important to remove the tops before curing. Garlic can also be cured by packing in jute sacks in a well-ventilated shady place [97]. The processing of ginger in Nigeria has however not been standardized. Ginger rhizomes are sorted, washed and splitted or peeled. The traditional methods vary considerably resulting in mold growth and loss of important volatile oils [133]. Curing enhances the valued yellow color and aroma of tumeric which is attributed to its curcumin content. The rhizomes must however be separated from the fingers before curing

#### *7.1.2 Peeling*

The skin of whole ginger must be peeled to allow for removal of the water content as it constitutes a barrier for evaporation during drying. Peeling is usually done with knives by scraping off the skin. The use of mechanical peeling machine is gaining popularity among ginger processors; however, the abrasion of machine is similar to hand-scrapping. The time spent during peeling is very critical as it contributes to the loss of volatile oils. In Nigeria, peeling is majorly carried out when the ginger is meant for culinary purposes [134]. Peeled ginger does not attract high prices compared to the splitted ginger and farmers consider it time consuming.

#### *7.1.3 Splitting*

Ginger splitting is the most common and widely acceptable processing operation that is carried out before drying. Flavor components are concentrated under the peel, thus splitting helps to retain as much as 20% of the flavor that could be lost due to peeling.

#### *7.1.4 Blanching*

Blanching is not desirable for ginger and garlic. Although blanching is generally helpful in inactivating spoilage causing organisms and enzymes, ginger, onions and garlic take exceptions as the specific flavors they are valued for are reduced by blanching. Tumeric's quality, on the other hand, is enhanced by blanching in the presence of an alikali [135]. In a study conducted by [136] blanching at 100°C for 10 min before drying was shown to yield tumeric powder with higher color value and curcumin content compared to those that were not blanched or those that were boiled before drying.

#### *7.1.5 Sun drying*

Sun drying still remain the predominant dehydration methods for ginger, garlic and onions in Nigeria. The processing of the Nigerian ginger has not been standardized resulting in low rating in international market and loss of foreign exchange earnings. The microbial load, organoleptic properties and chemical composition usually fall short of specifications. There have been reports of the presence of salmonella, aflatoxin and molds in dried split ginger and ground ginger from Nigeria which have been attributed to poor drying and occurrence of fungal infection. This limits Nigeria's access to the international market thereby creating the need for serious efforts in improving food safety and quality of dried root vegetables.

#### *7.1.6 Storage*

Cool environment with temperature of between 10 and 15°C is suitable for storage of dried rhizomes, slices and splits. Higher temperatures above 23°C can cause losses of up to 20% of its oleoresin. These postharvest operations are usually carried out on ginger from November to January in the northern states of Nigeria, but middlemen can do further sorting and cleaning after getting them from the farmers. The price of these root spices is determined by the extent of sorting, drying and packing done at individual farmers' level. Ginger that have been thoroughly washed and dried hygienically to produce a white-light cream slices is referred to as "American standard". The lower grade is usually called Mozo and the lowest grade of ginger are the unsorted dried ginger rhizomes.

### 7.2 Culinary uses

Garlic, onions and ginger are consumed fresh or dried in form of spice and also as ingredient to flavor various dishes. Tasty spice powders are produced from garlic cloves, ginger and tumeric rhizomes and onion bulbs by drying mainly in the sun. Garlic is also used to cover the odor and flavor of salted meat and fish. Dry root spices in powdered form are used in many homes and restaurants in Nigeria. It is a common seasoning applied to pasta, pizza, and grilled chicken. They are common components of spice mix which are usually seasoned with salt.

### 7.3 Traditional/medicinal uses

In addition to their nutritional benefits, root vegetables also have medicinal value, and their consumption has been used traditionally to treat digestive tract problems, morning sickness, arthritis, high level of cholesterol, etc. They are also good for healthy eyes and good-looking skin. Garlic posseses digestive, carminative and anti-rheumatic properties. Since acient times, garlic is used in formulation of Ayurveda curing muscular pain, giddiness, lungs, heating intestinal ulcer, etc. Different phyto-remedies including ginger, turmeric, onion, and garlic have achieved foremost importance in prevention of metabolic syndrome. Ginger is well known for its medicinal properties, ginger can help with alleviating cough, flu, rheumatoid arthritis and travel sickness based on reports. It aids digestion and said to be an inflammation fighter. It is also said to reduce cholesterol and blood pressure level and provides better blood circulation. Ginger is recognized as one of the most important constituents in herbal medicine. Herbal medicine practitioners use ginger as medicine for treating many diseases.

## 7.4 Value added products

Despite the volume of ginger production in Nigeria, the country is yet to tap into the global market of value-added products; the level of value addition to ginger, garlic, tumeric and onions in Nigeria is very low. Some of the processed products include dried slices, powder, paste, ginger beer, spice mixes, teas, ginger honey, pickle, oil, oleoresin, candy, soft drinks and juice. Increased storage life, reduced transportation cost and foreign exchange earnings are the main advantages of preparing value-added products from these root vegetables.

### *7.4.1 Dehydrated products*

Garlic and ginger are exported either as dried flakes or powder. These spice powders are gaining considerable recognition globally and locally. They are popular in household, restaurants, eateries, caterers, and clubs. Ease of handling and usage has made the demand for these processed products increase everyday.

### *7.4.2 Enhanced Ogi paste*

*Ogi* is a cheap and readily available health-sustaining fermented porridge commonly consumed in West Africa. It has been documented as the most popular fermented health food in many countries in West Africa. It is economical, easy to process and the raw materials are readily available, and a choice food for weaning children [137, 138]. The inclusion of ginger and garlic to *ogi* has been explored as a means of enhancing its taste and acceptability. Inclusion of garlic has been shown to improve the color and texture, while the use of ginger improves the aroma. Inclusion of both contributes to improved nutritional qualities of *ogi*.

### *7.4.3 Spiced jam*

Fruit jams are important in the diet and eaten by people of all ages. In order to improve the nutritional content and the health functionality of jams, produced watermelon, apple and pineapple jams spiced with garlic, ginger and/or turmeric. A good jam which has a soft even consistency and exhibits a high nutritional and antioxidant quality was produced. The total soluble solid and total titratable acidity were found to be within recommended range by Codex Alimentarius [139].

#### *7.4.4 Oleoresin*

Oleoresins are commonly used in many industries because of their strong aroma and flavor. They possess some advantages over raw or dried spices as flavoring agents [140]; their flavors are 5–20 times stronger than that of the corresponding spices and they are commonly sold as spice drop. Oleoresins are essential oils, their microbiological advantages, uniformity in flavor, pungency, and ease of storage and conveyance made them to be preferred. For a good yield of oleoresin, ginger should be harvested at 7–8 months after planting [141]. Through the collaboration of the Raw Materials Research and Development Council (RMRDC) and Belphins Nigeria Ltd., Kaduna State ginger oleoresin was produced in Nigeria which showed favorable physico-chemical and microbial properties [142]. The traditional method is usually a manual process which entails preliminary processing and hand pressing while the improved method comprise of chemical extraction and mechanical expression. Locally, the demand for ginger oleoresin is on the increase from both food and pharmaceutical industries but local production has not been able to meet the demand; there is still a dependence on importation. Therefore, the country has huge potentials for expanding oleoresin production to take advantage of the ever expanding local and global oleoresin market.

### 7.5 Nutraceuticals

The awareness and concerns for the physical wellbeing of humans has been a major driving force for the nutraceutical market in Africa. The inclusion of nutraceuticals in the diet is gaining popularity because they are considered to be natural, safe with less side effects compared to products that are chemically derived. The processing of nutraceuticals has become a means of livelihood especially in the South-west and South-east of Nigeria. Based on industry reports, nutraceuticals market value of Africa is projected to be over USD 430 billion by 2025 and it's expected to register a good Compound Annual Growth Rate (CAGR) of 6.05% during the projected period of 2020–2025 [143]. Nigeria, Namibia, Morocco and Egypt are markets said to drive the sales of pet nutraceuticals. Neutraceuticals in Nigeria are mainly herbs, medicinal and aromatics. The industry is active but small in value involving traditional productions.

### 7.6 Chemical and nutritional composition of root vegetables

The place of cultivation and postharvest treatments influence variation in the chemical and nutrient components of root vegetables. **Tables 6** and 7 shows the nutritional composition and chemical components of some root vegetables in Nigeira; with the highlights of nutrients, antioxidant and antimicrobial properties presented below.

#### *7.6.1 Nutrients*

Root vegetables like ginger, onions, garlic and turmeric are known for their rich mineral and very low fat content. Many Root vegetables have absolutely no calorie content thereby making it a weight loss diet. Root vegetables contain significant quantities of minerals such as magnesium, iron, calcium, and potassium.

| | Garlic (DW) | Ginger (DW) | Turmeric (DW) | White onions (WW) | Red onions (WW) |
|---|---|---|---|---|---|
| Moisture (%) | 4.55 | 6.37 | 8.92 | 89.62 | 88.48 |
| Ash (%) | 4.08 | 6.30 | 2.85 | 3.33 | 3.17 |
| Crude protein (%) | 15.33 | 8.58 | 9.40 | 3.22 | 3.02 |
| Fat (%) | 8.58 | 5.35 | 6.85 | 2.17 | 6.50 |
| Crude fiber (%) | 2.10 | 3.25 | 4.60 | 3.83 | 2.83 |
| Carbohydrate (%) | 73.22 | 68.15 | 67.38 | 87.44 | 84.48 |
| Sodium (mg/100 g) | 4.10 | 4.10 | NA | 0.40 | 0.37 |
| Calcium (mg/100 g) | 26.30 | 25.76 | 21.00 | 1.36 | 1.32 |
| Iron (mg/100 g) | 5.29 | 3.46 | 4.50 | 0.09 | 0.12 |
| Phosphorous (mg/100 g) | 10.19 | 12.56 | 63.00 | 3.09 | 3.68 |
| Potassium (mg/100 g) | 54.00 | 215.00 | 46.00 | 14.29 | 16.02 |
| Zinc (mg/100 g) | 0.34 | 0.04 | NA | 0.026 | 0.03 |
| Manganese (mg/100 g) | 0.001 | 0.002 | NA | 0.01 | 0.01 |
| Magnesium (mg/100 g) | 4.10 | 5.00 | 0.92 (ppm) | 0.86 | 0.97 |
| Vitamin C (mg/100 g) | 8.00 | 4.80 | 8.40 | 14.67 | 18.00 |

*DW: Dry weight, WW: Weight weight, NA: Not Available. Source: [144–148].*

**Table 6.**
*Nutritional composition of selected root vegetables in Nigeria.*

| Selected root vegetables | Some important antioxidant and chemical components | Health benefits and common uses |
|---|---|---|
| Garlic | Allicin, diallyl sulfide, diallyl disulfide, diallyl trisulfide, allyl isothiocyanate, *S*-allyl cysteine | Antioxidants, immunostimulants, antineoplastic, anti-inflammatory, antihypertensive, antidiabetic, antithrombotic, antihyperlipidemia, antibacterial, antifungal, antiviral, neuroprotective, anticarcinogenic |
| Ginger | Gingerol, turmeric, paradol, geraniol, geranial, borneol, linalool, camphene, zingerol, zingiberon | Antioxidant, anti-inflammatory, neuroprotective, antinausea, antiobesity, antilhyperlipidemia, antimicrobial |
| Tumeric | Curcumins, essential oils, eugenol, carotene, ascorbic acid, caffeic, *p*-coumaric, protocatechuic, syringic, vanillic acid | Antioxidant, antidiabetic, antibacterial, anticarcinogenic, antispasmodic, antitussive, antihelmintic |
| Onions | Quercetin, apigenin, dipyridyl disulfide, rutin, quercetin-4-glucoside | Antioxidants, antihyperlipidemia, anticarcinogenic, antidiabetic, antibacterial, antiplatelet activity |

*Source: [149–151].*

**Table 7.**
*Chemical components and health benefits of selected root vegetables.*

### *7.6.2 Antioxidants*

Root vegetables which are also used as spices are excellent sources of natural antioxidants; they contain antioxidant enzymes like catalase, glutathione peroxidase and superoxide and non-enzymatic antioxidants such as ascorbic acid, polyphenols, carotenoids and chelating agents. Ginger contains polyphenol compounds like gingerol and its derivatives like zingiberone, bisabolene, camphene, geranial, linalool, borneol and oleoresin. These bioactive compounds are responsible for the sensory attributes and capacity to delay food spoilage or inhibit disease causing reactions or agents that have been reported. Diverse bioactive compounds have been reported in garlic [152]. Flavonoids (flavones and quercetins) and sulfur-containing compounds (allyl-cysteine, diallyl sulfide, and allyl trisulfide are found in garlic, and [153] reported the antioxidant properties of the garlic compounds allyl cysteine, alliin, allicin, and ally disulfide. The S- allyl-L-cysteine sulfoxide is converted into allicin (which produces the unique odor associated with garlic). Consumption of garlic preparations have been shown to reduce blood lipid peroxidation while increasing the vitamin E concentration [154]. The curcumin in turmeric is a natural phenolic compound. Apart from being a coloring pigment, it has the capacity to neutralize free radicals [155].

#### *7.6.3 Antimicrobial properties*

The biological and antimicrobial properties of spices are also linked to the presence of bioactive compounds. Allicin present in garlic exhibits a broad spectrum of effects on a variety of fungal species. Phenolic compounds in ginger have high antimicrobial activities and effectiveness for the control of certain viral, bacterial and fungal diseases. These plants are used in many countries for food preservation. Gingerols and gingerdiol are the main anti-fungal components of ginger. Some antiviral properties have been reported for ginger. Ginger is reported to be effective in management of hepatitis C virus infection where viral clearance is affected [156, 157].

## 8. Opportunities and constraints

Vegetables plays pivotal role in the Nigerian economy by providing food, nutritional and economic security to households and higher returns to producers. In addition, vegetable crops have high productivity and short maturity cycle leading to higher returns per unit area and time. Globally, without significant increase in yield of vegetables, the strategy to reduce poverty is impossible. In Nigeria with diverse agro-ecological conditions that favor the cultivation of several types of fruit, vegetables and other classes of crops, diversification into NRV cultivation provides interesting and profitable opportunities. A very promising premise in the production of NRVs is the ease of cultivation, they do not need large area of land for profitable production. Successful cultivation could be achieved without excessive investments as they could be cultivated with little inputs like cash and land. Their production, though more labour intensive can provide twice the amount of employment in one hectare of land used for production compared to staple food crop production. Their value chain is longer with more complex stages than staple crops as a result of available job opportunities [158]. This is beneficial for generating additional employment opportunities in rural areas where labour abounds for attaining widespread and equitable growth.

NRVs are particularly viable enterprise for women, the landless and youths because they can be suitably produced in gardens, in and around homesteads, likewise

| Constraints | Mean | Rank |
|---|---|---|
| High cost of inputs (seeds, seedlings, fertilizers) | 1.46 | 1st |
| High transportation costs | 1.42 | 2nd |
| High labor costs | 1.27 | 3rd |
| Non availability of inputs | 1.16 | 4th |
| Non availability of labor for production | 1.10 | 5th |
| Pest and diseases | 0.60 | 6th |
| Seasonality of seeds and seedlings | 0.57 | 7th |
| Knowledge of production | 0.55 | 8th |
| Technical advice/ Extension service | 0.55 | 8th |
| Poor market information | 0.39 | 10th |

*Source: [56].*

**Table 8.**
*Constraints to production.*

provide opportunities for profit-oriented enterprises and opportunity to contribute to the local economy [64]. There are good potentials for entrepreneurship development by small-scale on-farm and off-farm processing that provides higher income from value-adding activities and sales of processed spices. The growing demand for high-value crops world-wide is an opportunity for rural households to diversify towards spices enterprises considering the strong potential for higher returns to land, labor and capital.

There is the need for breeding programmes to develop improved cultivars of existing cultivated NRVs. This will encourage farmers into more cultivation and commercialization with the attendant benefits inherent in cultivation and potential impacts of NRVs. In spite of the diverse opportunities inherent in production of NRVs, farmers still operate under high cost of production such as high transportation costs which reduce their productivity to a very large extent. In Nigeria, other factors that affect high productivity in small scale farming include input availability and use. As opined by [159], a continuous cycle of low productivity, income, input availability and use is prevalent among farmers in Nigeria as yields involve combination of education by extension services, access to appropriate and timely inputs along with ability to access finance to purchase inputs. It also further buttress the findings of [160] who posited that poor road network affect the quality of life of producers, their ability to transfer produce to markets which increases spoilage of harvest stalling enterprise sustainability, productivity and income.

As shown in **Table 8**, there are diverse constraints militating against optimum production of root vegetables. These constraints include challenges occasioned by infrastructural factors like high cost of inputs such as seeds, seedlings and other requirements for a successful production enterprise. This ranks as a prime constraint in the production of root vegetables in Kano state, Nigeria. Other infrastructural constraints are high transportation costs, high labor costs, non- availability of inputs and labor for farming activities as well as pest and diseases. Institutional constraints that challenge optimum production of root vegetables include inadequate technical advice, extension support, knowledge of production of root vegetables and poor market information.

## 9. Conclusion

Production of root vegetables should be popularized especially among rural, smallholder farmers who have limited land resources and agronomic inputs. Their short production cycles compared to other field crops and their high-value, low-volume, high-yields should be promoted as added advantages. The general public needs to be well-informed about accessible and affordable sources of micronutrients and antioxidants which impacts highly on human health and wellbeing. More researches on integrated conservation techniques, promotional campaigns to enhance consumption and utilization as well as promotion and development of value chains are required for a better understanding of benefits inherent in nutrient-dense NRVs. Reviews such as this that provide insights into the great benefits of root vegetables grown in Nigeria including perspectives on agronomy, genetics and breeding, biotechnology, nutraceutics and socio-economics, would ultimately enhance their production and utilization for healthy lives.

## Author details

Emmanuel O. Ajayi[1†], Pamela E. Akin-Idowu[1†], Olaide R. Aderibigbe[1], Dorcas O. Ibitoye[1], Gloria Afolayan[2], Oluyemisi M. Adewale[1], Esther A. Adesegun[1] and Benjamin E. Ubi[3*]

1 National Horticutural Research Institute (NIHORT), Ibadan, Nigeria

2 National Centre for Genetic Resources and Biotechnology, Ibadan, Nigeria

3 Deptartment of Biotechnology, Ebonyi State University, Abakaliki, Ebonyi State, Nigeria

*Address all correspondence to: ben.ubi@ebsu.edu.ng

† Co-first authors.

## References

[1] Alabi O, Banwo OO, Alabi SO. Crop pests management and food security in Nigeria agriculture. Archives of Phytopathology and Plant Protection. 2006;**39**(6):457-464

[2] Godfray HCJ, Beddington JR, Crute IR, Haddad L, Lawrence D, Muir JF, et al. Food security: The challenge of feeding 9 billion people. Science. 2010;**327**(5967):812-818. DOI: 10.1126/science.1185383

[3] Foley JA, Ramankutty N, Brauman KA, Cassidy ES, Gerber JS, Johnston M, et al. Solutions for a cultivated planet. Nature. 2011;**478**(7369):337-342. DOI: 10.1038/nature10452

[4] Tilman D, Balzer C, Hill J, Befort BL. Global food demand and the sustainable intensification of agriculture. Proceedings of the National Academy of Sciences. 2011;**108**(50):20260-20264. DOI: 10.1073/pnas.1116437108

[5] Valin H, Sands RD, Van der Mensbrugghe D, Nelson GC, Ahammad H, Blanc E, et al. The future of food demand: Understanding differences in global economic models. Agricultural Economics. 2014;**45**(1):51-67. DOI: 10.1111/agec.12089

[6] Frankel OH. Genetic perspective of germplasm conservation. In: Arber W, Limensee K, Peacock WJ, Stralinger P, editors. Genetic Manipulations: Impact on Man and Society. Cambridge, UK: Cambridge University Press; 1984. pp. 161-170. DOI: 102.100.100/281332

[7] Brush SB. In situ conservation of landraces in centers of crop diversity. Crop Science. 1995;**35**(2):346-354. DOI: 10.2135/cropsci1995.0011183X003500020009x

[8] Aladele SE, Afolayan G, Jamaleddine ZO, Oduoye OT, Bolatito OA, Afolayan AO, et al. Biotechnological approaches for genetic resources conservation and use. In: Agricultural Biotechnology, Biodiversity and Bioresources Conservation and Utilization. Boca Raton: CRC Press; 2022. pp. 427-450. DOI: 10.1201/9781003178880

[9] Ebert AW. The role of vegetable genetic resources in nutrition security and vegetable breeding. Plants. 2020;**9**(6):736. DOI: 10.3390/plants9060736

[10] FAO-Food and Agriculture Organization of the United Nations. The Second Report on the State of the World'sPlant Genetic Resources for Food and Agriculture. Rome, Italy: FAO; 2010. p. 399

[11] Antonelli A, Smith RJ, Fry C, Simmonds MS, Kersey PJ, Pritchard HW, et al. *State of the World's Plants and Fungi* (doctoral dissertation). Royal Botanic Gardens (Kew) Sfumato Foundation; 2020. DOI: 10.34885/172

[12] Adam IK, Mustapha AA. Omics and biotechnology: Tools to improve onion crop production and quality in developing counties. Bio/Technology. 2014;**76**:28390-28396

[13] Adam IK. The role of biotechnology in agricultural sustainability. International Journal of Chemical Sciences. 2009;**2**(**2**):255-261

[14] Zheng SJ, Henken B, Sofiari E, Keizer P, Jacobsen E, Kik C, et al.

Effect of cytokinin and lines on plant regeneration from long-term callus and suspension cultures of *Allium cepa* L. Euphytica. 1999;**108**(2):83-90

[15] Zheng SJ, Henken B, de Maagd RA, Purwito A, Krens FA, Kic C. Two different *Bacillus thuringiensis* toxin genes confer resistance to beet armyworm (*Spodopteraexigna* Hubner) in transgenic Bt-shallots (*Allium cepa* L.). Transgenic Research. 2005;**14**(3):261-272

[16] Ponce M, Martinez L, Galmarini C. Influence of CCC, putrescine and gellam gum concentration on gynogenic embryo induction in *Allium cepa*. Biologia Plantarum. 2006;**50**:425-428

[17] Ramming DW. The use of embryo culture in fruit breeding. HortScience. 1990;**25**:393-398

[18] Shimonaka M, Hosoki T, Tomita M, Yasumuro Y. Production of somatic hybrid plants between Japanese bunching onion (*Allium fistulosum* L.) and bulb onion (*Allium cepa* L.) via electrofusion. Journal of the Japanese Society for Horticultural Science. 2002;**71**(5):623-631

[19] Brown CR, Mojtahedi H, James S, Novy RG, Love S. Development and evaluation of potato breeding lines with introgressed resistance to Columbia root-knot nematode (*Meloidogyne chitwoodi*). American Journal of Potato Research. 2006;**83**:1-8

[20] Barrell PJ, Meiyalaghan S, Jacobs JME, Conner AJ. Applications of biotechnology and genomics in potato improvement. Plant Biotechnology Journal. 2013;**11**:907-920

[21] Wang H, Li X, Liu X, Oiu Y, Song J, Zhang X. Genetic diversity of garlic (*Allium sativum* L.) germplasm from China by fluorescent based AFLP, SSR and In Del markers. Plant Breeding. 2016;**135**:743-750

[22] Ijeomah C, Amuda O, Babatunde B, Abutu P. Evaluation of genetic diversity of spring onions based on DNA markers. Journal of Experimental Agriculture International. 2020;**42**(3):23-33

[23] Datta AK, Biswas AK. EMS induced mitotic consequences in three rhizomatous, spice yielding plants. Chromatography Information Services. 1985;**38**:25-26

[24] Ramachandran K, Chandrasekharan NPN. Induced tetraploids of ginger (*Zingiber officinale* Rosc.). Journal of Spices and Aromatic Crops. 1992;**1**:39-42

[25] Zheng YQ, Liu YM, Ma M, Xu K. Increasing *in vitro* microrhizome production of ginger (*Zingiber officinale* roscoe). Acta Physiologiae Plantarum. 2008;**30**:513-519

[26] Kavyashree R. An efficient *in vitro* protocol for clonal multiplication of ginger-var, Varada. Indian Journal of Biotechnology. 2009;**8**:328-331

[27] Kambaska KB, Santilata S. Effect of plant growth regulator on micropropagtion of ginger (*Zingiber officinale* Rosc.) cv-Suprava and Suruchi. Journal of Agricultural Technology. 2009;**5**:271-280

[28] Sathyagowri S, Seran TH. *In vitro* plant regeneration of ginger (*Zingiber officinale* Rosc.) with emphasis on initial culture establishment. International Journal of Medicinal and Aromatic Plants. 2011;**1**:195-202

[29] Rahman MM, Amin MN, Jahan HS, Ahmed R. *In vitro* regeneration of plantlets of *Curcuma longa* Linn. A valuable spice plant in Bangladesh.

Asian Journal of Plant Sciences. 2004;**3**:306-309

[30] Rout GR, Palai SK, Samantaray S, Das P. Effect of growth regulator and culture condition on shoot multiplication and rhizome formation in ginger (*Zingiber officinale* Rosc.) *in vitro*. In Vitro Cellular & Developmental Biology: Plant. 2001;**37**:814-819

[31] Murashige T, Skoog F. A revised medium for rapid growth and bio assays with tobacco tissue cultures. Physiologia Plantarum. 1962;**15**:473-497

[32] Zahid NA, Jaafar HZE, Hakiman M. Micropropagation of ginger (*Zingiber offificinale* roscoe) 'Bentong' and evaluation of its secondary metabolites and antioxidant activities compared with the conventionally propagated plant. Plants. 2021;**10**:630. DOI: 10.3390/plants10040630

[33] Nkere CK, Mbanaso ENA. Optimizing concentrations of growth regulators for in-vitro ginger propagation. Journal of Agrobiology. 2010;**27**(2):61-65. DOI: 10.2478/s10146-009-0009-9

[34] Sharma TR, Singh BM. High frequency *in vitro* multiplication of disease free *Zingiber officinale* Rosc. Plant Cell Reports. 1997;**17**:68-72

[35] Balachandran SM, Bhat SR, Chandel KPS. *In vitro* clonal multiplication of turmeric (*Curcuma* spp.) and ginger (*Zingiber officinale* Rosc.). Plant Cell Reports. 1990;**8**:521-524

[36] Hirata S, Abdelrahman M, Yamauchi N, Shigyo M. Diversity evaluation based on morphological, physiological and isozyme variation in genetic resources of garlic (*Allium sativum* L.) collected worldwide. Genes & Genetic Systems. 2016;**91**(3):161-173. DOI: 10.1266/ggs.15-00004

[37] Tesfu K, Aschale N, Regassa E, Gebremdhen H, Desalegn O, Deres D, et al. Molecular characterization of garlic (*Allium sativum* L.) using SSR markers in Ethiopia. African Journal of Plant Science. 2021;**15**(2):59-63

[38] Etoh T. Studies on the sterility in garlic *Allium sativum* L. Memoirs of the Faculty of Agriculture Kagoshima University. 1985;**21**:7-132

[39] Verbeek M, Van Dijk P, Van Well EMA. Efficiency of eradication of four viruses from garlic (*Allium sativum* L.) by meristem-tip culture. European Journal of Plant Pathology. 1995;**101**:231-239

[40] Lot H, Chovelon V, Souche S, Delecolle B. Effects of onion dwarf and leek yellow stripe viruses on symptomatology and yield loss of three French garlic cultivars. Phytopathology. 1998;**82**:1381-1385

[41] Haque MS, Wada T, Hattori K. Shoot regeneration and bulblets formation from shoot and root meristem of garlic cv Bangladesh local. Asian Journal of Plant Sciences. 2003;**2**:23-27

[42] Keller ERJ, Senula A. Micropropagation and cryopreservation of garlic (Allium sativum L.). In: Lambardi M et al., editors. Protocols for Micropropagation of Selected Economically-Important Horticultural Plants, Methods in Molecular Biology. Vol. **994**. NY, USA: Springer; 2013. pp. 353-368

[43] Barandiaran X, Martin N, Rodriguez M, Di Pietro A, Martin J. Genetic variability in the callogenesis and regeneration of garlic. Plant Cell Reports. 1999;**18**:434-437

[44] Metwally EI, El-Denary ME, Dewir YH, Naidoo Y. *Invitro* propagation of garlic (*Allium sativum* L.) through adventitious shoot organogenesis. African Journal of Biotechnology. 2014;**13**(38):3892-3900

[45] Olojede AO, Nwokocha CC. A decade of research on minor root and tuber crops at NRCRI: The contribution towards food sufficiency and economic empowerment in Nigeria. Amadi CO, Ekwe KC, Chukwu GO, Olojede AO, Egesi CN, editors. Root and Tuber Crops Research for Food Security and Empowerment. Umudike, Nigeria: NRCRI; 2011. pp. 387-396

[46] Malamug JF, Inden A, Asahira T. Plantlet regeneration and propagation from ginger callus. Scientia Horticulturae. 1991;**48**:89-97

[47] Babu KN, Samsudeen K, Ravindran PN. Direct regeneration of plantlets from immature inflorescences of ginger (*Zingiber officinale* Rosc.) by tissue culture. Journal of Spices and Aromatic Crops. 1992;**1**:43-48

[48] Ugochukwu SC, Bob SE, Ozioma O, Odii EB, Ijeoma IC, Olanike O. Shoot proliferation of invitro tumeric (*Curcuma longa* L.) affected by different concentrations of Benzylaminopurine (BAP). World Journal of Agricultural Sciences. 2013;**9**(3):227-230

[49] Pandey YR, Sagwansupyakorn C, Sahavacharin O, Thaveechai N. *In vitro* propagation of ginger (*Zingiber officinale* roscoe). Kasetsart Journal (Natural Science). 1997;**31**:81-86

[50] Nasirujjaman K, Salah Uddin M, Zaman S, Reza MA. Micropropagation of turmeric (*Curcuma longa* Linn.) through *in vitro* rhizome bud culture. Journal of Biological Sciences. 2005;5:490-492

[51] Naz S, Ilyas S, Javad S, Ali A. *In vitro* clonal multiplication and acclimatization of different varieties of turmeric (*Curcuma longa* L.). Pakistan Journal of Botany. 2009;**41**:2807-2816

[52] Dipti T, Ghorade RB, Swati M, Pawar BV, Ekta S. Rapid multiplication of turmeric by micropropagation. Annual Review of Plant Physiology. 2005;**19**:35-37

[53] Sit AK, Tiwari RS. Micropropagation of turmeric (*Curcuma longa* L.). Recreational Horticulture. 1998;**4**:145-148

[54] Adewale OM, Fariyike TA, Akinpelu CA, Adebisi-Adelani O, Muazu YG, Shuaibu SM. Assessment of constraints to spices production in Kano State, Nigeria. In: Adesina JM, Iwala OS, Kekere O, Ajayi AJ, Ajayi EO, Idowu-Agida OO, editors. Horticultural production for sustainable economic growth and Development. In: Book of Proceedings of the 37th Annual Conference of Horticultural Society of Nigeria, held at Faculty of Agricultural Technology. Rufus Giwa Polytechnic, Owo, Ondo State Nigeria. 18th - 22nd November, 2019. pp. 31-34

[55] Akinpelu CA, Adebayo OS, Adelaja BA, Adesegun EA, Aminu-Taiwo BR, Adeoye IB. An economic analysis of spices marketing in three selected states of Nigeria. The Research on Crops. 2011;**12**(3):817-821

[56] Adewale OM, Oladeji JO. Explorative analysis of benefits derived from spices production enterprise in Kano State, Nigeria. Journal of Sustainable Development. 2021;**19**(1):45-52

[57] Akinpelu CA, Adebayo OS, Fariyike TA, Adewale OM. Challenges in horticultural crops marketing in Nigeria: The spices experience. In: Proceedings of

the 2nd All Africa Horticulture Congress in South Africa Acta Hort 1007. ISHS; 2013. pp. 915-921

[58] Oladeji JO, Oyesola OB. Use of information and communication technologies among private agricultural organization workers in Oyo state, Nigeria. The Journal of Agriculture and Food Information. 2011;**12**:258-269

[59] Eurostat. Quality of life indicators, statistics explained-education. 2013. Available from: https://ec.europa.eu/eurostat/statistics-explained/index.php?title=Quality_of_life_indicators_-_education

[60] Welch C, Alemu B, Msaki T, Sengendo M, Kigutha H, Wolff A. Improving Household Food Security: Institutions, Gender and Integrated Approaches. USA: BASIS Management Entity; 2000

[61] National Bureau of Statistics. General Household Survey-Panel Wave 3 (Post Planting) Abuja. The NBS Publication; 2016

[62] Balogun OL, Akinyemi BE. Land fragmentation effects on technical efficiency of cassava farmers in south-west geopolitical zone, Nigeria. Cogent Social Sciences. 2017;**3**:1-10

[63] National Bureau of Statistics. In: Statistical News: Labor Force Statistics No. 476 Abuja. The NBS Publication; 2014

[64] Food and Agriculture Organization of the UN FAO. Spices and Herbs for Home and Market. Rural Infrastructure and Agro-allied Industries Division. Rome: Food and Agriculture Organisation of the United Nations; 2011 Diversification Booklet # 20

[65] News Agency of Nigeria (NAN) 2021. https://www.vanguardngr.com/2021/11/nigeria-produces-2m-metric-tonnes-of-onions-annually-%E2%80%95-agric-minister/

[66] Stan Edom (2016). https://startuptipsdaily.com/onion-farming-in-nigeria/

[67] National Horticultural Research Institute (NIHORT). Production Guide for Onions. Production Guide No. 1.

[68] Ibrahim ND. Growth and yield of onion (Allium cepa L.) in Sokoto, Nigeria. *Agriculture and Biology Journal of North America*. 2010;**1**(4):556-564 ISSN Print: 2151-7517, ISSN Online: 2151-7525

[69] Finelib.com. https://www.finelib.com/about/nigeria-cash-crops/varieties-of-carrots-and-the-producing-states-in-nigeria/147

[70] Maxipharo (2021). https://agricincome.com/carrot-farming-in-nigeria/#:~:text=However%2C%20commercial%20carrot%20farming's%20suitable,to%20get%20the%20desired%20yield

[71] Amadi POE, Adiele-Ezekiel C, Ibe MN, Nzeakor FC. Turmeric production, processing and marketing for sustainable agricultural development in Nigeria. Journal of Community & Communication Research (JCCR). 2018 ISSN: 2635-3318. www.jccr.org.ng

[72] Nwaekpe J, Anyaegbulam H, Okoye BC, Asumugha GN. Promotion of turmeric for the food/pharmaceutical industry in Nigeria. American Journal of Experimental Agriculture. 2015;**8**(6):335-341. Article no AJEA 2015.117. ISSSN: 2231-0606. www.sciencedomain.org

[73] Yusuf VA. Turmeric: 'Golden crop' neglected by Nigerian farmers. Daily Trust, Thursday, June 25. 2016:2020

[74] Amadi POE, Ibe MN, Kalu CI. Turmeric research in Nigeria: Challenges and Prospects for Economic Growth. In: Proceeding of the 51st Annual Conference of the Agricultural Society of Nigeria. Abuja. 2017. pp. 19-21

[75] Nwakor FN, Asumugha GN, Nwokocha CC, Ekedo TO. Guide to Turmeric Production, Processing and Marketing in Nigeria. Extension Guide No. 28. Extension Services Programme. Umudike, Nigeria: National Root Crops Research Institute; 2014. pp. 1-11

[76] Olojede AO, Iluebbey P, Dixon AGO. IITA/NRCRI collaborative germplasm and data collection on minor root and tuber crops in Nigeria. In: Annual Report 2005. Umudike: National Root Crops Research Institute; 2005. pp. 77-78

[77] Dialy Trust (2018). https://dailytrust.com/reap-the-benefits-of-planting-turmeric

[78] Nwokocha CC, Olojede AO, Ano AO. Effect of planting depth and mulching of turmeric on Ultisol. NRCRI Annual Report. 2005;**2005**:94-97

[79] Olojede AO, Nwakocha CC, Akinpelu AO. Effects of planting date and sett size on growth and yield of turmeric at Umudike. NRCRI Annual Report. 2007:71-73

[80] Njoku SC, Olojede AO, Mlifonwu AA, Nwadili C. Determination of the period of weed interference for optimum performance and productivity of rizga, turmeric and Hausa Poato at Umudike. NRCRI Annual Report. 2009;**2009**:70-73

[81] Nwubiko O, Olojede AO, Nwokocha CC. Effect of time of fertilizer application on the growth and yield of Livingstone potato (rizga), turmeric and Hausa potato under Umudike condition. NRCRI Annual Report. 2007;**2007**:83-86

[82] Olojede AO, Nwokocha CC. A Decade of Research on Minor Root and Tuber Crops at NRCRI: The Contribution towards Food Sufficiency and Economic Empowerment in Nigeria. 2011. pp. 387-395

[83] Nwokocha CC, Olojede AO, Ano AO, Chukwu GO, Korieocha DS. NPK requirement for turmeric production on an Arenic Hapludult in Southern Nigeria. In: Proceedings of the 33rd Annual Conference of the Soil Science Society of Nigeria. Held at University of Ado-Ekiti. 2009. p. 166

[84] Ohaeri JE, Olojede AO, Nwokocha CC. Effect of Lime on Growth and Yield of Turmeric on An Ultisol of South Eastern Nigeria NRCRI Umudike Annual Report. 2008. pp. 69-72

[85] Ohaeri JE, Olojede AO. Effect of lime and NPK fertilizer on growth and yield of turmeric on an ultisol in South-Eastern Nigeria. In: Proceedings of the 45th Annual Conference of the Agricultural Society of Nigeria held at Usman Danfodio University Sokoto, Nigeria, October 24-28, 2011. pp. 77-82

[86] Kaduna State Agricultural Development Project—KADP. Production of Ginger. An Extension Guide. Kaduna State, Agricultural Development Project, Kaduna; 2000

[87] BluePrint. 2021. https://www.blueprint.ng/nigerias-contribution-to-global-ginger-output-low-rmrdc-2/

[88] Adegboye MA. Evaluation of farmers' response to extension services on ginger production in Kagarko local government area of Kaduna State. Scientific Research and Essays.

2011;**6**(6):1166-1171, http://www.academicjournals.org/SRE. DOI: 10.5897/SRE09.447 ISSN 1992-2248

[89] Ahmed Maigari Ibrahim. Ginger farming practice in Jaba region, Kaduna state, Nigeria. Dutse Journal of Pure and Applied Sciences (DUJOPAS). 2018;**4**(2):155-168

[90] Phillips-Howard KD. Soil and fertilizer use among small-scale farmers on the Jos Plateau, Nigeria. In: Tony Binns Edited: People and Environment in Africa. Chichester New York: John Wiley & Sons; 1995

[91] National Horticultural Research and Development Foundation (NHRDF), 2008, http://nhrdf.com/

[92] FAO—Food and Agricultural Organization. Rome. the Origin and Distribution of Garlic. Washington, DC, USA: USDA Agricultural Research Unit; 2001

[93] Inuwa BBD. A study of issues arising from the production of garlic (*Allium sativum* L.) in Nigeria. In: Proceedings of the Training Workshop on Improving and Accelerated Garlic Production for Local and Export Needs in Nigeria at Food Crops Production Technology Transfer Station (FDA). Dan Hassan, Kano, Nigeria; 2001. p. 16 Google Scholar

[94] Aliyu BS. Ethno-Medicinal Plants of Semi- arid Region of West African States. Vol. 1. Gidan Sa'adu zungur, Kano, Nigeria: Triumph Publishing Company; 2006. pp. 83-84

[95] FAO. Food and Agricultural Organization Rome: The Origin and Distribution of Garlic. USA: USDA Agric. Res. Unit; 2001. pp. 98-108

[96] IAR. Report on horticultural crops research programme. In: Proceedings of the 1996 Cropping Scheme Meeting. Zaria, Nigeria: Institute for Agricultural Research Ahmadu Bello University; 1996 Google Scholar

[97] NAERLS (National Agricultural Extension and Research Liaison Services) Garlic Production Under Irrigation. Extension Bulletin No. 205 Agric. Engineering Series No. 6

[98] Purseglove JW. *Tropical Crops Monocotyledons*. London, UK: Longman; 1972

[99] Magaji GU, Fagam AS, Rilwanu UD, Zaharaddeen I. Phenotypic response of two garlic varieties to different nitrogen fertilization grown under irrigation in Sudan Savannah ecological zone of Nigeria. International Journal of Agronomy. 2016;**2016**, Article ID 2495828 . DOI: 10.1155/2016/24958289 pages

[100] Ahmed HG, Magaji MD, Yakutu AI, Aliyu L, Singh A. Response of garlic (*Allium sativum* L.) to irrigation interval and clove size in semi-arid, Nigeria. *Journal of Plant Sciences*. 2007;**2**:202-208. DOI: 10.3923/jps.2007.202.208 https://scialert.net/abstract/?doi=jps.2007.202.208

[101] Buwalda, J.G. (1986). Nitrogen nutrition of garlic (*Allium sativum* L.) under irrigation, components of yield and indices of crop nitrogen status. *Scientia Horticulturae*, vol. 29, no. 1-2, pp. 69-76, Publisher Site | Google Scholar

[102] Kakar AA, Abdullahzai MK, Salim M, Qaim Shah SA. Effect of nitrogenous fertilizer on growth and yield of garlic. Agricultural Research Institute, Sariab, Quetta, Pakistan. Asian Journal of Plant Science. 2002;**1**(5): 544-545 Google Scholar

[103] Farooqui MA, Naruka IS, Singh PP, Rathore SS, Shaktawat RPS. Effect of

Nitrogen and sulphur levels on growth and yield of garlic (*Allium sativum* L.). Asian Journal of Food and Agro-Industry. 2008;**2**:18-23. Google Scholar

[104] Babaji BA. Effect of plant spacing and nitrogen fertilization on growth and yield of garlic [M.S. thesis]. Zaria, Nigeria: ABU Zaria; 1994

[105] Bichi AM. Response of garlic *(Allium sativum L.)* to varying levels of phosphorus and nitrogen [B.Sc thesis], Department of Agronomy, ABU, Zaria, Nigeria. 1997

[106] Amadi P, Daniel-Ogbonna C, Nzeakor F. Identifying the potentials of turmeric for food security in Nigeria. Direct Research Journal of Agriculture and Food Technology. 2018;**6**(10):281-286

[107] Shanmugam PS, Indhumathi K, Sangeetha M, Tamilselvan N. Evaluation of different pest management modules against major insect pests and diseases of turmeric. Current Biotica. 2015;**9**(1):17-24

[108] Nirmal Babu K, Samsuddin K, Ratnambal MJ. *In vitro* plant regeneration from leaf derived callus in ginger (*Zingiber officinale* Rosc.). Plant Cell Tissue and Organ Culture. 1992;**29**:71-74

[109] Jagtap GP, Mali AK, Dey U. Bioefficacy of fungicides, bio_control agents and botanicals against leaf spot of turmeric incited by *Colletortricum capsici*. African Journal of Microbiological Research. 2013;**7**(18):1865-1873

[110] Anoop K, Suseela Bhai R. Evaluation of antagonistic potential of indigenous *Trichoderma* isolates against *Pythium aphanidermatum* (Edson) fitz. causing rhizome rot in turmeric (*Curcuma longa* L.). Journal of Science. 2014;**4**(2):99-105

[111] Sarathi I, Senthil Kumar V, Senthil Kumar R, Panneerselvam A. Studies on rhizome rot pathogen in *Curcuma longa*. International Journal of Current Microbiology and Applied Sciences. 2014;**3**(8):296-302

[112] Adesegun EA, Fajinmi OB, Olaleye O, Fariyike TA. Antifungal activity of turmeric (*Curcuma longa*) extract on*fusarium oxysporum* f. sp *lycopersici* causal agent of fusarium wilt of tomato. In: The Proceeding of the 33rd Conference of the Horticultural Society of Nigeria, Theme: "The Role of Horticulture in Food Security and Sustainable Development in Recessed Economy" held at Agricultural Research council of Nigeria, Abuja. 2015. pp. 401-407

[113] Paret ML, Cabos R, Kratky BA, Alvarez AM. Effect of plant essential oils on *Ralstonia solanacearum* race 4 and bacterial wilt of edible ginger. Plant Diseases. 2010;**94**:521-527

[114] Sharma BR, Dutta S, Roy S, Debnath A, Roy MD. The effect of soil physicochemical properties on rhizome rot and wilt disease complex incidence of ginger under hill agro climatic region of West Bengal. Journal of Plant Pathology. 2010;**26**:198-202

[115] Zakka U, Lale NES, Okereke VC. A survey of pests of stored ginger (*Zingiber officinale* (Rosc.)) in some selected markets in River state, Nigeria. *African Journal of Agricultural Research*. 2010;**5**(18):2529-2534

[116] Ewuziem JE, Alleluyanatha E. Economic evaluation of the incidence of yellow leaf spot disease on the yield of ginger in the rainforest agro-ecology of Nigeria. Nigerian Agricultural Journal. 2019;**50**(1):53-58

[117] Ibrahim N. Important insect pests of onions in Sokoto. Journal of Agricultural Research and Policies. 2010;**5**(2):1-4

[118] Emechebe AM, Erinle ID, Bos WS, Tyagi PD, Surdaram NV, Manzo SK. Check-list of Diseases of Crops in the Savanna and Semi-arid Areas of Nigeria. Samaru, Zaria, Nigeria: Institute of Agricultural Research (IAR). p. 6

[119] Abubakar L, Ado GO. Variability pattern for resistance to purple blotch (*Alternaria porri*) disease of onions (*Allium cepa* L.) in North western Nigeria. Nigerian Journal of Basic and Applied Sciences. 2013;**21**(2):109-115

[120] Dauda WP, Alao SEL, Zarafi AB, Alabi O. First report of die-back disease of onion (*Allium cepa* L.) induced by *Fusarium equiseti* (Mart) Sacc in Nigeria. International Journal of Plant and Soil Science. 2018;**21**(3):1-8

[121] Agri Farming. Carrot diseases and pests, symptoms, control methods. 2018. https://www.agrifarming.in/carrot-diseases-and-pests-symptoms-control. [Accessed: June 17, 2022]

[122] Hill DS, Waller JM. Pests and diseases of tropical crops. Longman Scientific and Technical, U.S.A. 1990

[123] Jiskani MM. Plant viruses: Major threat to crops. Plant Disease. 1993;**79**(360):369

[124] Balogun OS. Patterns of disease manifestation in tomato seedlings, singly or doubly infected with potato x-potexvirus and tobacco mosaic tobamovirus. Biokemistri. 2003;**14**(59):68

[125] Steven TK. Vegetable Dseases Caused by Soil Borne Pathogens. University of California, Division of Agriculture and Natural Resources; 2003. p. 13

[126] Wittwer SH. The Encyclopedia. Americana International Edition. Vol. 26. 1998. pp. 832-833

[127] Villareal RL. Carrot in the tropics. Vol. 97. Colorado: Westview Press; 1992. pp. 101-104

[128] Ohlendorf B, Moratorio BM, Yao G. Yolo County Master Garden. Cooperating: University of California and USDA; 2004

[129] Douglas SM. Powdery Mildew of Tomato. The Connecticut Agricultural Experimental Station: Department of plant pathology and ecology; 2005

[130] Wagini NH, Abubakar S. Incidence and severity of common diseases of carrot and their pathogenic agents in northern Nigeria. Katsina Journal of Natural and Applied Sciences. 2015;**4**(1):212-219

[131] Ajav EA, Ogunlade CA. Physical Properties of Ginger (*Zingiber officinale*). Global Journal of Science Frontier Research: D Agriculture and Veterinary. 2014;**14**(8) Version 1.0: 1-8 Online ISSN: 2249-4626 & Print ISSN: 0975-5896

[132] CBI, Ministry of Foreign Affairs. 2021. Food Loss in Nigeria Value Chain Analysis (VCA) of Tomato, Onion, Chilli Value Chains. https://www.rvo.nl/sites/default/files/2021/02/Report-Food-Loss-in-Nigeria.pdf

[133] Ebewele RO, Jimoh AA. Feasibility Study of Kaduna State Ginger Processing Industry. Nigeria: Ahmadu Bello University Chemical Engineering Consultant; 1981. pp. 1-45, 50-56, 63-80

[134] Fumen GA, Yiljep YD, Ajisegiri ESA. Survey of ginger processing and drying methods in Nigeria: A case of southern Kaduna of Kaduna state. In: Fourth International Conference and 25th AGM of the Nigeria Institution of Agricultural Engineers, Damaturu, Nigeria, September 8th–12th. 2003

[135] Suresh P, Hanumanthappa M, Srinivasan K. Effect of heat processing of spices on the concentrations of their bioactive principles: Tumeric (*Curcuma longa*), red pepper (*Capsicum annum*) and black pepper (*Piper nigrum*). Journal of Food Composition and Analysis. 2007;**20**(3-4):346-351

[136] Emelike NT. Functional and physicochemical properties of turmeric powder as affected by processing methods. Asian Food Science Journal. 2020;**19**(2):1-10

[137] Bolaji OT, Adepoju PA, Olalusi AP. Economic implication of industrialization of a popular weaning food "Ogi" production in Nigeria: A review. African Journal of Food Science. 2015;**2015**(9):495-503

[138] Olaniran AF, Abiose SH. Proximate and antioxidant activities of bio-preserved Ogi flour with garlic and ginger. F1000Research. 2019, 2019;7:1936

[139] Adeoti OA, Alabi AO, Ogunjobi EO, Elutilo OO, Adeodokun SO. Comparative study on the proximate, physicochemical and sensory properties of jams from selected tropical fruits spiced with ginger, garlic and turmeric. Asian Food Science Journal. 2021;20(3):86-99

[140] Sharma MM, Sharma RK. Chapter 12 Coriander. In: Peter KV, editor. Handbook of Herbs and Spices. 2nd ed. Vol. 2. Cambridge, England: Woodhead Publishing Ltd; 2004. pp. 216-249

[141] Kaushal M, Gupta A, Vaidya D, Gupta M. Postharvest management and value addition of ginger (Zingiber officinale Roscoe): A review. International Journal of Environment, Agriculture and Biotechnology. 2017;2(1):397-412

[142] Olife IC. Promoting ginger oleoresin production in Nigeria for economic growth and sustainable supply to user industries. Journal of Natural Sciences Research. 2021;**12**(22):2021. www.iiste.org

[143] Mundor Intelligence. Africa Nutraceutical Market—Growth, Trends, COVID-19 Impact, and Forecast (2022-2027); https://www.mordorintelligence.com/industry-reports/africa-nutraceutical-market. [Accessed: May 10, 2022]

[144] Otunla GA, Oloyede OB, Oladiji AT, Afolayan AJ. Comparative analysis of the chemical composition of three spices—Allium sativum L. Zingiber officinale Rosc. and Capsicum frutescens L. commonly consumed in Nigeria. African Journal of Biotechnology. 2010;9(41):6927-6931

[145] Ikpeama A, Onwuka GI, Nwankwo C. Nutritional composition of tumeric (*Curcuma longa*) and its antimicrobial properties. International Journal of Scientific and Engineering Research. 2014;**5**(10):1085-1090

[146] Imoru A, Onibi GE, Osho IB. Nutritional and biochemical compositions of turmeric (*Curcuma longa* Linn) rhizome powder—A promising animal feed additive. International Journal of Scientific and Engineering Research. 2018;**9**(1):424-429

[147] Lawal A, Matazu SS. Comparative studies of white and red Allium cepa cultivated in Sokoto, Nigeria. ChemSearch Journal. 2015;**6**(2):14-20

[148] Akinwande BA, Olatunde SJ. Comparative evaluation of the mineral profile and other selected components of onion and garlic. International Food Research Journal. 2015;**22**(1): 332-336

[149] Karakol P, Kapi E. Chapter 19: Use of selected antioxidant-rich spices

and herbs in foods. In: Waisundara V, editor. Antioxidants—Benefits, Sources, Mechanisms of Action. Intech Open; 2021. DOI: 10.5772/intechopen.92918

[150] Suleria HAR, Butt MS, Anjum FM, Saeed F, Khalid N. Onion: Nature protection against physiological threats. Critical Reviews in Food Science and Nutrition. 2015;**55**(1):50-66. DOI: 10.1080/10408398.2011.646364

[151] Yashin A, Yashin Y, Xia X, Nemzer B. Antioxidant activity of spices and their impact on human health: A review. *Antioxidants*. 2017;**6**(70):1-18

[152] Shang A, Cao S-Y, Xu X-Y, Gan R-Y, Tang G-Y, Corke H, et al. Bioactive compounds and biological functions of garlic (*Allium sativum* L.). Food. 2019;**8**:246. DOI: 10.3390/foods8070246

[153] Chung LY. The antioxidant properties of garlic compounds: Allyl cysteine, alliin, allicin, and allyl disulfide. Journal of Medicinal Food. 2006;**9**(2):205-213. DOI: 10.1089/jmf.2006.9.205

[154] Duda G, Suliburska J, Pupek-Musialik D. Effects of short-term garlic supplementation on lipid metabolism and antioxidant status in hypertensive adults. Pharmacological Reports. 2008;**60**(2):163-170

[155] Masuda T, Maekawa T, Hidaka K, Bando H, Takeda Y, Yamaguchi H. Chemical studies on antioxidant mechanism of curcumin: Analysis of oxidative coupling products from curcumin and linoleate. Journal of Agricultural and Food Chemistry. 2001;**49**(5):2539-2547

[156] Ozgoli G, Goli M. Effects of ginger capsules on pregnancy, nausea, and vomiting. Journal of Alternative and Complementary Medicine. 2009;**15**(3):243-246

[157] Mehdizadeh M, Dabaghian F, Nejhadi A, Fallah-Huseini H, Choopani S, et al. Zingiber officinale alters 3,4-methylenedioxymethamphetamine-induced neurotoxicity in rat brain. Cell Journal. 2012;**14**(3):177-184

[158] AVRDC. 2007. The World Vegetable Center. AVRDC Publication # 05-613, Working Paper # 15, p. 20

[159] Mgbenka RN, Mbah EN. A review of smallholder farming in Nigeria: Need for transformation. International Journal of Agriculture Extension and Rural Development Studies. 2016;**2**(3):43-54

[160] Lokesha MN, Mahesha M. Impact of road infrastructure on agricultural development and rural road infrastructure development tprogramms in India. International Journal of Humanities and Social Science Inventions. 2016;**11**(5):1-7

Chapter 12

# Sweet Potato (*Ipomoea batatas* (L.) Lam): A Review of Modern Varieties and Production Guidelines for Enhanced Food and Nutrition Security

*Vethaiya Balasubramanian*


## Abstract

Sweet potato is an important root crop that feeds millions of people, mostly the poor. Being a versatile crop, it is generally adapted to varying environments. The potential of sweet potato as food, feed, and industrial raw material has not been fully realized due to the: (a) dominance of subsistence farming with local varieties and poor-quality vine cuttings; (b) low or no knowledge and awareness of the new high-yielding yellow-fleshed sweet potato (YFSP) varieties rich in beta-carotene and micronutrients that could alleviate hunger and malnutrition globally; (c) high soil nutrient depletion by the crop under continuous cultivation with low or no nutrient inputs; (d) huge (40–80%) losses of roots after harvest due to poor postharvest management; and (e) inadequate farmers' access to sweet potato value chain. This review shows how to increase farmers' productivity and income and simultaneously sustain soil health by using improved, drought-tolerant varieties and climate-smart integrated crop and resource management technologies; reduce harvest and post-harvest losses through improved postharvest management; reduce malnutrition by producing and consuming YFSP varieties; and increase sweet potato product lines to boost market demand and farmers' income, which in turn will encourage farmers to intensify sweet potato production with adequate inputs.


**Keywords:** sweet potato, improved varieties, disease-free vine cuttings, soil nutrients depletion, integrated crop and resource management, postharvest management, value addition, food energy, nutrition, beta-carotene, anthocyanin, antioxidants

## 1. Introduction

Sweet potato (*Ipomoea batatas* (L.) Lam) is an important root crop that produces edible roots, like cassava. It's the seventh most important food crop after rice, wheat, potato, maize, barley, and cassava [1]. It is a hardy, versatile crop growing in marginal to fertile soils, and at altitudes ranging from sea level to 2500 metres (m). It produces the highest amount of edible energy per unit area and time and a cheap source of beta-carotene—a precursor of vitamin A [2]. Sweet potato is considered as a life saver in times of famines in China in 1594 and 1959–1961, or natural calamities such as typhoons that destroyed the principal food crop of rice in Japan or virus outbreaks on cassava in Uganda in the 1990s [1].

There is a huge yield gap between farmers' current yields of 4–6 tonnes per hectare (t/ha) to reported potential yields of 80–100 t/ha [3, 4], and the attainable yields of 40–60 t/ha in commercial farms of South Africa [2, 5]. The reasons for this large yield gaps are: use of traditional varieties and poor-quality planting materials due to lack of awareness of improved varieties and healthy planting materials [2]; progressive depletion of soil nutrients due to soil mining by planting sweet potato with low or no fertilizer inputs [6]; random distribution of rainfall and increased occurrence of droughts and heat waves due to climate change [7]; and non-harvesting of rainwater for supplemental irrigation. This review shows how to minimize this yield gap and increase farmers' productivity and profitability by promoting improved varieties and integrated crop and resource management; reducing crop losses through improved postharvest management; enhancing market access to small-scale farmers; and increasing market demand by using sweet potato as stock for developing diverse food and industrial products.

## 2. Origin and history of sweet potato

Sweet potato is believed to have originated from either Central or South America [8]. *Camote and batatas* lines came from the Caribbean and Central America (Yucatan Peninsula of Mexico), while *Kumara* clones could have been domesticated in western parts of South America (along Orinoco River in Venezuela). Sweet potato lines and clones were introduced to New Guinea which formed the secondary centre of diversity [9]. Later, it spread to Asia, Africa, Europe, and other parts of the world [8].

Sweet potato has a long history as a life saver in times of food scarcity: during famines in 1594 [10] and later in 1959–1961 in China; natural calamities such as typhoons destroying rice crops in Japan; virus devastation of the cassava crops in Uganda in the 1990s [1].

## 3. Sweet potato production environments and productivity

Sweet potato is a versatile crop that can be cultivated in a wide range of environments from humid tropics to cool subtropics to warm temperate zones and from sea level to 2500 m altitude. Although it prefers a warm climate (21–26° C) with an annual rainfall of 750–1500 millimetres (mm), it has fair tolerance to drought as well as to low temperatures of high-altitude areas [2]. Cool nights and sunny days are conducive for better root development. It grows well in well-drained sandy loam to silt loam soils with an optimum pH of 5.5–6.5, but it doesn't tolerate waterlogging.

Although farmers' yields are low at 4–6 t/ha in many developing countries, well-managed irrigated crops can attain yields of 40–60 t/ha [5], while the potential yields can be as high as 80–100 t/ha [3, 4]. The most productive sweet potato farms in the world are found in China, Ethiopia, Malawi and USA where the national average yields ranged between 21 and 27 t/ha in 2020 [11].

## 4. Sweet potato production statistics

Globally, in 2020, sweet potato was cultivated on 7.4 million hectares (mha) that produced 89.49 million tonnes (mt) of edible roots, with a mean yield of 12.1 t/ha [11]; about 95% of the global sweet potato production occurs in developing countries. In 2020, Asia stood first by producing 55.98 mt of roots from 2.785 mha, followed by Africa with 28.80 mt from 4.214 mha, Americas 3.81 mt from 0.253 mha and Oceania 0.90 mt from 0.149 mha. In 2020, mean yield of sweet potato was the highest in Asia (20.1 t/ha), followed by Americas (15.1 t/ha), Africa (6.8 t/ha) and Oceania (6.1 t/ha) (**Figure 1**).

China, Indonesia, Vietnam, Angola, Ethiopia, Malawi, Nigeria, Tanzania, Uganda, and USA are the top 10 sweet potato producing countries. The area harvested, yield, and production of roots for 2010, 2015 and 2020 are provided in **Table 1**. China is the largest producer and consumer of sweet potato in the world. Sweet potato yields increased steadily from 2010 to 2020 in most countries. During the period of

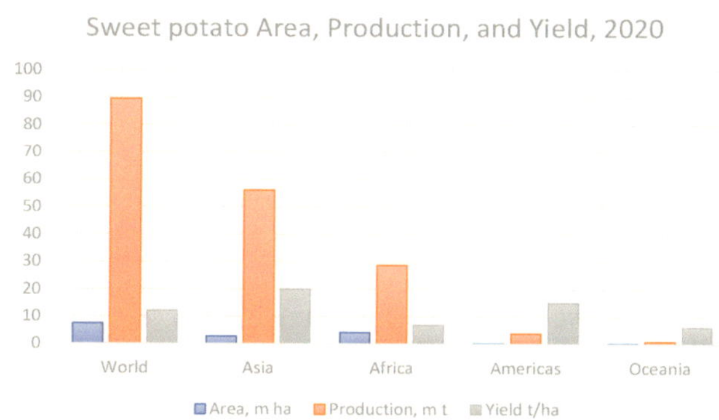


**Figure 1.**
*Global and region-wise sweet potato production statistics, 2020 [11].*

| Country | Area, million ha | | | Yield: t/ha | | | Prod., million tonnes | | |
|---|---|---|---|---|---|---|---|---|---|
| | 2010 | 2015 | 2020 | 2010 | 2015 | 2020 | 2010 | 2015 | 2020 |
| China | 3.135 | 2.519 | 2.240 | 20.925 | 21.515 | 21.851 | 65.605 | 54.201 | 48.950 |
| Indonesia | 0.181. | 0.143 | 0.070 | 11.327 | 16.053 | 21.166 | 2.051 | 2.298 | 1.487 |
| Vietnam | 0.151 | 0.128 | 0.110 | 8.743 | 10.481 | 12.533 | 1.319 | 1.336 | 1.373 |
| Angola | 0.157 | 0.168 | 0.183 | 6.300 | 11.517 | 9.457 | 0.987 | 1.933 | 1.738 |
| Ethiopia | 0.080 | 0.055 | 0.062 | 9.061 | 29.523 | 25.739 | 0.724 | 1.635 | 1.599 |
| Malawi | — | 0.213 | 0.303 | — | 20.289 | 22.800 | — | 4.325 | 6.918 |
| Nigeria | 1.299 | 1.553 | 0.151 | 2.670 | 2.484 | 2.559 | 3.467 | 3.857 | 3.868 |
| Tanzania | 0.576 | 0.747 | 0.612 | 4.207 | 4.627 | 7.249 | 2.424 | 3.455 | 4.435 |
| Uganda | 0.442 | 0.455 | 0.366 | 4.496 | 4.500 | 4.197 | 1.987 | 2.045 | 1.536 |
| USA | 0.047 | 0.062 | 0.063 | 22.863 | 22.706 | 24.553 | 1.0812 | 1.407 | 1.558 |
| World | 7.930 | 7.768 | 7.401 | 11.830 | 11.790 | 12.092 | 93.891 | 91.581 | 89.488 |

*Source: [11]*

**Table 1.**
*Area harvested, yield, and total production of sweet potato in selected countries.*

2015–2020, national average sweet potato yields stabilized around 27 t/ha in Ethiopia, 23 t/ha in the USA, 21 t/ha in China and Malawi, 18 t/ha in Indonesia, and 11 t/ha in Vietnam and Angola; sweet potato productivity was very low at 4–6 t/ha in Nigeria, Tanzania, and Uganda [11].

Globally and in Asia, the sweet potato area and root production have been going down steadily from 1994 to 2020. Africa is the only region where sweet potato area and production have been steadily increasing with time [11].

## 5. Sweet potato: nutrients profile

Sweet potato is rich in carbohydrates, fibre, vitamins, micronutrients, and plant compounds that have antioxidant properties.

### 5.1 Edible roots

The average composition of fresh sweet potato roots is water (77%), carbohydrates (20.1%), sugar (4.2%), fibre (3%), protein (1.6%), and fat (0.1%) (**Table 2**). Most of the carbohydrates come from starch (17%) and fibre (3%). The soluble fibres (15–23%) are mostly pectin which may increase fullness while eating, decrease food intake, and reduce blood sugar spikes; the insoluble fibres (77–85%) are composed of cellulose, hemicellulose, and lignin. Although protein content is low (1.6%), 80% of the total proteins are composed of sporamin which may have antioxidant properties. All sweet potato varieties also contain variable amounts of vitamins A; B1, B2, B3, B5, B6 and B9; C and E; and low levels of vitamin K, and high levels of potassium and medium levels of calcium, magnesium, phosphorus, and trace elements such as iron, manganese, zinc, copper, and selenium [12].

It provides 86 calories per 100 g of boiled roots. Glycaemic index (GI) of cooked sweet potato is medium to high (44–96). Boiling is linked with lower GI values than baking, frying or roasting [13, 14]. Therefore, people with type 2 diabetes must be careful not to consume too much sweet potato.

Sweet potato is rich in many plant compounds: beta carotene, anthocyanins, chlorogenic acid. The purple-skinned sweet potato contains anthocyanins and chlorogenic acid that have antioxidant properties. The OFSP varieties are rich in beta-carotene—the precursor of vitamin A. Breeding, distribution, and promotion of high-yielding OFSP varieties is critical to reduce the widespread deficiency of vitamin A and minerals such as iron and zinc in most developing countries [15]. A daily intake of 125 g of OFSP will satisfy the daily vitamin A needs of below 5-year-old children [16].

Purple-fleshed sweet potatoes rich in anthocyanins [17] are used in juices, alcoholic beverages, jams, confectionaries, bread, snacks, and noodles.

### 5.2 Vines and leaves

Sweet potato vines with leaves are equally valuable as human food. Young vines with tender leaves are boiled or steamed with salt and condiments and consumed as a vegetable. They are low in carbohydrates (7.4%), beta carotene [1725 micrograms per 100 grams (mcg/100g)], sodium, vitamin B5, and micronutrients such as copper and selenium, but richer in protein, fat, all B vitamins except B5, and vitamins E and K than the roots (**Table 2**). Leaves also contain some plant compounds that exhibit antioxidant properties [18].

| Nutrients | Unit | Sweet potato edible roots (fresh) | Sweet potato vines + leaves (steamed) |
|---|---|---|---|
| Water | % | 77.0 | 86.8 |
| Carbohydrates | % | 20.0 | 7.4 |
| Starch | % | −17.0 | −5.5 |
| Fibre | % | −3.0 | −1.9 |
| Sugar | % | 4.2 | 5.5 |
| Energy | Cal/100 g | 86.0 | 42.2 |
| Glycaemic index | — | 44–96 | — |
| Protein | % | 1.6 | 2.2 |
| Fat | % | 0.1 | 0.3 |
| Beta carotene | mcg/100 g | 8509 | 1725 |
| Vitamins | | | |
| A | mcg/100 g | 709 | 147 |
| B1 (Thiamine) | mcg/100 g | 78 | 112 |
| B2(Riboflavin) | mcg/100 g | 61 | 267 |
| B3 (Niacin) | mcg/100 g | 557 | 1003 |
| B5 (Pantothenic acid) | mcg/100 g | 800 | 200 |
| B6 (Pyridoxine) | mcg/100 g | 209 | 160 |
| B9 (Folate) | mcg/100 g | 11.0 | 49 |
| C | mcg/100 g | 2400 | 1500 |
| E | mcg/100 g | 260 | 960 |
| K | mcg/100 g | 1.80 | 109 |
| Minerals | | | |
| Calcium | mg/100 g | 30 | 33 |
| Magnesium | mg/100 g | 25 | 48 |
| Potassium | mg/100 g | 337 | 312 |
| Phosphorus | mg/100 g | 47 | 40 |
| Sodium | mg/100 g | 55 | 7 |
| Iron | mg/100 g | 0.61 | 0.63 |
| Zinc | mg/100 g | 0.30 | 0.26 |
| Manganese | mg/100 g | 0.26 | 0.23 |
| Copper | mg/100 g | 0.15 | 0.03 |
| Selenium | mg/100 g | 0.60 | 0.009 |

*Source: [12].*

**Table 2.**
*Nutrient profiles of fresh sweet potato roots and steamed vines + leaves.*

Sweet potato vines are excellent fodder for animals. Vines are periodically cut and fed to animals such as cattle, goats, sheep, and pigs. The vines can also be preserved as silage or hay for feeding animals during dry periods or summer when other fodders are scarce. An added advantage is that animals fed with sweet potato vines produce less methane [19], a potent greenhouse gas that cause climate change.

## 6. Constraints to sweet potato production

Farmers' low productivity in sweet potato is attributable to many constraints: abiotic, biotic, management, and socio-economic constraints.

### 6.1 Abiotic constraints

Low and declining soil fertility, excessive soil moisture due to poor drainage, soil salinity, poor soil aeration due to soil compaction, inadequate rainfall during the growing season, floods, drought, and high temperature [2].

### 6.2 Biotic constraints

Biotic constraints include insect pests and diseases: Insect pests of sweet potato are weevils, nematodes, sweet potato butterfly, millipedes, and rats [20, 21]; diseases include stem blight [22], Fusarium wilt, bacterial wilt, black spot, root rot, and virus diseases [21].

### 6.3 Management constraints

They are: poor awareness about improved sweet potato varieties and the production of healthy, disease-free planting materials [2]; absence of rainwater harvesting in rain-fed areas and poor management of irrigation water in irrigated areas; and other agronomic constraints such as poor land preparation, use of local varieties and poor-quality vine cuttings, late planting, inadequate pest and disease control, and inefficient postharvest management.

### 6.4 Socio-economic constraints and opportunities

Smallholders face several problems in terms of agro-ecological, economic, institutional, and social contexts. These are also called as commodity value chain challenges, constraints and opportunities by [23]. Sweet potato production is seasonal as dictated by local weather; farmers' poor access to capital and markets; poor agricultural extension support to farmers, particularly with respect to timely distribution of improved varieties and good-quality planting materials at the start of the season; farmers' inability to buy and apply fertilizers; high postharvest losses (40–80%) linked inefficient handling, storage, and transport of bulky sweet potato roots from farms to markets or processing centres [23–25]; low profit due to increasing production costs and decreasing revenue due to poorly developed markets for sweet potato and low farmgate price at peak harvest periods [26]; and low or no availability of safe storage and processing industries in rural production zones.

## 7. Breeding and distribution of improved sweet potato varieties

New high-yielding and early maturing sweet potato varieties that are resistant to local insect pests and diseases and resilient to climate change; varieties rich in nutrients; and varieties that have a long shelf life are needed to increase sweet potato farmers' productivity and income [2].

### 7.1 Factors affecting farmers' choice of sweet potato varieties

Understanding farmers' perceptions on improved sweet potato varieties is crucial to develop new varieties that will suit farmers' different needs in different countries and regions. In addition to high yields, farmers consider crop duration, labour requirement, drought and pest tolerance, cooking time, taste, firmness of roots, storability and shelf life, and profitability. While most farmers prefer short duration varieties with stable yield to suit shortening rainy seasons, others favour varieties with good tolerance to drought and insect pests and diseases; still others choose dual purpose varieties with high biomass yield for fodder and some roots [15, 24]. Commercial farmers go for varieties with high starch yield for industrial processing [2]. In addition, nutrient-rich sweet potato varieties such as the YFSP varieties are urgently needed to address the problem of malnutrition among the poor [2, 27].

### 7.2 Sweet potato breeding in Asia and Africa

There is a high diversity of sweet potato varieties as exhibited by the colour, width, thickness, and shape of leaves [28]; skin and flesh colours, size, shape, texture, and taste of the edible roots [28, 29]. This huge diversity of sweet potato can be used to breed improved varieties with desirable traits.

The International Potato Centre (CIP) has been collaborating with national sweet potato breeders in Africa and Asia to develop high-yielding, climate-resilient, and pest- and disease-tolerant sweet potato varieties suitable for different growing environments, farming systems, and market demands. These collaborative breeding programs placed higher priority to developing beta-carotene-rich YFSP varieties to tackle the problem of vitamin A and micronutrient (iron and zinc) deficiency, specifically in women and children. As a result, more than 100 pro-vitamin-A–rich sweet potato varieties suitable for local agro-ecologies and consumer preferences have been developed and released in more than 20 countries of Africa and Asia [30].

Most popular sweet potato varieties and their key traits are given in **Table 3**. Some of the varieties are climate-resilient by their tolerance to drought and high temperatures [2, 31], some are tolerant to salinity [32, 33], and others have wide adaptability to varying environments [2]. Among the 19 OFSP and three other sweet potato varieties released in Mozambique in 2011 and 2016, Alisha, Irene, and Sumaia are the three best varieties with high vine survival rates under drought; they yielded 18–25 t/ha in rain-fed fields with low or no fertilizer application [34]. The variety Irene with wide adaptability has been released in four African countries, and it performed well in saline soils under irrigation in Abu Dhabi [2]. The South African varieties A15 and Resisto are tolerant to deficit irrigation or restricted water supply [35]. The Vietnamese varieties such as Khoai ruot vang, Khoai cao san, and Khoai voi were selected for their tolerance to salinity [36].

Most OFSP varieties are richer in beta-carotene than some common carotenoid-rich vegetables and fruits such as carrot, mango, and tomato [37]. Thus, OFSP varieties are an inexpensive source of dietary vitamin A and essential micronutrients to fight the widespread malnutrition [38]. It is estimated that more than 6.8 million farm families in Africa and South Asia are now growing and eating vitamin-A–rich OFSP varieties [30].

The purple-fleshed sweet potato (PFSP) roots are rich in anthocyanin and chlorogenic acid which are excellent antioxidants [39]; anthocyanin is also used as a natural

| Countries | Varieties with different flesh colours | | |
|---|---|---|---|
| | **White, cream, or pale yellow** | **Purple or reddish purple** | **Orange or bright orange** |
| |  |  |  |
| | **Low in beta-carotene, anthocyanin, chlorogenic acid** | **Rich in anthocyanin & chlorogenic acid, but low in beta-carotene** | **High in beta-carotene, medium in chlorogenic acid, low in anthocyanin** |
| USA | Hannah, O'Henry, Jersey, Oriental, White Delight, White Hamon, Murasaki, Satsuma-Imo | Okinawa, Kotobuki, Purple, Brown Horse, Porto Rico | Jewel, Garnet, Beauregard, Centennial, Regal, Apache, Carolina, Darby, Excel, Hernandez |
| China | Xuzhou 18, Nanshu 18, Ji-03314, Ji-03468, Nanzi-8, CX-1 (Industrial use) | Okinawa, Kotobuki | Benihayato |
| India | Varsha, Pusa Safed, Sree Bhadra, Sourin, Sree Arun, C0-1, Co-2, Co-3 | Purple flesh ST-13 | Kamala Sundri, Gouri, Shankar, ST-14, Sree Kanaka |
| Indonesia | CIP-204, CIP-LSQ, CIP-WH15 | CIP-BDG, MSU01017-16, MSU01022-12, MSU01002-7 | Antin-1, CIP-B9 |
| Vietnam | Khoai ruot vang, Khoai cao san, Khoai voi, H12, K51, TV1 | Okinawa | — |
| Japan | Murasaki, Oriental, Satsuma-Imo | Okinawa, Kotobuki, Ayamurasaki | Benihayato, Kyushu No. 114 |
| Ethiopia | Awasa-83, Koka-6, Temesgen, Beletech, Berkume | — | Tulla (CIP 420027) |
| Mozambique | Amelia, Irene, Sumaia | Bie, Bita, Caelan | Ivone, Alisha, Victoria, Lawrence |
| Nigeria | Barth, Arrotwtip, Benue, Shaba | — | UMUSP01, UMUSP03, UMUSP04 |
| Tanzania, Kenya | Polista, Sinia, and SPN/O | | Salyboro, Tainung 65, Japon Tresmeniso, Jonathan, Zapallo, Excel and Kandee |
| Uganda | Tororo-3, New Kawogo, Tanzania, Kakamega | — | NASPOT-8, 11, 12, & 13; NAROSPOT-1, 2, 3, & 5 |
| South Africa | Blesbok, Bosbok, Ribbok, Mafutha | — | W-119, A59, A-15, Resisto, Tainung 64 |

*Sources: [34, 40, 41].*

**Table 3.**
*Most popular improved high-yielding white-, purple-, and orange-fleshed sweet potato varieties in selected countries.*

food colour. They are less popular in most countries except China, Indonesia, Japan, and USA. PFSP roots are used to prepare vegetable dishes, bakery products, beverages, etc.

Some sweet potato varieties (e.g. CX-1 in China) are rich in complex carbohydrates and bioactive compounds, and they are used in industries.

## 8. Multiplication and supply of healthy planting materials

Traditional farmers multiply sweet potato vines by growing them on residual moisture in rice fields after rice harvest or under fruit trees [42, 43]. The danger is that this traditional method of vines multiplication could transmit pests and diseases from one crop to the next. Therefore, improved methods using healthy mother plants are needed to produce healthy, disease-free vines for planting. Vines are multiplied in primary and secondary nursery as shown below.

### 8.1 Primary nursery

Primary nursery is prepared 3 months prior to planting in the main field. An area of 100 square metres ($m^2$) is required for primary nursery. About 100–125 kg of medium-size roots are needed to produce vine cuttings enough to plant in the secondary nursery. Seed roots taken only from healthy, disease-free mother plants are planted at 20-cm spacing on ridges formed 60 cm apart. The nursery is irrigated as required. About 15 days after planting (DAP), 1.5 kg urea per 100 $m^2$ is top-dressed to boost the growth of vines. At about 40–45 DAP, vines are harvested and cut into 20–30 cm long cuttings for multiplication in secondary nursery [44].

### 8.2 Secondary multiplication

A secondary nursery area of 500 $m^2$ is required to produce vines for planting one hectare of main field. The nursery area is ploughed, harrowed, and levelled before forming the beds or ridges. About 500 kg of organic manure or compost is incorporated into the soil during first ploughing. Vine cuttings taken from primary nursery are planted at 60 cm x 20 cm spacing on raised beds or ridges. To enhance vegetative growth, 5 kg urea is applied in two splits at 15 and 30 DAP. The nursery is irrigated as required. Vines are harvested at 45 DAP and cut into 20–40 cm long cuttings for planting in the main field [44].

### 8.3 Preparing and hardening of vine cuttings

Only the middle and top parts of the vines are used for getting vine cuttings. The length of cuttings ranges from 20 to 40 cm. The 20–40 cm long cuttings produce high yields [45]. Vine cuttings from the top produce higher root yields than those from mid-stems [46].

Cut vines can be planted soon after cutting, or they can be hardened by keeping them in a shady place for 1–3 days. During this hardening time, roots will begin to grow on the nodes, and the cuttings will become tougher and more resistant to the transplanting shock. Vine cuttings stored for 3 days established well and produced higher marketable root yield than fresh cuttings [47]. When stored for longer than 3 days, vine cuttings start drying, and planting of such dried cuttings results in poor crop establishment and low yields.

## 9. Land preparation

Sweet potato is planted on well-prepared flat land, on raised beds, on ridges or on mounds depending on the local soil and agroclimatic conditions. On sloping lands, contour ridges and furrows are formed to slow down the movement of rainwater and to minimize soil erosion. In the ridge-furrow system, the furrows are made gently sloping along the natural slope of the field to facilitate easy drainage of water during heavy rains. The highest yields were reported for crops planted on mounds in India [48] and in trenches in alluvial soil in Bangladesh [49]. Whatever be the system, it should enable good drainage and soil aeration and less soil compaction so as to facilitate the formation and bulking of storage roots.

## 10. Crop establishment

Proper establishment of the crops is critical for successful crop production. The aim is to get a fast and uniform sprouting of the planted vine cuttings and a vigorous early growth of the young sweet potato plants—all in order to develop an early and full canopy cover over the soil to suppress weeds, conserve soil moisture, minimize soil erosion, and to use all resources efficiently and evenly over the entire field. Once a uniform crop stand is established, further management of the crop becomes easier.

### 10.1 Spacing and plant population

Optimum planting density varies with soil type and other environmental conditions. The most commonly used spacing for sweet potato is 60 cm between rows or ridges and 20–30 cm between hills, giving a planting density of 83,333/ha for 20 cm spacing and 55,555/ha for 30 cm spacing [48]. The mounds are spaced at 70 cm x 70 cm and 3–6 vine cuttings are planted in each mound, giving a population density of 61,224–122,449 plants/ha. Increasing planting density beyond the optimum decreases plant vigour, increases the number of roots but decreases the root size, and increases the infestation of weevils [50]. The highest marketable root yields were obtained with a planting density per hectare of 66,666 in South Korea; 74,074 on alluvial soil with two plants per hill and 55,555 on flat bed in Bangladesh; 83,333 in Ethiopia [48]. Thus, a plant population range of 50,000–80,000 per hectare appears to be optimum for root production in most situations.

### 10.2 Methods of planting

Sweet potato is established mostly by planting vine cuttings, while the high-tech method of plug transplanting is practised in South Korea [48].

#### *10.2.1 Planting vine cuttings*

Prior to planting, vine cuttings can be dipped in fungicide solutions for 15 minutes to kill the pathogens, if any. The field is flush irrigated before planting for irrigated crops, while the vine cuttings are planted on a rainy day or just after rains for rain-fed crops.

The most common method planting is by burying the middle part of the vine cutting into the soil, leaving the two cut ends exposed to the air [45]. A notched stick can be used to push the cuttings into the soil. Some farmers plant by burying half of the cutting into the soil in an inclined position and exposing the other half to the air. Still others bury horizontally into the soil long cuttings with 5–6 nodes [48]. Depth of planting varies between 2.5 cm and 10.0 cm.

*10.2.2 Plug transplanting*

High-quality sweet potato plug seedlings are produced in bubble trays by using single node cuttings or shoot tip cuttings taken from virus-free mother plants under controlled environment [48]. This method is highly efficient in terms of time and labour use. Plug transplants are transplanted directly into the main field by a transplanter. High yields of storage roots are obtained by this method [51].

## 11. Water conservation and management

An adequate and well-distributed rainfall and/or irrigation water is needed to get high sweet potato yields.

### 11.1 Water requirement of sweet potato

Like any other crop, water requirements of sweet potato are location-specific due to variations in climate, soil type, crop varieties, and other crop-growing conditions [52]. Water use efficiency (WUE) of a crop is defined as the total biomass produced per unit area and per unit of water used up in evapotranspiration of the crop (ETc) [53]. WUE is used to quantify agricultural productivity of crops in any area, specifically in water-stressed areas.

Being a drought-tolerant crop, sweet potato is more efficient in using limited available water than non-drought-tolerant crops [54, 55]. Drought tolerance in sweet potato is related more to its survival and fast recovery after dry spells than to yield potential under drought conditions [56]. Sweet potato's sensitivity to water stress varies with growth stages. Although, at establishment phase, sweet potato rooting is optimal at a soil moisture content of 80% of field capacity, substantial rooting occurs even at 40% of field capacity [57]. Prolonged drought at storage roots initiation and bulking stages reduces crop yields drastically.

### 11.2 Irrigation management

Proper scheduling of irrigation as per crop demand at different growth stages is critical to maintain high WUE [58]. In-field water conservation practices such as soil cover to reduce evaporation losses, precise supply of water at the root zone as in drip irrigation and crop-need-based nutrient management to enhance yields will all help increase WUE. In South Africa and Ethiopia, irrigation at 60% and 100% ETc produced root yields that are not significantly different; it implies that deficit irrigation to meet the 60% ETc of the crop can be recommended under water-scarce conditions [59, 60]. As a thumb rule, sweet potato crops require 2 millimetres (mm) of water per day during the establishment phase, and it can be gradually increased to 5–6 mm per day during the critical storage root initiation and bulking stages [61].

In practice, soil is moistened with a flush irrigation before planting so as to ensure sufficient soil moisture for proper sprouting and establishment of the vine cuttings after planting. The second irrigation is scheduled at 3 DAP, and then the crop is irrigated at 7–10 days intervals. The irrigation is stopped 7–10 days before harvesting. Excess moisture must be avoided to prevent excessive vegetative growth at the expense of storage roots formation [62, 63], while water stress must be prevented through regular or supplemental irrigation during the most critical stages of storage root initiation, early bulking and late bulking in order to maximize root yields [64].

### 11.3 Rainfall management

Adequate amount and good distribution of rainfall during the entire crop growing period are critical for high yields of rain-fed crops. If the soil is dry during planting, the planted vine cuttings will dry up, resulting in poor crop establishment. Water stress or drought at storage roots initiation and bulking stages decreases leaf area index, increases metabolic toxicity and reduces yields [65]. Supplemental irrigation will save crops from droughts at critical crop growth stages.

Developing rainwater-harvesting structures such as farm ponds, small community reservoirs or check dams can help mitigate both drought and flooding impacts on rain-fed sweet potato crops by collecting and storing rainwater from peak floods and using it for supplemental irrigation of crops during periods of drought [66]. Providing supplemental irrigation from in-field water harvesting structures increased sweet potato yields significantly [67]. In India, rainwater harvesting structures, which are refilled during monsoons, reduce runoff by 40% and soil losses by 50%, and increase cropping intensity by 180% [68, 69]

## 12. Soil fertility and nutrient management of sweet potato crops

Sweet potato is a nutrient-intensive crop, and it depletes soil nutrients when cultivated continuously without application of adequate nutrients [6]. A crop yielding 35 t/ha of edible roots will remove from one hectare of land 151 kg nitrogen (N), 28 kg phosphorus or 64 kg $P_2O_5$, 263 kg potassium (K) or 328 kg $K_2O$, 46 kg calcium (Ca) and 18 kg magnesium (Mg) in vines and roots [4]. In another case, for a tuber yield of 10 t/ha, the crop removes 50 kg N, 22 kg $P_2O_5$ and 100 kg $K_2O$ [2]. These data help us to determine the ratio of major nutrients required for sweet potato as 2.5 N–1 $P_2O_5$–5 $K_2O$.

Potassium is essential for the synthesis and translocation of carbohydrates. For sustainable high yields, it is important to maintain soil K level at 200 parts per million (ppm) and plant tissue K concentration at 3.8% [70]. The next important nutrient is N which improves crop growth and development and root yields. However, excess N application promotes excessive vegetative growth at the expense of storage roots development [62, 63]. Sweet potato crops can tolerate low soil P levels because their roots are highly efficient in absorbing P from soil due to their association with vesicular-arbuscular mycorrhizae [4]. As Ca deficiency can lead to reduced root growth, lime or gypsum is applied to increase soil Ca levels to 1500 ppm minimum [70]. Among the micronutrients, boron (B) and manganese (Mn) are important; application B prevents blistering disorder in sweet potato [71].

Balanced nutrient use is important because deficiency of any one nutrient can depress the uptake of other nutrients. For example, crop response to N is poor when K is deficient in soils [72]. Calculated nutrient requirements will be 5.0 kg N, 2.2 kg $P_2O_5$ and 10 kg $K_2O$ for every tonne of root yield. The entire P and half of the N and K are incorporated into the soil at planting, and the second half of N and K is applied and covered with soil at 35–40DAP during the second weeding and earthing-up operations. Deficiency of Ca and micronutrients, if any, must be corrected as per local recommendations. Point placement or banding of inorganic fertilizers increases nutrients uptake and reduces fertilizer requirements. The combined application of organic amendments and supplemental fertilizers is called integrated nutrient management (INM), and it helps to increase soil carbon stocks, which in turn will improve the water and nutrients retention capacity of soils and at the same time enhance nutrients uptake by crops [73, 74]. Organic inputs help neutralize soil acidity and replenish K, Ca, and Mg in soils [75].

In practice, only moderate amounts of fertilizers recommended for sweet potato in most countries: 5–15 t/ha of cattle manure plus 50–60 kg N, 30–60 kg P2O5, and 75–120 kg K2O per hectare [41]. In Africa, where fertilizers are scarce and expensive, micro-dosing of fertilizers by applying small amounts of fertilizers to individual planting holes reduces fertilizer rates and increases crop uptake of nutrients [76].

## 13. Weeds and their management

Weeds compete with sweet potato crops for space, water, nutrients and light, thereby inhibiting crop growth and reducing crop yields. Weeds can also serve as alternate hosts for insect pests, pathogens, and vectors of diseases and provide a safe habitat for rats which feed on storage roots.

Weeds are a problem in sweet potato during the slow early growth stages (up to 40–45 DAP). Reduction in yields could be as high as 20–80% [77]. Farmers should adopt integrated weed management (IWM) strategies by using a combination of cultural, manual, mechanical and/or chemical weed control methods. Cultural weed control methods include good land preparation and levelling; selection of sweet potato varieties with vigorous early growth and long trailing vines; practising crop rotation and intercropping; surface soil mulch with crop residues; and reducing weed "seed bank" in soils. For manual weed control, two times digging the soil and earthing up around the base of plants at 15–20 DAP and 30–35 DAP are recommended. Mechanical weeding involves inter-row or inter-ridge cultivation and earthing up that uproots and buries young weeds. Finally, for chemical weed control, farmers can use herbicides [78].

## 14. Pruning

Under favourable rainfall and temperature and/or high N application, sweet potato grows vigorously and produces large amounts of vines at the cost of roots formation [62, 63]. If this happens, the top parts of the vines are cut from 75 DAP and used as greens in cooking, as fodder for animals, or as planting materials for a new crop.

## 15. Management of insect pests and diseases of sweet potato

Biotic constraints include insect pests and diseases. Sweet potato weevils are the most important insect pest, stem wilt and black rot are the two key fungal diseases, and sweet potato virus disease is the most destructive virus disease.

For control of insect pests and diseases, integrated pest management (IPM) is recommended. The three cardinal principles of IPM are (i) growing a healthy crop by adopting resistant varieties and disease-free planting materials as well as best crop management practices; (ii) field sanitation and maintaining a healthy agro-ecosystem; and (iii) the strategic use of external pest control inputs that are known to have a minimal impact on the agro-ecosystem and the surrounding environments and communities. IPM is thus a knowledge-intensive approach, one which requires systematic learning through continuous observation and practice (e.g. via Farmer Field Schools) [79].

Strict enforcement of phytosanitary measures at national borders is crucial to stem the spread of sweet potato virus diseases from one region to another or from one country to another.

Farmers currently can use any or any combination of the six IPM tactics: (i) genetic-host plant resistance; (ii) cultural control; (iii) mechanical methods; (iv) biological control; (v) use of safe botanicals and microbial bio-pesticides; and (vi) judicious chemical control with environmentally safe pesticides [80]. Specific IPM tactics are given below for individual insect pests and diseases.

### 15.1 Key insect pests and their management

*Sweet potato weevils (Cylas spp.):* The species *C. formicarius* is important in Asia, Oceania, the Caribbean and the USA, while *C. puncticollis* and *C. brunneus* are common in Africa. Adult weevils feed on epidermis of vines and leaves as well as surfaces of roots. The larvae of the weevils are the most damaging by their tunnelling of vines and roots. In response to attack by larvae, storage roots produce toxic terpenes, which render storage roots inedible even at low concentrations and low levels of physical damage.

Cultural control involves the use of weevil- and virus-free planting material; good field sanitation by removing volunteer plants and crop residues as well as alternate hosts in nearby areas; timely planting and prompt harvesting at maturity; flooding the field for 24 h soon after harvest; minimizing soil cracking by proper irrigation; and crop rotation. For biological control, fungi *Beauveria bassiana* and *Metarrhizium anisopliae* and the nematodes *Heterorhabditis* spp. and *Steinernema* spp. are used. The fungi attack and kill the adult weevils, whereas the nematodes kill the larvae. Finally, in severe cases, malathion 0.1% or carbaryl 0.2% is sprayed on the crop twice at 10-day intervals [41].

*Leaf-eating caterpillars*: The moths hover over the flowers around sunset time. The caterpillars feed gregariously on leaves. Deep ploughing after harvest and spraying Endosulfan @ 0.05% are recommended to control this pest.

### 15.2 Important diseases and their management

*Stem rot or wilt*: is caused by *Fusarium oxysporum* f. sp. *batatas*. This soil-borne fungus can survive in soil and in debris for several years. In affected plants, leaves turn yellow and wilt and then the vines die. It can be controlled by using disease-free vine cuttings and/or dipping vine cuttings in fungicide solutions [41].

*Black rot*: is caused by the soil-inhabiting fungus *Ceratocystis fimbriata*. Infected plants show dark to black sunken cankers in lower stems, and then the leaves turn yellow and the plants wilt and die. Transmission occurs through wounds made by sweet potato weevils, wireworms, crickets, and mice. Control measures include field sanitation; use of resistant varieties and disease-free vines; and dipping the vine cuttings in 0.2% Aretan or Agallol solution before planting [41].

*Foot rot*: is a minor fungal disease caused by *Plenodomus destruens*. Brown lesions form on the stem at or below the soil surface. Plants wilt and die in severe cases. It spreads through infected roots and stems.

*Charcoal rot*: is the fungal disease of the fleshy roots in storage, and it is caused by *Macrophomina phaseolina*. Good aeration and sanitation of the warehouse will prevent this disease.

*Sweet potato virus disease (SPVD)*: Plants attacked by this virus are stunted with small and narrow chlorotic leaves with mottled pale spots. Using resistant varieties and SPVD-free planting materials will help control this disease [35] and also increase yields by 40% [81].

## 16. Harvesting and postharvest processing

### 16.1 Harvesting

Sweet potato is ready for harvest when leaves turn yellow and drop. After maturity, storage roots can be kept in the ground for 1–2 months and harvested progressively as and when required. Its wide harvesting window allows it to act as a famine reserve food crop and is invaluable in managing labour schedules and in improving family's cash flow by selling the roots in local market over an extended period. Although root yields can increase with delayed or progressive harvesting, root quality declines and attack of roots by weevils and fungi increases [78]. The field is wet by a flush irrigation 2–3 days before harvest to facilitate easy lifting of the roots; the vines are cut and the storage roots are lifted carefully without causing any injury to the roots [45].

### 16.2 Postharvest management

Huge losses (40–80%) of storage roots occur during and after harvest due to high temperatures of 32–35°C and high RH of 80–95% prevailing at harvest time in the tropics [25] and inefficient handling, storage and transport of the harvested roots [23–25]. Being bulky and tender, losses in storage roots occur in both quality and quantity due to physical injury, sprouting, attack by weevils and fungal diseases, and loss of weight [25]. Therefore, sweet potato storage roots must be handled carefully at all stages.

*Curing*: It is intended to heal injuries, if any, caused during harvesting and to keep the roots in good condition for marketing or to preserve 'seed' roots for the next crop. Soon after harvest, the roots must be cured in well-ventilated curing rooms which provide adequate circulation of air for breathing of roots and to prevent excessive condensation of moisture [41]. Curing efficiency varied with dry matter content of storage roots and curing periods; roots with low dry matter content have high curing efficiency [25]. Curing methods such as fungicide treatment, bio-control, gamma irradiation, hydro-warming, and storage in sand and saw dust helped limit spoilage of roots and enhanced shelf life of sweet potato roots [25].

*Storage*: Red-skinned varieties store better than white-skinned varieties. An ideal storage room must be maintained at 15 °C and 85% relative humidity (RH). Two commonly used, inexpensive storage methods are pit storage and indoor storage. In pit storage, a pit covered with straw is used to store roots for 2–3 months [41]. Storage in sand and saw dust helped limit spoilage and enhanced shelf life of sweet potato roots [25]. Some farmers store the roots inside their houses, while others keep the roots in soil and harvest as and when required. In developed countries, refrigerated storage and transport are used to keep the roots fresh for long periods of time [41].

*Processing and value addition*: In addition to simple boiling and eating, sweet potato roots can be used to prepare several convenient foods: flour, chips, crisps, breads, biscuits, pies, candies, and juices. Coloured sweet potato flour is used in various bakery and noodles preparations.

Sweet potato roots are used to produce starch, glucose, sugar syrup, and industrial alcohol [82]. Sweet potato starch is used in textiles, paper and food manufacturing industries, preparation of liquid glucose and adhesives. Yellow- and violet-fleshed sweet potatoes are used to extract beta carotene and anthocyanins that are used as natural food colours and antioxidants [82]. Enzymes such as sporamin and beta-amylase are also produced from storage roots.

## 17. Conclusions

Sweet potato is an important root crop that feeds millions of people in developing countries. It is also a valuable raw material for food, feed and other industries. Use of improved varieties and healthy planting materials and improved crop production and postharvest management technologies will help close the yield gap, increase farmers' productivity and income, and build climate resilience for sweet potato farming. Increased production and consumption of nutrient-rich yellow-fleshed sweet potato (YFSP) varieties will address the widespread problem of vitamin A and micronutrients deficiency in children and women of developing countries. Being a nutrient intensive crop, it is important to apply enough nutrients through organic amendments and fertilizers to maintain soil health for sustainable high crop yields over long-term. Installing in-field rainwater harvesting systems and using the water for supplemental irrigation at critical stages of storage roots initiation and bulking will help stabilize the yields of rain-fed crops under changing climate. Market demand for sweet potato roots must be increased significantly by developing processing industries near production centres and diversifying sweet potato product lines. Finally, farmers should get greater access to sweet potato value chain so that they are encouraged to produce and sell more sweet potatoes in local, national, and international markets, as well as to food processing industries.

## Acknowledgements

I am thankful to my family for supporting me in all aspects while writing this manuscript.

## Author details

Vethaiya Balasubramanian
Ramya Nursery Illam, Coimbatore, India

*Address all correspondence to: vethaiyab@gmail.com

## References


[1] American Society for Horticultural Science. Sweet potato promises hunger relief in developing countries. Science Daily. 2007. Available from: www.sciencedaily.com/releases/2007/11/071102084811.htm

[2] Low JW, Ortiz R, Vandamme E, Andrade M, Biazin B, Grüneberg WJ. Nutrient-dense orange-fleshed sweetpotato: Advances in drought-tolerance breeding and understanding of management practices for sustainable next-generation cropping systems in Sub-Saharan Africa. Frontier in Sustainability. 2020;**4**:50. DOI: 10.3389/fsufs.2020.00050

[3] Karan YB, Şanli ÖG. The assessment of yield and quality traits of sweet potato (*Ipomoea batatas* L.) genotypes in middle Black Sea region, Turkey. PLoS One. 2021;**16**(9):e0257703

[4] O'Sullivan JN, Asher CJ, Blamey FPC. Nutrient Disorders Sweet Potato. Canberra, Australia: Australian Centre for International Agricultural Research; 1997

[5] Niederwieser JG. Guide to Sweet Potato Production in South Africa. Pretoria (South Africa): ARC-Roodeplant Vegetable and Ornamental Plant Institute; 2004

[6] Hartshorn AS, Chadwick OA, Vitousek PM, Kirch PV. Prehistoric agricultural depletion of soil nutrients in Hawaii. Proceedings of the National Academy Science USA. 2006;**103**(29):11092-11097. DOI: 10.1073/pnas.0604594103

[7] Zimmermann A, Webber H, Zhao G, Ewert F, Kros J, Wolf J, et al. 2017: Climate change impacts on crop yields, land use and environment in response to crop sowing dates and thermal time requirements. Agricultural Systems. 2017;**157**:81-92. DOI: 10.1016/j.agsy.2017.07.007

[8] O'Brien P. The sweet potato: Its origin and dispersal. American Anthropologist. 1972;**74**:342-365

[9] Roullier C, Kambouo R, Paofa J, McKey D, Lebot V. On the origin of sweet potato (*Ipomea batatas* (L.) Lam.) genetic diversity in New Guinea, a secondary center of diversity. Heredity. 2013;**110**:594-604

[10] Jia R. Weather shocks, sweet potatoes and peasant revolts in historical China. The Econometrics Journal. 2013;**124**: 92-118. DOI: 10.1111/ecoj.12037

[11] FAOSTAT. 2022. Available from: https://www.fao.org/faostat/en/#data/QCL [Accessed May 11, 2022]

[12] Available from: Nutritionvalue.org, https://www.nutritionvalue.org/Sweet_potato_leaves%2C_raw_nutritional_value.html#:~:text=Sweet%20potato%20leaves%2C%20raw%20contains%200%20g%20of%20saturated%20fat,and%20178%20mg%20of%20potassium

[13] Sweet potatoes 101: Nutrition facts and health benefits. Available from: https://www.healthline.com/nutrition/foods/sweet-potatoes

[14] Allen JC, Corbitt AD, Maloney KP, Butt MS, Truong VD. Glycemic index of sweet potato as affected by cooking methods. Open Nutrition Journal. 2012;**6**:1-11. DOI: 10.2174/1874288201206010001

[15] Low J, Ball LA, Magezi S, Njoku J, Mwanga R, Andrade M, et al. Sweet

potato development and delivery in sub-Saharan Africa. African Journal of Food and Agriculture. 2017;**17**:11955-11972

[16] Hotz C, Loechl C, Lubowa A, Tumwine JK, Ndeezi G, Nandutu Masawi A, et al. Introduction of β-carotene-rich orange sweet potato in rural Uganda resulted in increased vitamin A intakes among children and women and improved vitamin A status among children. The Journal of Nutrition. 2012;**142**:1871-1880. DOI: 10.3945/jn.111.151829

[17] Montilla EC, Hillebrand S, Winterhalter P. Anthocyanins in purple sweet potato (Ipomoea batatas L.) varieties. Fruit. Vegetable and Cereal Science and Biotechnology. 2011;**5**:19-23

[18] Teow CC, Truong VD, McFeeters RF, Thompson RL, Pecota KV, Yencho GC. Antioxidant activities, phenolic and β-carotene contents of sweet potato genotypes with varying flesh colours. Food Chemistry. 2007;**103**(2007):829-838

[19] Facts and figures about sweet potato. Available from: https://cipotato.org/site/worldfoodprize/pdf/FactsandFiguresSweetpotato.pdf

[20] Ebregt E, Struik PC, Abidin PE, Odongo B. Farmers' information on sweet potato production and millipede infestation in north-eastern Uganda. I. Associations between spatial and temporal crop diversity and the level of pest infestation. Journal of Life Science. 2004;**52**:47-68. DOI: 10.1016/s1573-5214(04)80029-0

[21] Ames T, Smit N, Braun AR, O'Sullivan JN, Skoglund LG. Sweetpotato: Major Pests, Diseases, and Nutritional Disorders. Lima: International Potato Center (CIP); 1996

[22] Osiru M, Adipala E, Olanya OM, Lemaga B, Kapinga R. Occurrence and distribution of alternaria leaf petiole and stem blight on sweet potato in Uganda. Plant Pathology Journal. 2007;**6**:112-119. DOI: 10.3923/ppj.2007.112.119

[23] Parmar A, Hensel O, Sturm B. Post-harvest handling practices and associated food losses and limitations in the sweetpotato value chain of southern Ethiopia. Journal of Life Science. 2017;**80**:65-74. DOI: 10.1016/j.njas.2016.12.002

[24] Tomlins K, Ndunguru G, Kimenya F, Ngendello T, Rwiza E, Amour R, et al. On-farm evaluation of methods for storing fresh sweet potato roots in East Africa. Tropical Science. 2007;**47**:197-210. DOI: 10.1002/ts.214

[25] Ray R, Ravi V. Post-harvest spoilage of sweetpotato in tropics and control measures. Critical Reviews in Food Science and Nutrition. 2005;**45**:623-644

[26] Biazin BT, Low J, Brouwer R, Feyissa TA, WoldeGiorgis M. Sweetpotato marketing in southern Ethiopia: Examining market chains, seasonal availability, volume and price fluctuations at district and provincial markets. In: Ethiopian Economics Association Conference. Hawassa: Ethiopian Economics Association; 2019. pp. 1-11

[27] Ssebuliba JM, Muyonga JH, Ekere W. Performance and acceptability of orange fleshed sweetpotato cultivars in Eastern Uganda. Africa Crop Science. 2006;**14**:231-240

[28] Hue S, Chandran S, Boyce A. Variations of leaf and storage roots morphology in Ipomoea batatas L. (Sweet potato) cultivars in Proceedings of Asia Pacific Symposium on Postharvest Research, Education and Extension. Acta Horticulturae. 2010;**943**:73-79

[29] Shekhar S, Mishra D, Buragohain AK, Chakraborty S, Chakraborty N. Comparative analysis of phytochemicals and nutrient availability in two contrasting cultivars of sweet potato (Ipomoea batatas L.). Food Chemistry. 2015;**173**:957-965

[30] Innovation: Biofortified orange-fleshed sweet potato. Available from: https://www.cgiar.org/innovations/biofortified-orange-fleshed-sweetpotato/

[31] International Potato Center (CIP). SASHA Brief 02. A breakthrough: Hybrid population breeding validated in Peru, Uganda, and Mozambique. CIP. 6 p., Lima, Peru; 2019

[32] Chávez R, Mendoza H, Mahesh U, Espinoza J, Cabello R, Arévalo N, et al. Genetic improvement and adaptation of sweet potato (Ipomoea batatas) to dry and saline conditions. IDESIA. 2000;**18**:97

[33] Mukherjee A, Ray R, Pati K, Mukherjee A. Sweet potato and taro resilient to stresses: Sustainable livelihood in fragile zones vulnerable to climate changes. Journal of Environmental Society. 2015;**12**:53-64

[34] Andrade MI, Ricardo J, Naico A, Alvaro A, Makunde GS, Low JW, et al. Release of orange-fleshed sweetpotato (*Ipomoea batatas* [l.] Lam.) cultivars in Mozambique through an accelerated breeding scheme. Journal of Agricultural Science. 2017;**155**:919-929

[35] Van Heerden PDR, Laurie R. Effects of prolonged restriction in water supply on photosynthesis, shoot development and storage root yield in sweet potato. Physiologia Plantarum. 2008;**134**:99-109. DOI: 10.1111/j.1399-3054.2008.01111.x

[36] Nguyen VK, Mai TH, Nguyen TNH. Using salt-tolerant sweet potato varieties. Vietnam: SEARCA Agriculture & Development Discussion Paper No. 2013-3

[37] Gul K, Tak A, Singh A, Singh P, Yousu B, et al. Chemistry, encapsulation, and health benefits of β-carotene-a review. Cogent Food & Agriculture. 2015;**1**:1018696

[38] Kidane G, Abegaz K, Mulugeta A, Singh P. Nutritional analysis of vitamin A enriched bread from orange flesh sweet potato and locally available wheat flours at Samre Woreda, Northern Ethiopia. Current Research in Food and Nutrition Journal. 2013;**1**:49-57. Available from: http://www.foodandnutritionjournal.org/?p=283

[39] Zhao JG, Yan QQ, Lu LZ, Zhang YQ. In vivo antioxidant, hypoglycemic, and anti-tumor activities of anthocyanin extracts from purple sweet potato. Nutrition Research and Practice. 2013;**7**(5):359-365. DOI: 10.4162/nrp.2013.7.5.359

[40] Stratosphere: 35 different types of sweet potatoes. Available from: https://www.homestratosphere.com/types-of-sweet-potatoes/

[41] Chaudhary SK, Mahto DK, Yadav SK, Singh SK, Sinha N, Kumar M. Sweet potato (Ipomea batatas L.) A storage root crop—An overview. An International Journal Society for Scientific Development in Agriculture. 2014;**9**:943-950

[42] Gurmu F, Hussein S, Laing M. Diagnostic assessment of sweetpotato production in Ethiopia: Constraints, post-harvest handling and farmers' preferences. Research Crops. 2015;**16**:104-115

[43] Mudege NN, Biazin BT, Brouwer AR. Gender Situational Analysis of the Sweetpotato Value Chain in Selected

Districts in Sidama and Gedeo Zones. Southern Ethiopia: Social Science; 2019

[44] CTCRI. Cultural Practices of Sweet Potato. Thiruvananthapuram, India: Central Tuber Research Institute; 1987

[45] Nair GM. Agro-techniques and planting material production in sweet potato. In: Byju J, editor. Quality Planting Material Production in Tropical Tuber Crops. Thiruvananthapuram: Central Tuber Research Institute; 2006. pp. 55-58

[46] Choudhary SH, Ahmad SU, Sharafuddin AFM. Effect of number of nodes in different types of vine cuttings on the growth and yield of sweet potatoes. Bangladesh Hort. 1986;**14**:29-33

[47] Hammet HL. Effects of holding sweet potato cuttings. Lousiana Agriculture. 1983;**26**:6-7

[48] Nedeunchezhiyan M, Byju G, Jata SK. Sweet Potato Agronomy. SCIRP, Wuhan, China: Global Science Books; 2012

[49] Bhuiyan MKR, Hossain M, Hussain MAE, Attaluri S. Research and development activities of sweet potato in Bangladesh. In: 14th Triennial Symposium of International Society of Tropical Crops. Thiruvananthapuram, India: Central Tuber Research Institute; 2006

[50] Woolfe JA. Sweet Potato: An Untapped Food Resource. Cambridge, UK: Cambridge University Press; 2008

[51] He D, Chun C, Lok Y, Kozai T. Growth of sweet potato plug transplants in a closed system as affected by cell volume and type of cutting. In: Proceedings of 12th Symposium of International Society of Tropical Root Crops. Tsukuba, Japan; 2000. pp. 328-331

[52] Mbayaki CW, Kinama JM. More crop per drop: The magic of sweet potato (*Ipomea batatas* L.). Tropical and Subtropical Agroecosystems. 2022, 2022;**25**:18

[53] Molden D. Accounting for water use and productivity-SWIM Paper 1. 1997

[54] Shao H, Yong B, Xu P, Zheng H, Liao R, et al. Phytoene synthase gene (PSY) from sweet potato (Ipomoea batatas Lam.) enhances tolerance to abiotic stress. Brazilian Archives of Biology and Technology. 2018;**61**:1. DOI: 10.1590/1678-4324-2018160558

[55] Picotte JJ, Rosenthal DM, Rhode JM, Cruzan MB. Plastic responses to temporal variation in moisture availability: Consequences for water use efficiency and plant performance. Oecologia. 2007;**153**(4):821-832

[56] Daryanto S, Wang L, Jacinthe PA. Global synthesis of drought effects on cereal, legume, root and root crops production: A review. Agricultural Water Management. 2017;**179**:18-33

[57] Belehu T. Agronomical and Physiological Factors Affecting Growth, Development and Yield of Sweet Potato in Ethiopia. Pretoria: University of Pretoria; 2003

[58] De Pascale S, Dalla Costa L, Vallone S, Barbieri G, Maggio A. Increasing water use efficiency in vegetable crop production: From plant to irrigation systems efficiency. Hort Technology. 2011;**21**(3):301-308

[59] Laurie RN, Laurie SM, du Plooy CP, Finnie JF, Van Staden J. Yield of drought-stressed sweet potato in relation to canopy cover, stem length and stomatal conductance. The Journal of Agricultural Science. 2014;7:201-214. DOI: 10.5539/jas.v7n1p201

[60] Alemayehu A. Growth, Yield and Physiological Response of Sweetpotato (Ipomoea Batatas (L.) LAM.) Genotypes to Different Deficit Irrigation Levels at Hawassa Zuriya Woreda, Southern Ethiopia. Hawassa: Hawassa University; 2019

[61] Gomes F, Carr MKV. Effects of water availability and vine harvesting frequency on the productivity of sweet potato in southern Mozambique. II. Crop water use. Experimental Agriculture. 2003;**39**:39-54

[62] Bourke RM. Recent research on sweet potato and cassava in Papua New Guinea. In: Concise Papers of the 2nd Int. Symposium on Sweet Potato and Cassava. Kuala Lumpur, Malaysia; 2005. pp. 125-126

[63] Biswal S. Response of Sweet Potato (Ipomea batatas L.) to Irrigation and Fertility Levels. Bhubaneswar, India: Orissa Agricultural University and Technology; 2008

[64] Goswami SB, Sen H, Jama PK. Rootization and yield potential of sweet potato cultivars as influenced by water management practices. Journal of Root Crops. 1995;**21**:77-81

[65] Obidiegwu JE, Bryan GJ, Jones HG, Prashar A. Coping with drought: Stress and adaptive responses in potato and perspectives for improvement. Frontiers in Plant Science. 2015;**6**:542

[66] Moudgil M. Ponds—Once a Lifeline of India's Agriculture—Are Being Revived by Some Punjab Farmers. India: India Water Portal; 2015

[67] Laurie S, Calitz F, Mtileni M, Mphela W, Tjale S. Performance of informal market sweet potato cultivars in on-farm trials in South Africa. Open Agriculture. 2017;**2**:431-441. DOI: 10.1515/opag-2017-0047

[68] Singh R, Garg KK, Wani SP, Tewari RK, Dhyani SK. Impact of water management interventions on hydrology and ecosystem services in Garhkundar Dabar watershed of Bundelkhand region, Central India. Journal of Hydrology. 2014;**509**:132-149

[69] Garg KK, Karlberg L, Barron J, Wani SP, Rockstrom J. Assessing impact of agricultural water interventions at the Kothapally watershed, Southern India. Hydrological Processes. 2012;**26**(3):387-404

[70] Vos C, Arancon N, Mathews B. Nutrient management on commercial Okinawan sweet potato farms on the Hamakua coast of Hawaii Island. Pacific Agricultural and Natural Resources. 2018;**18**:1-7

[71] Mathews BW, Senock RS. Sustainable soil management in the high rainfall areas of Hawai'i Island and the use of some conventional and alternative fertilizers. Journal of Hawaiian Pacific Agriculture. 1998;**2018**(1998):7-32

[72] Lebot V. Sweetpotato: Hand book of plant breeding. In: Bradshaw JE, editor. Root and Tuber Crops. Basel: Springer Nature; 2010. pp. 97-125

[73] Nedunchezhiyan M, Reddy DS. Growth, yield, and soil productivity as influenced by integrated nutrient management in rainfed sweet potato. Journal of Root Crops. 2004;**30**:41-45

[74] Chivenge P, Vanlauwe B, Gentile R, Wangechi H, Mugendi D, van Kessel C, et al. Organic and mineral input management to enhance crop productivity in Central Kenya. Agronomy Journal. 2009;**101**:1266-1275

[75] Batino A, Lampo F, Koala S. Research on nutrient flows and balances in West Africa: State-of-the-art. Agriculture, Ecosystems and Environment. 1998;**71**:19-35

[76] Kisinyo P, Opala P, Palapala V, Gudu S, Othieno C, Ouma E. Micro-dosing of lime, phosphorus and nitrogen fertilizers effect on maize performance on an acid soil in Kenya. Sustainable Agriculture Research. 2015;**4**:21-30

[77] Nedunchezhiyan M, Varma SP, Ray RC. Estimation of critical period crop-weed competition in sweet potato (*Ipomea batatas* L.). Advan. Horticultural Science. 1998;**12**:101-104

[78] Nedunchezhiyan M, Byju G, Jata SK. Sweet potato agronomy. Fruit, Vegetable and Cereal Science and Biotechnology. 2012;**6**:1-10

[79] IITA (International Institute of Tropical Agriculture). Biological control: A sustainable solution to crop pest problems in Africa. 1990

[80] Integrated pest management (IPM). 2022. Available from: https://content.ces.ncsu.edu/extension-gardener-handbook/8-integrated-pest-management-ipm

[81] Fuglie KO, Zhang L, Salazar L, Walker T. Economic impact of virus-free sweet potato planting materials in Shandong Province, China. CIP Program Report. 1997;**1997**:249-254

[82] Bach D, Bedin AC, Lacerda LG, Nogueira A, Demiate IM. Sweet potato (*Ipomea batata* L.): A versatile raw material for the food industry. Brazilian Archives of Biology and Technology. 2021;**64**:e21200568

Chapter 13

# State of the Art of Yam Production

*Abukari Wumbei, Sopkoutie Kengni Nerlus Gautier, Joseph Kwowura Kwodaga, Djeugap Fovo Joseph and Yamdeu Joseph Hubert Galani*

## Abstract

Yam is a labor-intensive and weed-sensitive food crop. The labor-intensive nature of the yam means that the production process requires the attention of the farmer all year round. However, the dwindling labor situation and the proliferation of weeds have forced farmers to think of modern ways of controlling weeds on their farms, that is, the adoption of chemical (herbicides) weed control. Even though the adoption of these chemicals has no doubt brought relief to the farmers and has resulted in increased yam production over the years, it has also brought in its wake, negative externalities of environmental pollution, human health effects, and food quality issues. The study was thus designed to investigate how yam is produced, the human and environmental health effects of how yam is produced, and food quality effects of how yam is produced. This was done through literature review, and field and laboratory experiments. It was revealed that, in recent years, new innovations have been introduced in yam production, the manner in which farmers handle herbicides in their yam production process exposes them to high doses of pesticides, thereby endangering their lives. The study findings also suggest that the use of herbicides in yam production does not affect the quality of the yam.

**Keywords:** yam production, food security, Ghana, chemical herbicides and West Africa

## 1. Introduction

Yam plants belong to the genus *Dioscorea* and produce tubers, bubils, or rhizomes that are of economic importance. They are monocotyledons in the family *Dioscoreaceae* within the order *Dioscoreales* which also includes the families *Stenomeridaceae, Trichoodaceae*, and *Stemonaceae*. In addition to the genus *Dioscorea*, the family also includes the genera *Stenomeris, Avetra, Trichopus, Rajana*, and *Tamus*. However, *Dioscorea* is by far the largest genus of the family [1].

Although more than 600 cultivars of the tubers have been recorded [2], only a few are important as staple food in the tropics. These include white yam (*D. rotundata*), yellow yam (*D. Cayenensis)*, water yam (*D. alata*), trifoliate yam (*D. dumetorum*), aerial yam (*D. bulbifera*), Chinese yam (*D. polystachya*), and Lesser yam (*D. esculenta*) [1, 3, 4]. *D. rotundata* is a native to West Africa, but it does not occur in the wild, and it was probably developed from *D. praehensilis* Benth. The extent of its cultivation parallels the preference of the people of West Africa for this type of yam over most other kinds. Its cultivation has also spread to other parts of the world as it is grown extensively in the Caribbean, Asia, and South America [5].

West Africa accounts for over 90% of world yam production with Nigeria, the largest single producer followed by Ghana and Cote d'Ivoire [5, 6]. In 2016, global yam production stood at 66 million metric tons (MT) with 86% of this coming from West Africa. In 2016, more than 90% (6.9 million ha) of the global area under yam cultivation was in West Africa, where the mean gross yield is 12 t/ha [5].

Worldwide annual consumption of yams is 18 million tons, with 15 million in West Africa. Annual consumption in West Africa is 61 kg/person/year [7]. Yams are consumed in the form of boiled, roasted, baked, or fried. Yam is an important staple food for many Ghanaians, accounting for 28% of total calorie sources in 2016 [8]. Per capita consumption of yam in Ghana increased from 83 kg/year 1995 to 160 kg/person/year in 2013, making yam the second most important calorie source after cassava in Ghana [8, 9]. Between 2005 and 2010, yam production in Ghana contributed about 16 percent to the country's agricultural gross domestic product [6].

In Ghana, as in many other West African countries, the yam species of economic importance include *D. rotundata*, *D. alata*, *D. cayenensis*, and *D. bulbifera* [3]. Among these economically important species, it is *D. rotundata*, popularly called white yam or white guinea yam, which is grown on a larger scale than any other yam species in the dominant yam production zone of West Africa and Central Africa [7]. Several varieties of yam are produced throughout Ghana. These include Pona (white yam), Dente, Asana, and Serwa. In recent years, Ghana's Crop Research Institute (CRI) introduced new high yield varieties, such as the Mankrong and Kukrupa. However, white yam/Pona (*D. rotundata*) remains the most preferred variety in both the domestic and export markets [6].

## 2. Results

### 2.1 Economic importance of Yam

In West Africa, yams are a major source of income and have high cultural value. They are used in fertility and marriage ceremonies, and a festival is held annually to celebrate its harvest in most cultures across West Africa. In West Africa, yam plays key roles in food security, income generation, and the sociocultural life of at least 60 million people [7]. It also serves as source of foreign exchange to government. Ghana has been exporting substantial quantities of yam to regions such as Europe, America, and African countries such as Mali, Burkina Faso, and Niger. Although the second largest yam producer after Nigeria, Ghana is the leading exporter of yam in the world [6, 7], exporting approximately 21,000 metric tons of yams annually, a number that has been increasing over the last decade. In addition to the food and market values, yams play vital roles in traditional sociocultural rituals and religions that the ethnocentric attachment to the crop remains strong for some ethnic groups in Africa [10].

In a typical Ghanaian urban center, household food budget formed about 51% of the total household budget [11]. Yam constituted about 12% of household at-home food budget and 13% of its away-from-home food budget. The shares of food budget that households allocate to yam generally increase during the peak harvest season (August to December) and drops during lean season (June to July) across all urban centers in Ghana [11].

| Rank | Country | Production (ha) | Production (tons) | Yield (tons/ha) |
|---|---|---|---|---|
| 1 | Nigeria | 6,307,232 | 50,052,977 | 8 |
| 2 | Ghana | 468,433 | 8,532,731 | 18 |
| 3 | Cote D'Ivoire | 1,200,405 | 7,654,617 | 6 |
| 4 | Benin | 228,998 | 3,150,248 | 14 |
| 5 | Togo | 98,547 | 868,677 | 9 |
| 6 | Cameroon | 62,008 | 707,576 | 11 |
| 7 | Central African Republic | 58,533 | 491,960 | 8 |
| 8 | Chad | 47,784 | 458,054 | 9.6 |
| 9 | Papua New Guinea | 21,185 | 363,387 | 17 |
| 10 | Haiti | 9983 | 63,358 | 6 |
| | **Total** | **8,503,108** | **72,343,585** | |

**Table 1.**
*List of top 10 yam producers in the World [5, 8]. Retrieved 24/05/2022.*

The top 10 yam-producing countries in the world are represented in **Table 1**. Together, they produced 72.3 million tons of yam in 2020, of which Africa accounted for about 99%. Most of the world's production comes from West Africa representing 97%, with Nigeria alone producing about 69%, equaling to more than 50 million tons [5].

Yam production is regarded as a source of food security and employment to a lot of people in many areas where it is cultivated [12]. Yam is among the major cash and most consumed food crops in West African countries like Nigeria, Cote D'Ivoire, Ghana, Benin, and Togo. Its cultivation is very profitable despite high costs of production and price fluctuations in the markets [13]. Over 60% of people grow yams as a primary source of livelihood [6].

In Ghana, yam is produced in commercial quantities in all the regions, except the Greater Accra region, Central region, and Upper east region. The highest concentration of yam production is found in the central and northern Savannah portions of the country [6]. The northern region is the second leading yam producer after the Brong Ahafo region (**Figure 1**). The region produces about 2.3 million metric tons of yam annually [14].

Even though, yam production is declining in some traditional yam-producing areas due to declining soil fertility, increasing pest pressures, and the high cost of labor [3, 15], production in Ghana has seen some steady increase over the past few years (**Figure 2**). Production grew by 19% between 2010 and 2020. This increase is mainly driven by growth in total hectares planted to yam, rather than growth in yam yields. The recent increases in production have often been attributed to the use of pesticides (herbicides) which allows small holder farmers to increase their area of cultivation [16]. The total yam production in Ghana in 2020 was 8,532,731 metric tons with mean gross yield of about 18 metric tons per ha [5].

## 2.2 Morphology and cultivation of yam

Yam is an annual or perennial vine and a climber with annual or perennial underground tubers. The life cycle of the yam plant consists of the following stages: propagules (true seed or tuber), emerging seedling or plantlet, mature plant, senescing

**Figure 1.**
*Five top yam-producing regions of Ghana [14].*

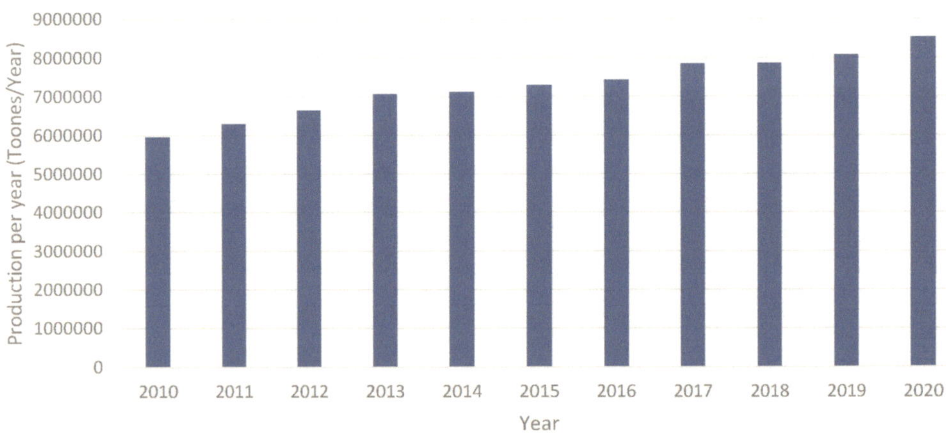


**Figure 2.**
*Yam production in Ghana, 2010–2020 [5, 8]. Retrieved on 24/05/2022.*

plant, and dormant tubers. Yam has an annual vegetative system (**Figure 3**) composed of a root system (some extend throughout the upper layers of the soil, others consist of root hairs), a stem system, a foliar system, and a reproductive system [17].

The root system generally consists of two categories of roots (**Figure 3**): the adventitious roots (appearing from the base of the stem) and roots arising from the skin of the tuber. The adventitious roots are usually about 3–6 mm thick and 1–3 m long and absorb nutrients and water. This type of roots extends throughout the upper layers of the soil and rarely branch out and emit few rootlets. The roots on the tuber are rarely more than few centimeters in length and are usually 1 or 2 mm thick [17].

**Figure 3.**
*Yam plants growing in the field, roots on tuber skin, and adventitious roots.*

These types of roots are not usually common on the tubers of the two yam varieties considered in this study (Laribako and Olodo).

The reproductive system consists of sexual components and a male or female inflorescence. The seed is flat, has a wing-like structure, and usually goes through a dormancy period of 3 to 4 months before germination can occur [3]. However, in Ghana and particularly in commercial production, yams are vegetatively propagated using the basal nodal region of the tuber, as flowering is rare.

The tuber, the economically important part of the yam plant, is rich in carbohydrates and contains modest amounts of mineral matter (calcium and iron), vitamin B, vitamin C, and crude fiber [3]. The plant usually produces a single annual tuber, which is 20–40 cm long and weighs from two to a dozen kilograms, depending on cultivar and growing conditions. The body can be elongated or spherical with a white, yellow, or purple flesh [18].

### 2.3 Temperature requirements

Yams are essentially tropical crops. Their growth is severely restricted at temperatures below 20°C. In general, they require temperatures of 25–30°C for normal growth to occur. Closely linked to temperature for optimal growth of the yam plant is light. It has been observed that the length of the day plays an important role in tuber formation. Short days tend to favor tuber formation and tuber growth, while long days favor vine growth. Though the influence of light intensity on yam growth and productivity has not been fully investigated, several points suggest that it is not a shade-loving plant. It requires and tolerates high intensities of sunlight to be maximally productive [1]. Because of this, in the Nanumba traditional area of Ghana, yam farmers after land preparation will usually devote substantial amount of their time to burn down or prune big trees on their farms before planting.

### 2.4 Water requirement

Yams thrive best when they receive sufficient moisture (about 1000 mm rainfall) well distributed throughout the growing cycle [3]. Since most yams require 7–9 months from planting to harvesting, they do best in areas where the rainy season is relatively long and where there are fewer than three or four rainless months in a year. Even though, the yam plant can survive droughts, they usually will give disappointing harvest. Hence, if yam is to be grown in an area where the dry season is longer, then supplementary irrigation must be provided [1]. In Ghana, the rain starts in April and ends in October, thereby providing the growing yam plants with 5 months continuous rain. A minimum of 1000 mm of water is required for the optimum growth [19].

### 2.5 Soil requirement

Yam like other crops requires soils of high fertility in order do well. For this reason, in traditional yam cultivation, it is grown on fertile forest soils as the first crop after clearing large trees or after a long term of fallowing [3, 17]. They are also cultivated in sandier savannah types of soils in the northern parts of West African [3]. Soils in the Nanumba traditional area are generally loamy in structure with good organic matter content and good water-holding capacity. This explains why the area is good for yam production. Soil structure has an impact on harvesting and ultimately on the quality of the tuber as loamy soils make harvesting easy with less bruises and clayey soils make harvesting difficult with lots of bruises on the yam (Rahaman, personal communication).

### 2.6 Agronomic practices

Yam is propagated from seed tubers or sections of the tuber. The use of true seeds as propagules is restricted to research stations, mainly in crop improvement programs. Traditionally, farmers obtain seed tubers in different ways. They may select small tubers (e.g. 300–500 g) from each harvest, use tubers from the second harvest of early maturing varieties, use small tubers from varieties that produce multiple tubers per stand, or cut ware tubers into pieces [3, 20]. Yam is sometimes planted alone, but it is more often intercropped with maize, cassava, rice, or other crops such as the legumes [3, 15]. Depending on the type of planting material, species, and location, effective duration of crop growth ranges from 6 to 12 months, calculated from the time of seed emergence until senescence of the leaves [19].

Yam is cultivated either on mounds or on ridges, and the growing plants can be staked or allowed to spread on the ground without stakes, especially in the savanna agro-ecological zone. Though staking may increase tuber yield, depending on the cultivar, the cost of stakes, the labor demand for their placement, and the environmental damage associated with their use may render this practice uneconomical [21, 22].

In Ghana, as in many other yam-producing countries in West Africa, yam production involves seven main processes. These include land preparation, planting, mulching, staking, weeding, harvesting, and storage. Fertilizer application on yam is rarely practiced at the farm level in Ghana.

### 2.7 Land preparation

Traditionally, land preparation in yam production is done by first clearing the land either manually with the use of a hoe and a cutlass, or the use of a tractor or bullocks [3]. However, with the modernization of yam farming, the use of herbicides has been adopted for land clearing in some yam-growing communities. In the case of manual land clearing and the use of herbicides, 2–4 weeks are allowed for the stubble to dry so as to enable the farmer to burn them and raise mounds. In the case of tractor ploughing, the farmer will usually go through the ploughed land and further clear grasses that are left under trees after which a second ploughing will be done before mounds are raised. Even though, it is said that yam can be grown on either mounds or ridges [3, 22], ridging is not yet practiced in the Nanumba traditional area of Ghana and mounding is done manually by the use of a hoe. In forested areas or in the savannah woodlands where there are substantial tree populations, the land clearing process also involves the burning or pruning of trees.

The use of herbicides for land clearing in yam farming is catching up very fast with farmers in the northern savannah areas of Ghana, especially farmers in the Nanumba traditional area, since it makes subsequent land preparation operations such as mounding easy and also gives them a better tuber yield (Rahaman, personal communication).

## 2.8 Planting

In the Nanumba traditional area of Ghana, yam is usually planted from December to February, the peak of the dry season. As a result of this, the planting operation is usually accompanied by capping (mulching) the yam mounds with some grass or leaves (**Figure 4**) to protect the planted sett from excessive heat and desiccation [1, 3, 17]. However, farmers who are not able to plant around this time would usually wait until the rain starts in April/May.

The time of planting yam depends on a number of factors, including the onset of the rains, the type of cultivars, the local ecological conditions, and the demand. For example, many *D. rotundata* and *D. cayenensis* cultivars are planted in the middle of the dry season, i.e. January–March [17]. These are generally cultivars with a short dormancy period whose long initial stems do not collapse on overheated soils and have a waxy coating that limits water loss.

**Figure 4.**
*Yam planted and mulched on manually prepared mounds [2].*

## 2.9 Staking

Staking is one of the yam farmer's main concerns [17, 21], since it is becoming increasingly difficult to obtain stakes throughout the yam-growing zones, especially in the dry savannah regions of West Africa. In its rudimentary form as it is practiced by yam farmers in the Nanumba traditional area of Ghana, stakes are obtained by cutting down young growing trees or cutting of the branches of bigger trees. Sometimes, trees that are burnt in the process of land preparation also serve as stakes. The stakes can be of different sizes (**Figure 5**) ranging from 1 to 4 m. The smaller ones are usually planted on the yam mound, while the bigger stakes are placed between yam mounds so that several plants can climb on it.

Tropical forests generally have high natural coppice rates, but these are reduced when trees are cut too close to the ground. It is suggested that ideally cuts should be made at least 45 cm from the ground to maximize regrowth. However, in an effort to get as much length from a sapling, yam stick cutters often cut sticks almost at ground level [23].

## 2.10 Weeding

Weeds are one of the serious challenges confronting yam production throughout the world [2]. The climbing growth habit of the plant coupled with its inability to shade the ground completely at any stage of its growth and development makes it very susceptible to weed interference [3]. In a study of Akobundu [24], it was observed that the critical period of weed interference in white yam is between the 8th and 16th week after planting. Proper weed management is necessary to obtain optimum yam yields. In general, a minimum of three to four weeding activities between planting and harvesting are necessary to minimize yield reduction. It is reported that weeds can cause yield loss in yam up to 90%. In most of the humid tropics, repeated hand weeding is a common feature of yam production. However, the commercialization of agriculture and high population growth with attendant increase in demand for yams,

**Figure 5.**
*Staked yam farm.*

in the absence of adequate labor [3], have created the need for the introduction of efficient and nonlabor-intensive methods of yam production [24].

In the yam-growing areas of Ghana, farmers usually go into the farm 1 month after planting with their hoe and cutlass to clear the farm of shrubs and other broad leaf weeds that have appeared. After this first weeding, the farmer can carry out three other weeding activities before harvesting in the case of manual weeding and one weeding in the case of chemical control.

In recent times, yam farmers in Ghana who are mostly located in rural areas adopt chemical weed control on yam. This is due to the dwindling labor force in the rural communities as a result of rural-urban migration of the youth in search of white collar jobs [25]. It can also be partly attributed to the expansion of farm sizes due to decreasing soil fertility and the commercialization of yam farming. However, whether a farmer uses chemical or manual weed control, the hoe is still employed, because the use of the hoe loosens the soil up to enhance aeration and water percolation.

### 2.11 Harvesting

Time for yam harvesting varies and may be spread out over several months in almost all regions due to the wide range of species and cultivars. For example, *D. rotundata* and *D. cayenensis* in West Africa are harvested twice [17], while *D. alata* is harvested only one time per season. This is similar to what pertains in the Nanumba traditional area in the northern region of Ghana (Rahaman, personal communication).

The first harvesting is done at a time when the plant has fully flowered (usually 6 months after planting) and the vines are about to cease growth with some of the bottom leaves turning yellow (**Figure 6**), that is usually around August and September. After this first harvest, the yam plant will regenerate and grow for some time and wither and finally dry up. This allows for the production of seed yam which is harvested in the second harvest period (December to January) together with late yam varieties like *D. alata* [26].

Early harvesting is done by the use of either a cutlass [2] or a sharp-ended stick where the mound is carefully cut open and the tuber is severed from the vine at the point below the base of the vine after which the mound is neatly covered to allow regenerative growth for seed tubers (**Figure 6**). On the other hand, late harvesting is done by the use of a hoe and cutlass, where the withered plant is cut off and half of the mound destroyed in order to remove the ware tuber or the seed tuber. In either

**Figure 6.**
*Harvesting of yam.*

situation, care is needed to minimize damage or bruises that lead to rot in tubers and a decrease in market value of the yam [6].

### 2.12 Storage

Yam as a tropical tuber crop has a relatively long shelf life (6–8 months) compared to other tropical fresh produce [3, 27]. This explains why, as a staple food, yam is available all year round for consumers. However, this long storability of yam notwithstanding, tubers are often damaged during harvesting and after harvesting, and this can lead to postharvest losses. After harvesting, yam can be stored either by adopting traditional or modern methods of storage [1, 12, 17]. The traditional system of yam storage varies among the different yam-producing countries of the world [27]. In Ghana, just as observed in other West African countries [27], yams can be stored by leaving them in the ground until they are needed for food or for sale. This system however exposes the yam tuber to attack by pests such as termites and rodents and harvesting also becomes difficult when the ground becomes hard during the dry season, resulting in tuber breakage and bruises which predispose the tubers to pathogens leading to loss of tubers in storage [28, 29]. Also, with this system, when there is heavy rain, the tubers may become rotten. The other traditional methods of yam storage include wooden platforms, cool and well-ventilated rooms, yam barns, heaping and covering with dry grass under trees, stored in a thatched shed, and Silo (burying in the soil) [27, 28, 30, 31]. The commonest traditional method of yam storage in Ghana is storage in the yam bam [32–34]. The modern methods on the other hand include chemical treatments (e.g. fungicides), storage in a cold room, and refrigeration; however, the method of cold storage is hardly practiced by farmers in West Africa [12, 28].

The efficacy of tuber storage structures for preserving yams until they are used is influenced by the cultivars, environmental conditions such as relative humidity and temperature, the physical condition of the tubers at the beginning of storage, and the effectiveness to exclude vermin such as rodents [27, 30, 32]. Traditional storage methods therefore vary according to ecology and the volume of yam produced [2].

Due to the economic circumstances in the yam-growing areas in Ghana and the Nanumba traditional area of Ghana in particular, farmers store yam through the traditional methods (**Figure 7**). In this traditional area, after harvest, farmers will usually burry the tubers in the soil or keep them under the shade of a tree and cover them with dry vines of yam or grass. They can also keep them in constructed barns on the farm or in the house or keep them in well-ventilated rooms. The latter practice is

**Figure 7.**
*Yam storage in the field.*

hardly done by farmers, but rather a common practice of yam traders (Imoro, personal communication).

Storage is an important element within the yam production chain which, when not properly done, can lead to high postharvest losses leading to low incomes for farmers and food insecurity. Onwueme [1] and Ravi et al. [27] respectively observed that shading, ventilation, and constant inspection are three essential elements for good yam storage in a barn. They asserted that ventilation serves two purposes, i.e. preventing the buildup of high humidity which favors rotting and preventing tubers from heating up owing to their own respiratory activities.

Postharvest losses for yam in Ghana are as high as 24 percent of production, despite the Ministry of Agriculture's goal to reduce these losses to only 12 percent [35, 36]. The major causes of postharvest losses are weight loss due to evapotranspiration intensified by sprouting, rotting due to fungal and bacterial pathogens, and insect infestation [35, 37].

### 2.13 Pests, diseases and weeds

In spite of the importance of yam as a food security and cash crop, yam farmers in Ghana suffer serious challenges in the production process. Among these production challenges are field and storage pests and diseases and weed burden [37–40].

#### *2.13.1 Pests*

Yams are infested by a broad taxonomic diversity of insect pests [41]. In a study of Braimah et al. [42], field and storage pests of yam in Ghana are identified. These pests (**Table 2**) cause significant losses to tubers both in the field and in storage.

#### *2.13.2 Diseases*

Yam in Ghana is attacked by a couple of diseases both in the field and in storage. However, the common diseases are those caused by fungi (**Table 3**). Other diseases affecting yam in Ghana are bacterial and viral diseases [37].

Yam is prone to infection right from the seedling stage through to harvesting and even after harvesting (in storage); hence, diseases of yam can be grouped into field and storage diseases. The field and storage diseases are many and varied. While Aboagye-Nuamah et al. [37] found that *B. theobromae, F. oxysporium*, and *R. stolonifer* were the most frequently encountered spoilage microorganisms in storage, Ripoche [43] observed anthracnose caused by *Colletotrichum gloeosporioides* to be the most severe foliar disease of water yam in the tropics and Amusa et al. [44] observed yam mosaic virus (YMV) disease to be causing the most severe losses in yams in Nigeria.

Tuber rots through microbial attack are an important cause of postharvest loss in yam throughout the tropics [33]. Rot from microbial infection of healthy tubers reduces their table quality and renders them unappealing to consumers, as it leads to total loss of tuber carbohydrates by transforming it into inedible colored mass. It is reported that postharvest losses in yam due to tuber rots can be as high as between 20 and 60% [31, 45, 46]. The rots are of different types, including soft rots, dry rots, and wet rots [29, 45, 47, 48].

Due to the importance of microbial tuber rots in yam, studies have been conducted to identify the causal organisms. In Ghana, Cornelius [45] in a survey identified some fungal, bacterial, and nematode species (**Table 3**) to be responsible for yam tuber rots.

| Pest | Scientific name | Order | Categorization |
|---|---|---|---|
| Termites | *Amitermes* spp | Isoptera | Field |
| Millipedes | *Deridontoyge* spp | Callipodida | Field |
| Tuber beetles | *Heterolygus meles* | Coleoptera | Field and storage |
| Leaf beetles | *Planococcus diocorea* | Coleoptera | Field |
| Mealybugs | *Pseudococcus brevipes* | Hemiptera | Storage |
| Scale insects | *Aspidiotus destructor* | Hemiptera | Field and storage |
| Crickets | *Gryllus campestris* | Orthoptera | Field |
| Nematodes | *Meloidogyna incognita* | Nematoda | Field |
| | *Pratylenchus coffeae* | | |
| | *Scutellonema bradys* | | |

**Table 2.**
*Field and storage pest of yam in Ghana.*

| Disease | Causal organism | Type | Categorization |
|---|---|---|---|
| Rots | Fungus | *Aspergillus flavus* | Storage |
| | | *Aspergillus niger* | Storage |
| | | *Aspergillus oryzae* | Storage |
| | | *Botryodiplodia theobromae* | Storage |
| | | *Colletotrichum gloeosporioides* | Field |
| | | *Fusarium culmorum* | Storage |
| | | *Fusarium oxysporium* | Storage |
| | | *Fusarium moniliforme* | Storage |
| | | *Penicilium* spp | Storage |
| | | *Rhizopus stolonifer* | Storage |
| | Bacteria | *Erwinia carotovora* | Field and Storage |
| | Nematode | *Pratylenchus coffeae* | Field |
| | | *Scutellonema bradys* | Field |
| Galling of tubers | Nematode | *Meloidogyne incognita* | Field |
| Yam mosaic | Virus | *Yam mosaic potyvirus* (YMV) | Field |
| | | *Dioscorea alata potyvirus* (DaV) | Field |

**Table 3.**
*Diseases affecting yam in Ghana.*

In a related study, Aboagye-Nuamah et al. [37] confirmed the fungal and bacterial species reported by Cornelius [45] to be the cause of yam tuber rots. Subsequently, Asare-Bediako et al. [49] in a study to identify microorganisms causing rot in white yam found *Aspergillus flavus*, *A. niger*, *A. ochraceus*, *Aspergillus* sp., *A. tamari*, *Sclerotium rolfsii*, *Cladosporium* sp. *Corynebacterium* sp. *Fusarium* sp. *Penicillium* sp.,

*Rhizopus stolonifer*, and *Trichoderma* sp. to be associated with the rots, with *Sclerotium rolfsii* causing the most severe rot, followed by *A. niger* and *Fusarium* sp. In Nigeria, Osai and Ikotun [50] in a study to identify the microbial causes of rot in yam minisetts observed that *Sclerotium rolfsii, Trichoderma longibrachiatum, Botryodiplodia theobromae*, and *Penicillium oxalicum* are the most pathogenic among nine fungal and two bacterial species, causing tuber losses between 40 and 60%. In a study of Ogaraku and Usman [48] in Nasarawa state of Nigeria, six fungal species *(Aspergillus niger, Aspergillus flavus, Rhizopus stolonifer, Sclerotium rolfsii, Fusarium oxysporum*, and *Rhizoctonia* spp*)* were isolated from rotted yam tubers. In Brazil, Muniz et al. [47] in a study identified three nematode species (*Scutellonema bradys*, *Pratylenchus coffeae*, and *Pratylenchus brachyurus*) to be associated with dry rots of yam. In China, Zhang et al. [51] reported for the first time that *Pythium ultimum* var. *ultimum* causes tuber rot in the Chinese yam (*Dioscorea polystachya*).

Even though there have been reports of the use of insecticides such as pirimiphos-methyl (Actellic 2% Dust) [52] leading to significant reduction in fungal infections and physical damage to yam tubers, yam farmers in the Nanumba traditional area currently do not treat yam on the field and after harvest with chemicals. They rather adopt crop rotation, bush fallow, and use of healthy planting materials. The use of chemical insecticides as practiced in the Delta State in Nigeria [53, 54] can indirectly reduce the incidence of rots in the yam tuber caused by fungi, because physical damage to the yam tuber, either mechanically or by insects attack, can serve as conduit for microbial infections [28, 30, 31].

### *2.13.3 Weeds*

The common weeds found on yam farms in Ghana can be seen in **Table 4**. They belong to three groups, i.e. the dicotyledons, monocotyledons, and the parasites [38].

In a study by Akobundu [24], it was observed that annual weeds caused a tuber yield reduction of about 54% and 90%, respectively, when the period of weed interference lasted 3 and 4 months after planting.

| Weed group | Name |
|---|---|
| Monocotyledons | *Axonopus compresus* (Sw.) P. Beauv. |
| | *Cyperus esculentus* L. |
| | *Cyperus rotundus* L. |
| | *Digitaria horizontalis* Willd. |
| | *Eragrostis tremula* Hochst. Ex Steud. |
| | *Hackelochloa granularis* (L.) O. Ktze. |
| | *Kyllinga erecta* Schumach. Var. |
| | *Kyllinga squamulata* Thonn. Ex Vahl. |
| | *Paspalum scrobiculatum* L. |
| | *Rottboellia cochinchinensis* (Lour.) Clayton |
| | *Setaria pallide-fusca* (Schum.) Stapf. & C.E. Hubbard |
| | *Imperata cylindrica* |

| Weed group | Name |
|---|---|
| Dicotyledons | *Corchorus olitorius* L. |
| | *Commelina benghalensis* L. |
| | *Commelina diffusa* Burm. |
| | *Desmodium scorpluras* (Sw.) Desv. |
| | *Hyptis suoveolens* Poit. |
| | *Mimosa inv*isa Mart. |
| | *Mimosa pigra* L. |
| | *Mitracarpus villosus* (Sw.) DC. |
| | *Oldenlandia corymbosa* L. |
| | *Phyllanthus amarus* Schum. & Thonn. |
| | *Scoparia dulcis* L. |
| | *Tridax procumbens* L. |
| | *Triumfeta cordiflora* A. Rich. |
| | *Vernonia galamensis* (Cass.) Less. |
| Parasites | *Striga hermonthica* (Del.) Benth. |

**Table 4.**
*Weeds associated with yam cultivation in Ghana.*

## 3. Yam consumption

Yam, sweet in flavor, is consumed as boiled yam (as cooked vegetable) or fufu or fried in oil (**Figure 8**) and then consumed. It is often pounded into a dough-like paste after boiling and is consumed with soup [3]. It is also processed into flour for use in the preparation of the paste. The tuber is the edible part of the yam plant with high carbohydrate content and low in fat and proteins and provides a good source of energy. Yam is an important source of minerals such as calcium, phosphorus, iron, carbohydrates, and vitamins such as riboflavin, thiamin, vitamin B, and vitamin C [55, 56]. Yam has some inherent characteristics which make it attractive and keep it in a high demand. First and foremost, it is rich in carbohydrates, especially starch;

**Figure 8.**
*Prepared yam dishes of fufu, boiled yam, and fried yam.*

consequently, it has a multiplicity of end uses. Secondly, it is available all year round, making it preferable to other seasonal crops [13].

Yam contains products such as alkaloids (saponin and sapogenin) and proteins (dioscorin and diosgenin) which can cause side effects in humans and animals such as inflammations, allergic reactions, kidney problems, and interference with the metabolic system [57, 58]. However, these same properties of the yam plant are exploited in medicine for the treatment and management of conditions such as allergies, metabolic disorders, hypertension, inflammations, and hormonal irregularities [58–60]. Yam has been used in traditional medicine in Africa and among the Chinese and other Asiatic people to treat diseases like diabetes, to increase coronary circulation, and to prevent hypercholesterolemia [55, 61, 62]. Although the industrial use potential of yam has not been fully exploited, its use as an industrial starch has been established as the quality of some of the species is able to provide as much starch as in cereals [11, 63, 64].

In recent times, efforts have been made by the scientific community to investigate the suitability of yam to be fried into chips like French fries. In a study to investigate the effect of blanching and frying on the textural profile and appearance of yam for French fries [9], it was found that yam fries with desirable texture and color attributes can be produced with *Dioscorea rotundata* by blanching that yam species at 75–80°C for 5–8 min and frying at 180°C for 3–3.5 min.

## 4. Marketing

The importance of yam in the economy of some producing areas appears to be declining due partly to competition from other crops like cassava in Nigeria and taro in the South Pacific. However, in Ghana, the contribution of yam to the economy by way of meeting household food needs and foreign exchange earnings through exports has been growing [11].

Harvested yams in most communities are mostly for household consumption with only 16.4% of farmers selling more than half of their harvested yam [26]. Yam tubers meant for sale are mostly sold to merchants in local markets and also directly to ordinary people and food vendors in towns. These merchants come from other regions and move from village to village to purchase tubers directly from farmers. In the Nanumba traditional area of the northern region of Ghana, merchants come from Accra and Kumasi in Southern Ghana to buy yam [65]. This practice conforms with what happens in the Niger State of Nigeria where merchants come from Ibadan and Ilorin in southern Nigeria to buy yam [26].

In Ghana, yam is generally traded in its original state and is not processed into a secondary product. Traders and chop bar owners (small restaurants) often buy yam to sell or prepare it for consumers directly. Yam for the export market is also not processed, but is treated, wrapped in paper, and packed in 20-kg boxes before it is shipped [37].

About a decade ago, the UK was the leading importer (49%) of Ghana yam export, followed by the Netherlands and the USA [6]. Since then, the trend has changed dramatically with the USA now leading the chat. In 2017, Ghana exported a total value of USD 38,393.00 worth of yam to 10 top importing countries of which the USA took 35%, followed by the UK and Belgium of 34% and 9%, respectively [66] (**Figure 9**). The large volumes of yam exported to these destinations are largely due to the high demand for yam by Ghanaians and other West Africans residing in these countries.

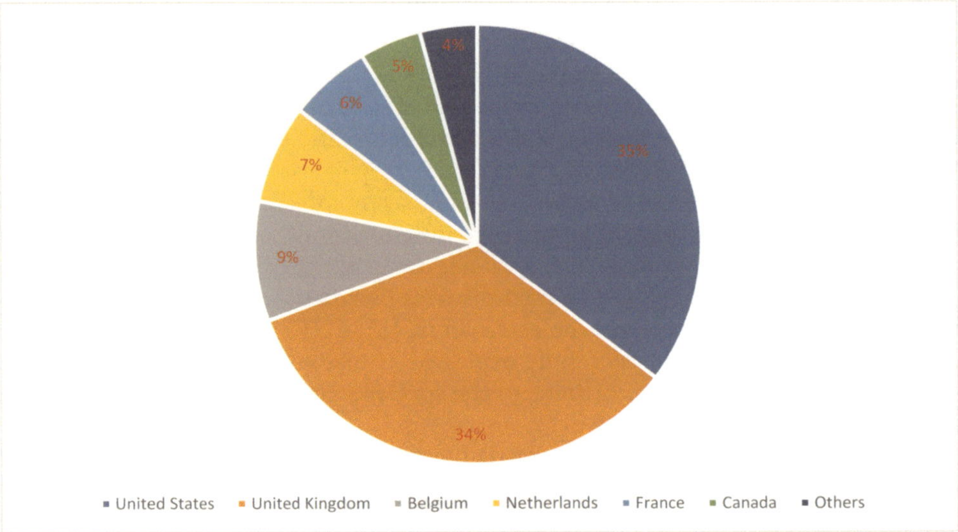


**Figure 9.**
*Share of Ghana yam export by importing countries.*

## 5. Monitoring of pesticides residues in yam

All over the world, there are concerns about pesticide residues resulting from the use of pesticides on crops [67]. For this reason, governments and especially in the EU allow pesticides to enter their respective countries as long as they are used in line with the law and the guidelines controlling their use.

In Ghana, public concerns are high about the use of pesticides by Ghanaian farmers and its attendant food safety and human health issues. This has led to the conduct of pesticides residue monitoring studies to assess the levels of pesticides in various food items. In a study by Asiedu [68] to determine pesticides residues in lettuce, garden egg, pineapple, and mango in three regions of Ghana, it was found that some market fruits and vegetables contained different types of pesticides of which chlorpyrifos (an organophosphate) and cypermethrin (a synthetic pyrethroid) were the most common ones. In a study by Bempah & Donkor [69] to assess the concentration of pesticides residues in fruits and vegetables from selected markets in Kumasi in the Ashanti region, it was found that 19% of the samples contained pesticides residues above the maximum residue level (MRL) with the health risk analysis further revealing that the pesticides endrin had exceeded the reference dose in vegetables, thereby suggesting a great potential for systematic poisoning in children that are considered as the most vulnerable population subgroup.

There is limited literature on pesticides residues in yam in contrast to other root and tuber crops in developing countries [70, 71]. This is partly because the crop is grown and consumed mostly in the developing world where there is limited scientific expertise and resources to set residue limits and to monitor them. In a study of Adeyeye and Osibanjo [70], 55% of yam samples from Nigerian markets were contaminated with one or more organochlorine pesticides (aldrin, dieldrin, HCH, and

DDT). In a monitoring program by the "UK pesticides residue committee," to check whether pesticides residues in food and drink are above the maximum residue levels (MRLs), it was found that yam was among the food commodities with pesticides residues exceeding the MRLs [72]. Out of the 52 yam samples analyzed, 9 samples contained residues at or below the MRL, 11 samples contained residues above the MRL, and 13 samples contained more than one residue. The report observed further that as in previous years relatively high numbers of samples with residues over the MRL were found in the specialty vegetables, okra and yam. The yam samples used in this monitoring study originated from Ghana, Brazil, and Jordan. The samples from Ghana had residue levels of 0.2–0.3 mg/kg carbendazim (MRL = 0.1 mg/kg) and 0.4 mg/kg tebuconazole (MRL = 0.02 mg/kg). Since MRLs are not safety limits, risk assessment were carried out with the monitoring results which showed that the residues found in the yam will be unlikely to have adverse effects on health [72]. Similarly, in the studies of Wumbei et al. [73, 74] to investigate pesticides residues

| | **Residues (mg/kg)** | | | | | |
|---|---|---|---|---|---|---|
| | **Fenitrothion** | **Fenpropimorph** | | | | |
| Mean | 0.0043 | 0.0003 | | | | |
| Median | 0.0023 | 0.0002 | | | | |
| P75 | 0.0069 | 0.0002 | | | | |
| P90 | 0.0069 | 0.0006 | | | | |
| P95 | 0.0079 | 0.0007 | | | | |
| P97.5 | 0.0097 | 0.0013 | | | | |
| P99 | 0.0144 | 0.0031 | | | | |
| | Yam consumption (kg/kgBW/day) | | | | | |
| Mean | Median | P75 | P90 | P95 | P97.5 | P99 |
| 0.006 | 0.006 | 0.008 | 0.009 | 0.01 | 0.011 | 0.013 |
| | Estimated daily intake (EDI) (mg/kgBW/day) | | | | | |
| | Fenitrothion | Fenpropimorph | | | | |
| Mean | 0.000026 | 0.000002 | | | | |
| Median | 0.000014 | 0.000001 | | | | |
| P75 | 0.000055 | 0.000002 | | | | |
| P90 | 0.000062 | 0.000006 | | | | |
| P95 | 0.000082 | 0.000007 | | | | |
| P97.5 | 0.000110 | 0.000014 | | | | |
| P99 | 0.000187 | 0.000040 | | | | |
| **Acceptable daily intake (ADI) (mg/kgBW/day)** | **0.005** | **0.003** | | | | |

**Table 5.**
*Estimated daily intake of fenpropimorph and fenitrothion through deterministic exposure assessment and corresponding ADIs. Adopted from Wumbei et al. [32, 74, 75].*

in yam, 12 pesticides, including five insecticides (cadusafos, fenitrothion, imidacloprid, profenofos, and propoxur), four fungicides (carbendazim, fenpropimorph, metalaxyl, and propiconazole), and three herbicides (bentazone, glyphosate, and pendimethalin) were detected. However, when consumption risk assessment was carried out, it was revealed that there was no risk of dietary intake of these pesticides in yam under the deterministic approach (**Table 5**) and simple distribution approach (**Table 6**), but there was intake risk in about 10% of the study population to fenpropimorph and fenitrothion under the probabilistic (upper bound scenario) approach (**Table 7**) [75].

| | | | **Residues (mg/kg)** | | |
|---|---|---|---|---|---|
| | **Cadusafos** | **Carbendazim** | **Glyphosate** | **Imidacloprid** | **Metalaxyl** |
| | 0.0005 | 0.0007 | 0.12 | 0.0007 | 0.0009 |
| Statistical dist. of yam consumption (kg/kgBW/day) = Loglogistic (−0.017192; 0.02288; 14,635) | | | | | |
| | | | EDI (mg/kgBW/day) | | |
| | Cadusafos | Carbendazim | Glyphosate | Imidacloprid | Metalaxyl |
| Mean | 0.0000029 | 0.0000041 | 0.00070 | 0.0000041 | 0.0000053 |
| Median | 0.0000028 | 0.0000039 | 0.00068 | 0.0000039 | 0.0000051 |
| P75 | 0.0000037 | 0.0000052 | 0.00089 | 0.0000052 | 0.0000067 |
| P90 | 0.0000047 | 0.0000066 | 0.00113 | 0.0000066 | 0.0000084 |
| P95 | 0.0000054 | 0.0000075 | 0.00129 | 0.0000075 | 0.0000097 |
| P97.5 | 0.0000061 | 0.0000085 | 0.00146 | 0.0000085 | 0.000011 |
| P99 | 0.0000071 | 0.00001 | 0.00169 | 0.00001 | 0.000013 |
| ADI (mg/kgBW/day) | 0.0004 | 0.02 | 0.5 | 0.06 | 0.08 |
| | | | Residues (mg/kg) | | |
| | Pendimethalin | Profenofos | Propiconazole | Propoxur | Bentazone |
| | 0.0003 | 0.0004 | 0.0002 | 0.0004 | 0.0007 |
| | | | EDI (mg/kgBW/day) | | |
| | Pendimethalin | Profenofos | Propiconazole | Propoxur | Bentazone |
| Mean | 0.0000018 | 0.0000023 | 0.0000012 | 0.0000023 | 0.0000041 |
| Median | 0.0000017 | 0.0000022 | 0.0000011 | 0.0000022 | 0.0000039 |
| P75 | 0.0000022 | 0.0000029 | 0.0000015 | 0.0000029 | 0.0000052 |
| P90 | 0.0000028 | 0.0000037 | 0.0000019 | 0.0000037 | 0.0000066 |
| P95 | 0.0000032 | 0.0000043 | 0.0000021 | 0.0000043 | 0.0000075 |
| P97.5 | 0.0000036 | 0.0000049 | 0.0000024 | 0.0000049 | 0.0000085 |
| P99 | 0.0000042 | 0.0000056 | 0.0000028 | 0.0000056 | 0.00001 |
| **ADI (mg/kgBW/day)** | **0.125** | **0.03** | **0.04** | **0.02** | |

**Table 6.**
*Estimated daily intake of cadusafos, carbendazim, glyphosate, imidacloprid, metalaxyl, pendimethalin, profenofos, propiconazole, and propoxur through simple distribution and corresponding ADIs. Adopted from Wumbei et al. [32, 74, 75].*

| Statistical distributions of residue and consumption data | | | | |
|---|---|---|---|---|
| **Fenpropimorph (mg/kg)** | **Fenitrothion (mg/kg)** | | **Yam consumption (kg/kgBW/day)** | |
| LB (0), UB (0.0035) | LB (0) UB (0.0115) | | Loglogistic (−0.017192; 0.02288; 14,635) | |
| | EDI (mg/kgBW/Day) | | | |
| Percentile | Fenpropimorph | | Fenitrothion | |
| | Lower bound | Upper bound | Lower bound | Upper bound |
| Mean | 0.000008 | 0.00058 | 0.000024 | 0.0012 |
| Median | 0.000000 | 0.00001 | 0.000000 | 0.0001 |
| P75 | 0.000000 | 0.00002 | 0.000043 | 0.0022 |
| P90 | 0.000006 | 0.0023 | 0.000076 | 0.0042 |
| P95 | 0.000011 | 0.0051 | 0.0001 | 0.0052 |
| P97.5 | 0.000019 | 0.0066 | 0.00013 | 0.0060 |
| P99 | 0.000044 | 0.0082 | 0.0002 | 0.0071 |
| **ADI (mg/kgBW/day)** | **0.003** | **0.003** | **0.005** | **0.005** |

*LB = Lower bound, UB = Upper bound.*

**Table 7.**
*Estimated daily intake of fenpropimorph and fenitrothion through probabilistic exposure assessment. Adopted from Wumbei et al. [32, 74, 75].*

**Author details**

Abukari Wumbei[1*], Sopkoutie Kengni Nerlus Gautier[2], Joseph Kwowura Kwodaga[3], Djeugap Fovo Joseph[2] and Yamdeu Joseph Hubert Galani[4]

1 Department of Planning, Evidence and Policy, Institute for Interdisciplinary Research, University for Development Studies, Ghana

2 Faculty of Agronomy and Agricultural Sciences, Department of Plant Protection, University of Dschang, Cameroon

3 Faculty of Agriculture, Department of Crop Science, Food and Consumer Sciences, University for Development Studies, Ghana

4 Section of Natural and Applied Sciences, School of Psychology and Life Sciences, Canterbury Christ University, Canterbury, UK

*Address all correspondence to: awumbei@uds.edu.gh

## References

[1] Onwueme IC. The Tropical Tuber Crops: Yams, Cassava, Sweet Potato, and Cocoyams. Wiley; 1978. Available from: https://books.google.be/books/about/The_Tropical_Tuber_Crops.html?id=WRFIAAAAYAAJ&redir_esc=y

[2] Asiedu R, Sart A. Crops that feed the World: Yams for income and food security. Food Security: Springer. 2010; **2**:305-315. DOI: 10.1007/s12571-010-0085-0

[3] Andres C, Adeoluwa OO, Bhullar GS. Yam (Dioscorea spp.). Encyclopedia of Applied Plant Sciences. 2016;**3** (September):435-441. DOI: 10.1016/B978-0-12-394807-6.00177-5

[4] Osunde Z. Minimizing postharvest losses in Yam (Dioscorea spp.): Treatments and techniques. International Union of Food Science and Technology. 2008;**2008**:1-12

[5] FAO. FAOSTAT Crop Production Data. 2020. Availale from: http://www.fao.org/faostat/en/#data/QC [Accessed: June 27, 2022]

[6] Anaadumba P. Analysis of Incentives and Disincentives for Yam in Ghana. Technical Notes Series. Rome: MAFAP, FAO; 2013. Available from: https://www.fao.org/in-action/mafap/resources/detail/en/c/394345/

[7] IITA. Maximizing profitability of yam production | IITA Annual Report. 2014. Available from: http://wpar12.iita.org/?p=1482. [Accessed: December 4, 2017]

[8] FAO. FAOSTAT: Food Balance—Food Supply—Crops Primary Equivalent. 2020. Available from: http://www.fao.org/faostat/en/#compare. [Accessed: September 11, 2022]

[9] Graham-Acquaah S, Ayernor GS, Bediako-Amoa B, Saalia FS, Afoakwa EO, Abbey L. Effect of blanching and frying on textural profile and appearance of yam (Dioscorea rotundata) French fries. Journal of Food Processing and Preservation. 2015;**39**(1): 19-29. DOI: 10.1111/jfpp.12204

[10] Nweke F, Aidoo R. Yam consumption patterns in West Africa. 2013 (February). Available from: https://agriknowledge.org/downloads/3x816m672

[11] Aidoo R. An analysis of yam consumption patterns in Ghanaian Urban communities [doctoral thesis]. 2009. p. 191

[12] Nahanga V, Vera B. Yam production as pillar of food security in logo local government area of benue state, Nigeria. European Scientific Journal. 2014; **10**(31):1857-7431

[13] Izekor OB, Olumese MI. Determinants of yam production and profitability in Edo State, Nigeria. African Studies on Population and Health. 2010;**00**(4):205-210. Available from: http://www.asopah.org

[14] Ministry of food and agriculture (MoFA). Agriculture in Ghana, facts and figures, 2016 final. Economic Journal of Development. 2016. Available from: http://mofa.gov.gh/site/wp-content/uploads/2018/05/Agric in Ghana F&F 2016_Final.pdf

[15] Maliki R, Toukourou M, Sinsin B, Vernier P. Productivity of yam-based systems with herbaceous legumes and short fallows in the Guinea-Sudan transition zone of Benin. Nutrient Cycling in Agroecosystems. 2012;**92**(1): 9-19. DOI: 10.1007/s10705-011-9468-7

[16] Wumbei A. The pesticides law and the attitudes of pesticides dealers in the northern region of Ghana: Implications for environmental security and human health. Journal of Environment and Earth Science. 2015b;**5**(12):106-114

[17] Degras L, Coste R. The Yam: A Tropical Root Crop. Macmillan; 1993. Available from: https://books.google.be/books/about/The_Yam.html?id=vCRIAAAAYAAJ&redir_esc=y

[18] Soibam H, Ayam VS, Mitra S. Morphological characteristics of yam cultivars at different stages of growth, development and harvesting. International Journal of Agricultural and Statistical Sciences. 2017;**13**(2):499-505. Available from: https://www.scopus.com/inward/record.uri?eid=2-s2.0-85039457944&partnerID=40&md5=75bd9caee10ce018559a1e5a0f564560

[19] Orkwor GC, Asiedu R, Ekanayake IJ. Food Yams. Advances in Research. Ibadan, Nigeria: International Institute of Tropical Agriculture (IITA) and NRCRI; 1998. Available from: https://biblio.iita.org/documents/U98BkOrkworFoodNothomNodev.pdf-80db79d0b1ecb5f20d691ddcfcf8fbd3.pdf

[20] Aighewi BA, Asiedu R, Maroya N, Balogun M. Improved propagation methods to raise the productivity of yam (Dioscorea rotundata Poir.). Food Security. 2015;**7**(4):823-834. DOI: 10.1007/s12571-015-0481-6

[21] Danquah EO, Ennin SA, Lamptey JNL, Acheampong PP. Staking options for sustainable yam production in Ghana. Sustainable Agriculture Research. 2014;**4**(1):2015. DOI: 10.5539/sar.v4n1p106

[22] Ennin SA, Otoo E, Tetteh FM. Ridging, a mechanized alternative to mounding for yam and cassava production. West African Journal of Applied Ecology. 2009;**15**:2009. DOI: 10.4314/wajae.v15i1.49424. Available from: https://journals.ug.edu.gh/index.php/wajae/article/view/830/512

[23] Beckford C, Campbell D, Barker D. Sustainable food production systems and food security: Economic and environmental imperatives in yam cultivation in trelawny, Jamaica. Sustainability. 2011;**3**(3):541-561. DOI: 10.3390/su3030541

[24] Akobundu IO. Weed interference and control in white yam (Dioscorea rotundata Poir). Weed Research. 2006; **21**(6):267-272. DOI: 10.1111/j.1365-3180.1981.tb00127.x

[25] Suleman MN, I. B., Majid, M. R. Rural-urban migration in Ghana: A menace or blessing? Current Politics and Economics of Africa. 2015;**8**(4):597-616. Available from: http://search.proquest.com/docview/1841074686?pq-origsite=gscholar

[26] Fu RHY, Kikuno H, Maruyama M. Research on yam production, marketing and consumption of Nupe farmers of Niger State, central Nigeria. African Journal of Agricultural Research. 2011; **6**(23):5301-5313. DOI: 10.5897/AJAR11.586

[27] Ravi V, Aked J, Balagopalan C. Review on tropical root and tuber crops I. Storage methods and quality changes. Critical Reviews in Food Science and Nutrition. 1996;**36**(7): 661-709. DOI: 10.1080/10408399609527744

[28] Falola A, Salami MF, Bello AA, Olaoye TA. Effect of yam storage techniques usage on farm income in Kwara State, Nigeria. Agrosearch. 2017; **17**(1):54. DOI: 10.4314/agrosh.v17i1.5

[29] Ogundana SK, Naqvi SHZ, Ekundayo JA. Fungi associated with soft rot of yams (Dioscorea spp.) in storage in Nigeria. Transactions of the British Mycological Society. 1970;**54**(3): 445-451. DOI: 10.1016/S0007-1536(70) 80159-0

[30] FAO. Post-Harvest Operations. Yams: Post-harvest Operations; 1999. Available from: http://www.fao.org/3/a-ax449e.pdf

[31] Nyadanu D, Dapaah H, Agyekum AD. Resistance to post-harvest microbial rot in yam: Integration of genotype and storage methods. African Crop Science Journal. 2014;**22**(2):89-95. Available from: https://www.ajol.info/index.php/acsj/article/download/104935/94985

[32] Wumbei A, Bawa JAK, Akudugu MA, Spanoghe P. Absence of effects of herbicides use on yam rots: A case study in Wulensi, Ghana. Agriculture. 2019;**2019**(9):95. DOI: 10.3390/agriculture9050095. Available from: www.mdpi.com/journal/agric ulture

[33] Aboagye EE. Patterns of pesticide use and residue levels in exportable pineapple (*Ananas Comosus* L. Merr.). 2002. Available from: http://ugspace.ug.edu.gh/handle/123456789/7339

[34] Cornelius EW, O. K. A. Storage diseases of white yam (Dioscorea rotundata, Poir): Causes, varietal susceptibility and control. Journal of the Ghana Science Association. 1999

[35] Ansah IGK, Tetteh BKD. Determinants of Yam postharvest management in the Zabzugu district of Northern Ghana. Advances in Agriculture. 2016;**2016**:1-9. DOI: 10.1155/2016/9274017

[36] Ministry of Food and Agriculture. FASDEP II | Ministry of Food & Agriculture. 2007. Available from: http://mofa.gov.gh/site/?page_id=598

[37] Aboagye-Nuamah F, Offei SK, Cornelius EW, Bancroft RD. Severity of spoilage storage rots of white yam (Dioscorea rotundata Poir.). Annals of Applied Biology. 2005;**147**(2):183-190. DOI: 10.1111/j.1744-7348.2005.00028.x

[38] Dzomeku IK, Abudulai M, Brandenburg RL, Jordan DL. Survey of weeds and management practices in peanut (*Arachis hypogaea* L.) in the Savanna ecology of Ghana. Peanut Science. 2009;**36**(2):165-173

[39] Kusi F, Asante SK, Adjebeng-Danquah J, Nutsugah SK, Buah SSJ, Owusu RK, et al. Participatory integrated pest management strategy for improving shelf-life of yam (Dioscorea spp). International Journal of Advance Agricultural Research. 2013;**1**:124-132

[40] Moro S. Trade potential factors influencing herbicides use of yam farm households in Kpandai District in northern region. Ghana. 2014;**1**(1):1-8. Available from: https://www.peakjourna ls.org/sub-journals-PJSRE.html

[41] Korada RR, Naskar SK, Edison S. Insect pests and their management in yam production and storage: A world review. International Journal of Pest Management. 2010;**56**(4):337-349. DOI: 10.1080/09670874.2010.500406

[42] Braimah H, Anchirinah VM, Adu-Mensah J. Yam pests in the Ashanti and Brong Ahafo regions of Ghana: A study of farmers\' indigenous technical knowledge and control practices. Ghana Journal of Agricultural Science. 2008; **40**(1):33-42. DOI: 10.4314/gjas. v40i1.2152

[43] Ripoche A, Jacqua G, Bussière F, Guyader S, Sierra J. Survival of Colletotrichum gloeosporioides (causal agent of yam anthracnose) on yam residues decomposing in soil. Applied Soil Ecology. 2008;**38**(3):270-278. DOI: 10.1016/J.APSOIL.2007.10.015

[44] Amusa N, Adigbite A, Muhammed S, Baiyewu R. Yam diseases and its management in Nigeria. African Journal of Biotechnology. 2003;**2**(12): 497-502. DOI: 10.4314/ajb.v2i12.14878

[45] Cornelius EW. Causes and Control of Tuber Rots of White Yam (*Dioscorea Rotundata Poir)* Varieties: Araba. Asana and Puna: University of Ghana; 1998. Available from: http://ugspace.ug.edu.gh

[46] Okigbo RN, Ikediugwu FEO. Studies on biological control of postharvest rot in yams (Dioscorea spp.) using Trichoderma viride. Journal of Phytopathology. 2000;**148**(6):351-355. DOI: 10.1046/j.1439-0434.2000.00515.x

[47] Muniz FM, Silva EJ, Castro JMC, Rocha FS, Alencar LMC, Gonzaga V. Intensity of dry rot disease of yam in the state of alagoas, Brazil. Nematropica. 2012;**42**(1989):198-200

[48] Ogaraku AO, Usman HO. Storage rot of some yam (Dioscorea Spp) in Keffi and Environs, Nasarawa State, Nigeria. Patnsuk Journal. 2008;**4**(2):22-27. Available from: http://www.pa tnsukjournal.net/Vol4No2/p3.pdf

[49] Asare-Bediako E, Showemimo FA, Opoku-Asiama Y. Microorganisms associated with rot of minisetts of white Yam (Dioscorea rotundata, Poir). Research Journal of Microbiology. 2007; **2**(3):278-283. DOI: 10.3923/ jm.2007.278.283

[50] Osai EO, Ikotun T. Microbial rot of yam minisette. Fitopatologia Brasileira. 1994;**19**(3):408-412. Available from: https://www.cabdirect.org/cabdirect/ abstract/19952306478

[51] Zhang YL, Li CS, Zhang B, Qi JS, Xu ZT, Ma LG. First report of root rot of chinese yam caused by *Pythium ultimum* var. *ultimum* in China. Plant Disease. 2018;**102**(3):687-687. DOI: 10.1094/ PDIS-06-17-0829-PDN

[52] Al Zadjali S, Morse S, Chenoweth J, Deadman M. Factors determining pesticide use practices by farmers in the Sultanate of Oman. The Science of the Total Environment. 2014;**476–477**: 505-512. DOI: 10.1016/j.scitotenv. 2013.12.040

[53] Ebewore S, Egho E, Enujeke E. Yam (Dioscorea spp.) farmers' coping strategies against yam beetle (Heteroligus spp.) in Delta State, Nigeria: Implication for food security. Asian Journal of Agricultural Extension, Economics & Sociology. 2015;**6**(4): 194-201. DOI: 10.9734/ajaees/2015/ 17041

[54] Tobih FO. Methods of insecticide application to control yam tuber beetles (heteroligus spp; coleoptera: Dynastidae) in yam monocrop system in delta state, Nigeria. 2014:259-264. DOI: 10.5251/abjna.2014.5.6.259.264

[55] Ijato J, Tedela P. Phytotoxic potentials of cold and hot aqueous extracts of chromolaena odorata against fungal deteriorating agents of yam tubers (Dioscorea rotundata, Poir) after harvest. American Journal of Experimental Agriculture. 2015;**5**(3): 262-266. DOI: 10.9734/AJEA/2015/12139

[56] Obilo O, Oguamanam K, Ogbedeh K, Onyia V, Ofor M, Ofor M. The use of plant extracts in the control of Aspergillus niger in the rot of yam (Dioscorea spp.) during storage. International Journal of Agriculture and

Rural Development. 2005;**6**(1):74-80. DOI: 10.4314/IJARD.V6I1.2591

[57] Diaz GJ. Toxicosis by plant alkaloids in humans and animals in Colombia. Toxins. 2015;7(12):5408-5416. DOI: 10.3390/toxins7124892

[58] Shih SL, Lin YS, Hou WC. Effects of yam dioscorin interventions on improvements of the metabolic syndrome in high-fat diet-induced obese rats. Botanical Studies. 2015;**56**(1):4. DOI: 10.1186/s40529-015-0084-8

[59] Naskar KK, Mishra O. Action of diosgenin and homoeopathic pathogenesis of Dioscorea villosa. Indian Journal of Research in Homoeopathy. Wolters Kluwer - Medknow. 2017;**11**: 5-11. DOI: 10.4103/0974-7168.200845

[60] Yi T, Fan L-L, Chen H-L, Zhu G-Y, Suen H-M, Tang Y-N, et al. Comparative analysis of diosgenin in Dioscorea species and related medicinal plants by UPLC-DAD-MS. BMC Biochemistry. 2014;**15**: 19. DOI: 10.1186/1471-2091-15-19

[61] McKoy M-L, Thomas P-G, Asemota H, Omoruyi F, Simon O. Effects of jamaican bitter yam (Dioscorea polygonoides) and diosgenin on blood and fecal cholesterol in rats. Journal of Medicinal Food. 2014;**17**(11): 1183-1188. DOI: 10.1089/jmf.2013.0140

[62] Okigbo RN, Nmeka Ia. Control of yam tuber rot with leaf extracts of Xylopia aethiopica and Zingiber officinale. African Journal of Biotechnology. 2005;**4**(August): 804-807. Available from: http://www.academicjournals.org/AJB/abstracts/abs2005/Aug/OkigboandNmeka.htm

[63] Chandrasekara A, Josheph Kumar T. Roots and tuber crops as functional foods: A review on phytochemical constituents and their potential health benefits. International Journal of Food Science. 2016;**2016**:3631647. DOI: 10.1155/2016/3631647

[64] Wesley T. Yam – sources, health benefits, nutrients, uses and constituents at NaturalPedia.com. 2017. Available from: http://www.superfoods.news/2017-07-06-yam-sources-health-benefits-nutrients-uses-and-constituents-at-naturalpedia-com.html. [Accessed: June 27, 2018]

[65] Ministry of Financce, Ghana. Composite budget of the Nanumba South District Assembly. 2016. Available from: https://www.mofep.gov.gh/sites/default/files/composite-budget/2016/NR/Nanumba-South.pdf

[66] Trade Map. Trade Map - List of importing markets for the product exported by Ghana in 2017 (Mirror). 2017. Available from: https://www.trademap.org/Country_SelProductCountry.aspx?nvpm=1%7C288%7C%7C%7C%7C07%7C%7C%7C2%7C1%7C2%7C2%7C1%7C%7C2%7C1%7C1. [Accessed: September 13, 2018]

[67] EFSA. The 2015 European Union report on pesticide residues in food. EFSA Journal. 2017;**15**(4):4791. DOI: 10.2903/j.efsa.2017.4791

[68] Asiedu E. Pesticide Contamination of Fruits and Vegetables: A Market-Basket Survey from Selected Regions in Ghana. University of Ghana. 2013. Available from: http://ugspace.ug.edu.gh

[69] Bempah CK, Donkor AK. Pesticide residues in fruits at the market level in Accra Metropolis, Ghana, a preliminary study. Environmental Monitoring and Assessment. 2011;**175**(1–4):551-561. DOI: 10.1007/s10661-010-1550-0

[70] Adeyeye A, Osibanjo O. Residues of organochlorine pesticides in fruits, vegetables and tubers from Nigerian markets. Science of the Total Environment. 1999;**231**(2–3, 233):227. DOI: 10.1016/S0048-9697(99)00067-4

[71] Khan Z, Kamble N, Bhongale A, Girme M, Bahadur Chauhan V, Banerjee K. Analysis of pesticide residues in tuber crops using pressurised liquid extraction and gas chromatography-tandem mass spectrometry. Food Chemistry. 2018;**241**:250-257. DOI: 10.1016/J.FOODCHEM.2017.08.091

[72] PRC, UK. UK Pesticide Residues Committee Annual Report. Residue Reviews. 2009;**35**. DOI: 10.1016/S0166-526X(08)00009-3

[73] Wumbei A, Senaeve D, Houbraken M, Spanoghe P. Pesticides residue analysis in yam from selected markets across Ghana and Belgium: An evaluation of the QUECHER method. International Journal of Food Contamination. 2018;(2018):5-4. DOI: 10.1186/s40550-018-0066-1

[74] Wumbei A, Goeteyn L, Lopez ED, Houbraken M, Spanoghe P. Glyphosate in yam from Ghana. Food Additives & Contaminants: Part B. 2019. DOI: 10.1080/19393210.2019.1609098

[75] Wumbei A, Abdul-Rahaman I, Abdulai A, Edelbis L, Spanoghe P. Consumption risk assessment of pesticides residues in yam. Tunisian Journal of Plant Protection. 2019;**14**(2): 49-64. Available from: http://www.tjpp.tn/SiteWeb/CurrentIssue/TJPP14_2/5Abukari.pdf